Apolipoproteins, Triglycerides and Cholesterol

Edited by Viduranga Y. Waisundara and Miljana Z. Jovandaric

Published in London, United Kingdom

IntechOpen

Supporting open minds since 2005

Apolipoproteins, Triglycerides and Cholesterol
http://dx.doi.org/10.5772/intechopen.88426
Edited by Viduranga Y. Waisundara and Miljana Z. Jovandaric

Contributors
David Coman, Suzanne Sindi, Mikahl Banwarth-Kuhn, Jie Lin, Gopi M, Vinayak Shashidhar Uppin, Ramaprasad Ravichandra Talahalli, Pooja Acharya, María A. Hidalgo, Marcelo Ratto, Rafael Burgos, Febri Odel Nitbani, Jumina Jumina, Miljana Z. Z Jovandaric, Svetlana J. Milenkovic, Muhammad Ayoub, Jie Wei Wan, Manzoor Ahmad, Ranitha Mathialagan, Sarah Farrukh, Mohammed Danish, Sami Ullah, Salman Raza Naqvi, Donghui Liu, Anh Le, Swati Sanjay Gaikwad, Jasmine Avari, Mansi Liladhar Patil, Viduranga Y. Waisundara, M.W. Kanchana Priyadarshani Samarasinghe

First published in London, United Kingdom, 2020 by IntechOpen
IntechOpen is the global imprint of INTECHOPEN LIMITED, registered in England and Wales, registration number: 11086078, 7th floor, 10 Lower Thames Street, London, EC3R 6AF, United Kingdom
Printed in Croatia

British Library Cataloguing-in-Publication Data
A catalogue record for this book is available from the British Library

Additional hard and PDF copies can be obtained from orders@intechopen.com

Apolipoproteins, Triglycerides and Cholesterol
Edited by Viduranga Y. Waisundara and Miljana Z. Jovandaric
p. cm.
Print ISBN 978-1-83962-519-0
Online ISBN 978-1-83962-520-6
eBook (PDF) ISBN 978-1-83962-521-3

For EU product safety concerns:
IN TECH d.o.o., Prolaz Marije Krucifikse Kozulić 3, 51000 Rijeka, Croatia, info@intechopen.com or visit our website at intechopen.com.

Meet the editors

Dr. Viduranga Waisundara obtained her Ph.D from the Department of Chemistry, National University of Singapore in Food Science & Technology in 2010. She was a lecturer at Temasek Polytechnic, Singapore from July 2009 to March 2013. She relocated to her motherland of Sri Lanka and spearheaded the Functional Food Product Development Project at the National Institute of Fundamental Studies from April 2013 to October 2016. She was a Senior Lecturer on a temporary basis, at the Department of Food Technology, Faculty of Technology, Rajarata University of Sri Lanka. She is currently the Deputy Principal of the Australian College of Business & Technology – Kandy Campus, in Kandy, Sri Lanka. She is also the present Global Harmonization Initiative (GHI) Ambassador to Sri Lanka.

Miljana Z. Jovandaric was born in Serbia. She graduated from the Faculty of Medicine, University of Belgrade. Her pediatric specialization ended in 1999 at the University Children's Hospital, Belgrade. She completed her specialization in neonatology in 2003. Her master's thesis, `` Analysis of lipid infants in women suffering from gestational diabetes mellitus (GDM)``, was defended in 2006 and her doctoral dissertation, `` Effect of hypoxia on electrolyte and lipid levels in term newborns``, was achieved in 2018 from the School of Medicine, University of Belgrade, Serbia. She has been the author and co-author of 74 scientific papers presented at national and international conferences and published in journals. She is the Head of the Department of Sick Newborns at the Clinic for Gynecology and Obstetrics, Department of Neonatology, Clinical Center of Serbia, Belgrade.

Contents

Preface

Lipid-related knowledge and research has been in the limelight for the past several decades. This is owing to the fact that this particular biomolecule has a deep-seated relationship with non-communicable diseases at large such as high blood pressure, elevated levels of blood cholesterol, obesity, and cardiovascular disease. Scientists and researchers across the globe have been working on demystifying the biochemical pathways and proposing remedial actions for lipid-related diseases, all with the good intention of improving the quality of life around the world.

Lipids as a whole cannot be avoided in their entirety. They exist practically everywhere, particularly in food. It is with this reflection that this book has been brought forward, to serve as a new perspective. As the title of the book suggests, the content focuses on apolipoproteins, triglycerides, and cholesterol. These are very broad terms, albeit in this book, they are given a new angle – biochemically and clinically. Some of the areas presented herein have been overlooked, and therefore, are given their due recognition.

My most sincere gratitude goes to my co-editor, Dr. Miljana Z. Jovandaric of the Department of Neonatology, Clinical Center of Serbia, Belgrade, who has worked very hard on editing the chapters and putting them together in line with the title of the book. I would also like to take this opportunity to extend my appreciation to the authors who have contributed so many wonderful chapters. Also, heartfelt appreciation to IntechOpen Publishers with whom I have worked with for many book projects. Last but not least, my appreciation goes to Ms. Sara Debeuc, the Publishing Process Manager assigned to this book, who has rendered her utmost support in putting the material together.

This book will be of value to academics who are conducting research on lipids. It is updated and current in terms of the content. There are novel perspectives brought forward through many of the chapters and therefore, it is without a doubt that, readers will find these fresh points of view insightful and astute in advancing the existing knowledge on apolipoproteins, triglycerides, and cholesterol.

Dr. Viduranga Waisundara
Australian College of Business and Technology—Kandy Campus,
Kandy, Sri Lanka

Section 1

Apolipoproteins

Chapter 1

Significance of Lipid and Lipoprotein in Organism

Miljana Z. Jovandaric and Svetlana J. Milenkovic

Abstract

Lipids are important energy and building compounds. Their decomposition provides a significant amount of energy required for various life processes. It can thus be deposited in triglycerides and adipocytes. Some of them, in conjunction with proteins, form the most important structural elements of cells and cellular organelles, while others are precursors for the synthesis of numerous active compounds such as some hormones or prostaglandins. Lipids are ingested but can also be synthesized in the body. In circulation, lipids are found packed in lipoprotein molecules because they are insoluble in water. Lipoproteins have a central lipid part (nucleus) containing triglycerides and cholesterol esters, and on the surface there is a sheath composed of certain proteins (apoproteins), phospholipids, and small amounts of free cholesterol. Thanks to this sheath, lipids can be transported via blood. It took a long time to determine the importance and role of lipids in the body, as well as their role in many metabolic disorders of various diseases. This field is still unexplored and is a challenge for many researchers to prevent and treat lipid metabolism disorders.

Keywords: lipids, lipoproteins, apolipoproteins

1. Introduction

There is evidence that Leonardo da Vinci first observed macroscopic changes in the arteries corresponding to atherosclerosis, noting that such lesions were later discovered on Egyptian mummies' aortic wall and large blood vessels. The presence of cholesterol in atherosclerotic plaque was first indicated by Vogel in 1847 [1].

An association between elevated blood cholesterol levels and atheroma was observed early this century. The credit goes to Alexander Ivanovsky, who in 1908 published a paper proving that high-energy and high-protein diets in rabbits lead to the development of atherosclerosis. Anichkov found that this resulted in increased fat infiltration into the arterial wall. The significance of this animal model was in the morphological and histochemical similarity of these changes with human atherosclerosis [2].

The landmark advances in lipidology were marked in 1985 by Michael Brown and Joseph Goldstein for the pioneering work on the role of LDL receptors, intracellular cholesterol metabolism, and its homeostasis in the body [3].

Undoubtedly of great importance for the better knowledge of the structure and metabolism of lipoprotein particles in the blood have been the great advances in the domain of learning about their protein component—apolipoprotein [4].

In recent years, hypertriglyceridemia has been found to be a risk factor for coronary artery disease and atherosclerosis [5].

IntechOpen

From the pioneering work of Ignatovsky and Anickov at the beginning of this century, a huge journey has been made, and great discoveries have been made that have expanded our knowledge of lipid and lipoprotein metabolism to unprecedented limits [6].

2. Definition of lipids

The term lipids refers to a group of organic chemicals that are found in all animal and plant organisms.

There are several types of lipids present in the human body, the most important of which are fatty acids, cholesterol, triglycerides, and phospholipids [7].

They are an integral part of all complex lipids (cholesterol esters, triglycerides, and phospholipids) and are present in the body in the form of the so-called free fatty acids (SMK), unbound for other lipids. The 7SMKs are composed of a straight chain of carbon atoms, the number of which is, as a rule, even and the molecule ends with a carboxyl group. According to the number of carbon atoms, they are divided into fatty acids of short (up to 8 carbon atoms), medium (8–12 carbon atoms), and long chain (more than 12 carbon atoms) [8].

Based on the presence or absence of double bonds, the fatty acids are divided into saturated and unsaturated. The unsaturated fatty acids by the number of double bonds are divided into monosaturated (having only one bond) and polyunsaturated, having from two to six double bonds. They can occur in two isomeric forms, the so-called cis and trans forms [9].

2.1 Fatty acids

Fatty acids are present in the blood in two forms, either free or nonesterified (SMK, FFA) and esterified. Esterified fatty acids account for the largest part in the circulation of about 95%, while 5% of free fatty acids are found in the blood [10].

Plasma free fatty acids are reversibly bound to proteins, primarily to albumins, and to a lesser extent to globulins and lipoproteins. They are present in a very low concentration (about 0.5 mmol/l) but have an extremely fast turnover, with a half-life of 1–3 minutes. Metabolically, the most active are the plasma lipids, and their oxidation is the main source of energy in the fasting state. The esterified fatty acids are esterically bound in the composition of triglycerides (45%), cholesterol esters (15%), and phospholipids (35%) [8].

A large number of fatty acids have been discovered in nature, many of which are present in the human body. Some of the fatty acids can be created in our body, and those fatty acids that the human body is unable to synthesize must be ingested through food and are called essential fatty acids. It has already been mentioned that most of the fatty acids are in bound form, in the composition of phospholipids, triglycerides, and cholesterol esters, and only 5% of free fatty acids are in free or unbound form (free fatty acids). It should be borne in mind that many properties of triglycerides, cholesterol, and phospholipids depend significantly on the type of fatty acids that make up their composition. It is also of great importance whether the fats in the diet are dominated by saturated, monounsaturated, or polyunsaturated fatty acids [11].

Cholesterol is best known among lipid fractions and plays the most important role in the formation of atherosclerosis. It is a special type of lipid, which differs significantly from other lipid substances in its chemical structure. In pure form, the whitish soft waxy substance is insoluble in water [11]. It is present in the body in free, nonesterified, and esterified form bound to a single fatty acid in the form of cholesterol esters [12].

Cholesterol esterification occurs in the plasma under the action of the enzyme lecithin-cholesterol acyltransferase (LCAT). Plasma contains about 75% of the cholesterol in the esterified state and most often esterifies with polyunsaturated linoleic fatty acid about 55% [13].

Most of the free (nonsterified) cholesterol is found in tissues. Cholesterol is a necessary component of the body. It is a structural element of all cellular and intracellular membranes and has specific roles in specific organs (e.g., in hepatocytes where it participates in the synthesis of bile acids, in the cortex, and in the synthesis of steroid hormones, and it plays the role of transporters of liposoluble vitamins A, D, E, and K).

The origin of cholesterol in the body is twofold (endogenous and exogenous). Most cells have the ability to synthesize it themselves, and its other source is the food it feeds on. It has been found that 2/3 of cholesterol is produced by synthesis in the body (in adults about 800–900 mg/day) and only 1/3 is ingested by food. Considering the ability of the organism to produce it in large quantities, it is quite sufficient to feed 150–300 mg daily with food. Excess cholesterol from the body is eliminated through the bile (by conversion to bile acids) and by skin peeling, and a very small amount is lost by urine. Breastfeeding women also lose some cholesterol through milk. From all of the above, it is obvious that cholesterol is essential to the body and is of great importance for the normal functioning of each cell. Its adverse effects are manifested when it is present in much higher concentrations in the blood [14].

2.2 Triglycerides

Triglycerides are esters of glycerol, a trihydroxy alcohol with fatty acids. It is possible to esterify only one, two, or all three hydroxyl groups of fatty acid glycerol to produce mono-, di-, or triglycerides. Triglycerides are most prevalent in the body, while diglycerides, and especially monoglycerides, are present in significant amounts only in the intestinal mucosa during fat absorption [12].

Their basic role in the body is the creation of energy depots from which, when necessary, they release fatty acids, whose oxidation provides the energy necessary for the life of all cells. The highest amount of triglyceride is found in the composition of adipose tissue (about 95%), while insignificant amounts are present in the blood. Triglycerides have also been reliably found today to have an increased amount in the blood in the process of atherosclerosis [15].

2.3 Phospholipids

The chemical structure of phospholipids is very complex. It is an ester-bound two-fatty acid molecule for one alcohol, which contains phosphoric acid as an integral part of its molecule. Depending on the alcohol they contain, they are divided into two groups, glycerophospholipids (containing glycerol) and sphingophospholipids (containing sphingosine). With the exception of fatty tissue dominated by triglycerides, phospholipids are the basic lipids of cell membranes and other cellular structural elements. It should be emphasized that the brain and nerve tissue are the richest in phospholipids [7].

3. Definition of lipoproteins

An important feature of lipids is their insolubility in water. In order to dissolve it in the blood and transport it to all cells of the body, all lipid substances are bound to certain proteins, thus forming particles called lipoproteins [8].

The lipid part of the lipoprotein particles consists of cholesterol, cholesterol esters, triglycerides, and phospholipids, and their protein parts are very different in structure and are called apolipoproteins. Depending on the type and amount of lipids, on the one hand, and the amount and type of protein part, or apolipoprotein, on the other, different types of lipoproteins are present in the blood [4].

The four basic types of lipoprotein particles present in the blood of all persons are chylomicrons, very low-density lipoproteins (VLDL), low-density lipoproteins (LDL), and high-density lipoproteins (HDL) (**Table 1**) [7].

Today, the labeling of lipoprotein particles with abbreviations deriving from English names, i.e., as VLDL, LDL, and HDL particles, is accepted worldwide. They should also be supplemented with chylomicrons, which have the largest circulating lipoprotein particle [7].

Certain lipids (cholesterol, triglycerides, phospholipids) are found in the blood in the composition of all lipoprotein particles, in different amounts, and in combination with other constituents of lipoprotein particles. Chylomicrons and VLDLs contain predominantly triglycerides, LDL mainly cholesterol, while HDL particles are most abundant in the protein moiety and lipids in phospholipids. One person's blood cholesterol represents the sum of cholesterol present in all lipoprotein particles, in chylomicrons, VLDL, LDL, and HDL, although its largest amount (about 70%) is found in LDL particles. Depending on the physiological role of individual lipoprotein particles, the increase in cholesterol in one of them will have a different significance for the organism. Cholesterol contained in the LDL particle leads to atherosclerosis, while the increase in cholesterol in HDL particles has a protective effect against atherosclerosis [6]. In addition to chylomicrons, VLDL, LDL, and HDL particles, some other types of lipoproteins are normally present in the blood. Among these lipoprotein species is the so-called lipoprotein Lp(a). It is a special type of lipoprotein, which has been shown to be present in the blood of every person at quite minimal concentrations (up to 0.25 g/l). Lipoprotein Lp(a) together with the LDL particle represents the most serious lipid risk factor for atherosclerosis [7].

3.1 Apolipoproteins

Lipids are mostly insoluble in water. In plasma, they can be in the form of stable complexes if they are attached to specific protein moieties called apolipoproteins or apoproteins (Apo).

The protein part of the lipoprotein particle was previously thought to be a lipid transport agent only. However, they have been found to have other, very significant roles in the body. Apoproteins provide stability of plasma lipid transport. They are cofactors of individual enzymes that participate in the metabolism of lipoprotein particles. Their most important function is to bind to specific receptors on cell membranes, thereby ensuring the entry of plasma lipoprotein particles into cells and their further catabolism [2, 15, 16].

• Hilomicrones
• Very low-density lipoproteins (VLDL) Pre-beta-lipoproteins
• Low-density lipoproteins (LDL) Beta-lipoproteins
• High-density lipoproteins (HDL) Alpha-lipoproteins

Table 1.
Basic blood lipoprotein particles.

The structure and concentration of individual apolipoproteins in a particular lipoprotein particle are under a direct genetic control, as opposed to serum lipid content, which is significantly more influenced by diet and metabolism [17].

3.1.1 Apolipoprotein A-I

Apolipoprotein A-I (Apo A-I) is the major apolipoprotein of HDL particles. About 90% of Apo A-I is in HDL fractions. In the HDL particle alone, it accounts for 65% of the total protein portion. It is present in small amounts in the chylomicrons and the VLDL fraction. It is synthesized in the liver and small intestine wall. Its most important role is the activator of the enzyme LCAT, which allows esterification of free cholesterol on the surface of HDL particles. In this way, small discoidal HDL3 particles accept excess cholesterol from the cells and accumulate it in the form of esterified cholesterol, thereby transforming it into larger, more soluble, lower-density HDL2 particles. Since it allows the uptake of free cholesterol and subsequently its esterification, it is logical that it has a protective effect on the process of atherosclerosis [18].

3.1.2 Apolipoprotein A-II

Apolipoprotein A-II (Apo A-II) represents about 20–30% of total HDL particle proteins. It forms in the liver and the small intestine wall. It is thought to have an inhibitory effect on LCAT and modulatory effect on the activity of lipoprotein lipase (LPL) and hepatic triglyceride lipase (HTGL) [19].

3.1.3 Apolipoprotein B

Apolipoprotein B (Apo B) is present in the human population in two forms, as apolipoprotein B-48 (Apo B-48) and apolipoprotein B-100 (Apo B-100).

Apo B-48, which is part of the chylomicron, is synthesized in the small intestinal wall. Far more significant is Apo B-100, which is synthesized predominantly in the liver, and is a constituent of VLDL, the intermediate-density lipoprotein (IDL) that circulates for a short time, LDL particles, and the lipoprotein Lp(a). This apoprotein is also characterized by containing certain amounts of carbohydrates (approximately 5% of the total mass), which is chemically classified as glycoproteins [20, 21].

Apo B-100 is secreted primarily from the liver in the form of VLDL particles. By further metabolizing VLDL via IDL, LDL particles are formed that contain exclusively Apo B-100 [4]. The most important role of Apo B-100 is to specifically bind LDL particles to cellular receptors in the liver and other tissues [22].

Given that Apo B-100 is the only apolipoprotein in LDL particles known to have the most heterogeneous effect, it is understandable that an increase in plasma Apo B-100 concentration increases atherogenic risk. Due to the fact that only one Apo B-100 molecule is present in the LDL particle, unlike the amount of cholesterol that can vary in them, the plasma concentration of Apo B-100 is a better indicator of the content of atherogenic lipoprotein particles in the blood than the determination of LDL cholesterol [23].

3.1.4 Apolipoprotein C

The apolipoprotein C (Apo C) group is represented by three different proteins that are synthesized in the liver [20]. There are chylomicrons, VLDLs, and newly formed HDL particles bound in the circulation. During intravascular metabolism of lipoprotein particles, there is an intense exchange of Apo C. From HDL particles, these apolipoproteins transfer to lipoprotein particles rich in triglycerides, namely, chylomicrons and VLDL. Apparently, Apo C plays a significant role in the catabolism of triglyceride-rich lipoproteins [24].

3.1.5 Apolipoprotein E

Apolipoprotein E (Apo E) is a protein molecule composed of 229 amino acids, which contains 10% of the amino acid arginine, and is therefore called the "arginine-rich protein" [16]. The main site of its synthesis is the liver, but it is also created in other organs: brain, spleen, lungs, adrenal gland, kidneys, ovaries, muscles. It plays a significant role in the metabolism of various lipoprotein particles: chylomicrons, VLDL, IDL, and HDL containing Apo E.

Apolipoprotein E participates in the reversible transport of excess cholesterol from peripheral tissues into the liver via HDL particles, which when incorporated into cholesterol incorporate Apo E [25].

Apo E also appears to be involved in reparative responses to tissue damage. An increase in its concentration is found in sites of peripheral nerve damage and regeneration [25]. It has significance in nerve regeneration, growth, and/or differentiation of neurons [26].

3.1.6 Apolipoprotein (a)

Apolipoprotein (Apo A) is a relatively large molecular weight protein that is incorporated into the lipoprotein composition of Lp(a).

If Apo A is isolated, the rest of Lp(a) in its composition is almost identical to the LDL particle because it has a similar lipid composition and contains a single molecule of Apo B-100. The gene that regulates the synthesis of this apolipoprotein is located on the sixth chromosome near the plasminogen gene. Apo A has a great structural similarity and shows immune cross-reactivity with plasminogen, which could be a link between atherosclerosis and fibrinolysis [27].

4. Conclusion

Lipid disorders are of fundamental importance for atherogenesis and even the occurrence of ischemic heart disease and other cardiovascular and cerebrovascular diseases. They are often associated with diabetes, obesity, and hypertension with which they interact synergistically, leading to arteriosclerotic changes. Atherosclerosis is caused by changes in the wall of blood vessels characterized by lipid deposition and cell proliferation. Deposited lipids in the blood vessel wall originate from plasma lipoproteins, and elevated cholesterol, especially LDL cholesterol, is a major risk factor. Atherogenic lipoproteins include, in addition to LDL particles, almost all classes of Apo B-containing lipoproteins (VLDL, VLDL residues, IDL, Lp(a), and oxidized LDL). A common feature of all atherogenic lipoproteins is that they contain different amounts of cholesterol esters and/or Apo B-100 or Apo B-48. Atherogenic effects of triglyceride-rich lipoproteins are associated with postprandial lipemia after fatty meal intake. Atherosclerosis is considered an inevitable process at the present stage of medical science development. In most people, around the age of 85, it is thought that about 60% of coronary circulation is covered by atherosclerotic plaques, provided that no risk factors are present during life. In the presence of risk factors such as hypercholesterolemia, such changes in the coronary vessels are reached sometime in the 42nd year of life. This early atherosclerosis is a global problem for humanity today. The major risk factors for cardiovascular disease are the values of total LDL and HDL cholesterol. Higher HDL cholesterol has also been shown to have a protective effect. Elevated triglyceride levels also increase the incidence of myocardial infarction, even when HDL cholesterol levels are normal. Based on the studies, it was concluded that the level of total cholesterol was important in the assessment of total individual risk but the value of LDL cholesterol was taken as the goal of therapy for lipid abnormalities.

Conflict of interest

No potential conflict of interest was reported by the authors.

Funding

None.

Author details

Miljana Z. Jovandaric* and Svetlana J. Milenkovic
Clinic for Gynecology and Obstetrics, Department of Neonatology, Clinical Center of Serbia, Belgrade, Serbia

*Address all correspondence to: rrebecca080@gmail.com

References

[1] Quiney JR, Wats GF, editors. Classic Papers in Hyperlipidaemia. London: MSD Science press; 1989. pp. 25-31

[2] Grundy SM. Use of emerging lipoprotein risk factors in assessment of cardiovascular risk. Journal of the American Medical Association. 2012;**307**:2540-2542

[3] Brown MS, Goldstein JL. Lipoprotein in the macrophage: Implication for cholesterol deposition in atherosclerosis. Annual Review of Biochemistry. 1983;**52**:223-261

[4] Sniderman AD, Cianflone K. Measurement of apolipoproteins: Time to improve the diagnosis and treatment of the atherogenic dyslipoproteinemias. Clinical Chemistry. 1996;**42**:489-491

[5] Ikeda Y, Ashida Y, Takagy A, et al. Mechanism of the production of small dense LDL(s LDL) in hypertriglyceridemia. In: Jacotot B, Mathe D, Fruchart J-C, editors. Atherosclerosis XI. Amsterdam-Lausanne-NewYork-Oxoford-Singapore-Tokyo: Elsevier; 1998. pp. 777-788

[6] Luc G, Lecerf J-M, Bard J-M, et al. Cholesterol et Atherosclerose. Paris-Milan-Barcelona-Bonn: Masson; 1991. pp. 15-20

[7] National Cholesterol Education Program (NCEP) Expert Panel on Detection, Evaluation, and Treatment of High Blood Cholesterol in Adults (Adult Treatment Panel III). Third report of the National Cholesterol Education Program (NCEP) expert panel on detection, evaluation, and treatment of high blood cholesterol in adults (adult treatment panel III) final report. Circulation. 2002;**106**:3143-3421

[8] De Buch H. Lipido –composition of lipoproteins. Annales de Biologie Clinique. 1973;**31**:65-67

[9] Maedler K, Spinas GA, Dyntar D, Moritz W, Kaiser N, Donath MY. Distinct effects of saturated and monounsaturated fatty acids on beta-cell turnover and function. Diabetes. 2001;**50**:69-76

[10] Frederikson DS, Gordon RS. Transport of fatty acids. Physiological Reviews. 1958;**38**:585-630

[11] Chapman MJ. LDL subfractions atherogenesis and coronary risk. The World of Lipids. 1995;**1**:4-7

[12] Ginsberg HN. New perspectives on atherogenesis: Role of abnormal triglyceride-rich lipoprotein metabolism. Circulation. 2002;**106**:2137-2142

[13] Olsson AG, Walldius G, Rössner S, Callmer E, Kaijser L. Studies on serum lipoproteins and lipid metabolism. Analysis of a random sample of 40 year old men. Acta Medica Scandinavica. 1980;**637**:1-47

[14] Kaluarachchi M, Boulangé CL, Karaman I, Lindon JC, Ebbels TMD, Elliott P, et al. A comparison of human serum and plasma metabolites using untargeted ^{1}H NMR spectroscopy and UPLC-MS. Metabolomics. 2018;**14**:32

[15] Rifai N, Dufour R, Cooper GR. Preanalytical variation in lipid, lipoprotein, apolipoprotein testing. In: Rifai N, Warnick RG, Dominikzak M, editors. Handbook of Lipoprotein Testing. Washington: AACC Press; 1977. pp. 75-77

[16] Marinetti GV. Disorders of Lipid Metabolism. New York and London: Plenum Press; 1990. pp. 35-49

[17] Li W-H, Tanimura M, Luo CC, et al. The apolipoprotein multigene family: Biosynthesis, structure-function

relation-ships, and evolution. Journal of Lipid Research. 1988;**29**:245-271

[18] Rader DJ, Schaefer JR, Lohse P, et al. Increased production of apolipoprotein A-I associated with elevated plasma levels of high-density lipoproteins, apolipoprotein A-I, and lipoprotein A-I in a patient with familial hyperalphalipoproteinemia. Metabolism. 1993;**42**:1429-1434

[19] Valimaki M, Taskinen M-R, Ylikahri R, et al. Comparison of the effects of two different doses of alcohol on serum lipoproteins A-I and A-II: A controlled study. European Journal of Clinical Investigation. 1988;**18**:472-480

[20] Breslow JL. Lipoprotein genetics and molecular biology. In: Gotto AM, editor. Plasma lipoproteins. Amsterdam-New Jork-Oxford: Elsevier; 1987. pp. 359-397

[21] Hoeg JM, Sviridov DD, Ge T, et al. Both apolipoprotein B-48 and B-100 are synthesized and secreted by the human intestine. Journal of Lipid Research. 1990;**31**:1761-1769

[22] Myant NB. Cholesterol Metabolism, LDL, and the LDL Receptor. San Diego-New York-Boston-London: Academic Press, Inc; 1990. pp. 124-129

[23] Rader DJ, Hoeg JM, Brewer HB Jr. Quantitation of plasma apolipoprotein in the primary and secondary prevention of coronary artery disease. Annals of Internal Medicine. 1994;**120**:1012-1025

[24] Baggio G, Monrato E, Gobelli C, et al. Apolipoprotein C-II deficiency syndrome. Clinical features, lipoprotein characterization, lipase activity and correction at hypertriglyceridemia after apolipoprotein C-II administration in two affected patients. The Journal of Clinical Investigation. 1986;**77**:520-527

[25] Mahley RW. Apolipoprotein C: Cholesterol transport protein with expanding role in cell biology. Science. 1988;**240**:622-630

[26] Rw M. Expanding roles for apolipoprotein E in health and disease. In: Jacotot B, Mathe D, Fruchort JC, editors. Atherosclerosis XI. Amsterdam-Lausanne-New York-Oxford-Singapore-Tokyo: Elsevier, Science; 1998. pp. 117-124

[27] Calabresi L, Sirtori CR, Paoletti R, Franceschini G. Recombinant apolipoprotein A-IMilano for the treatment of cardiovascular diseases. Current Atherosclerosis Reports. 2006;**8**:163-167

Chapter 2

High-Density Lipoprotein: From Biological Functions to Clinical Perspectives

Donghui Liu

Abstract

High-density lipoprotein (HDL) is a heterogeneous particle composed of apolipoproteins, enzymes, and lipids. Besides transporting cholesterol to the liver, HDL also exerts many protections on anti-oxidation, anti-inflammation, and anti-apoptosis. Initial understandings of HDL came from its protective roles against atherosclerosis and the observation that high plasma HDL cholesterol (HDL-C) levels seemed to decrease cardiovascular disease (CVD) attack. However, those patients either with cholesterol ester transfer protein (CETP) deficiency or taking CETP inhibitors substantially elevated HDL-C levels but did not necessarily decrease CVD risk. Thus, some researchers suggested that quantitative measurements of HDL particle (HDL-P) might be more valuable than traditional HDL-C measurements. What is more bewildering is that HDL from patients with systemic inflammation decreased its protective effects and even became a pro-inflammatory factor. Recently, synthesized HDL and apolipoprotein mimetic peptides showed biological functions similar to native ones. Expectedly, lots of novel measurement methods and therapeutic agents about HDL would be established soon.

Keywords: HDL, apolipoprotein, mimetic peptide, atherosclerosis, CVD

1. Introduction

Initial understandings of high-density lipoprotein (HDL) came from the epidemiological studies, which consistently showed that a low HDL cholesterol (HDL-C) level is regarded as an independent risk for the development of cardiovascular disease (CVD) [1, 2]. Inversely, elevated HDL-C concentration in plasma is correlated with reduced CVD risk [3]. Therefore, lots of strategies for raising HDL-C were considered to be the suitable targets for CVD prevention and treatment [4, 5]. Deficiency and inhibition of cholesterol ester transfer protein (CETP) increase plasma HDL-C levels; however, they do not necessarily reduce CVD risk as expected, which suggest that the compositions and functions of HDL are more complicated than we supposed before [6]. Besides reverse cholesterol transport (RCT), HDL possesses anti-oxidative, anti-inflammatory, and anti-apoptotic effects on endothelial cells, exerts anti-migrative and anti-proliferative functions on smooth muscle cells, and presents anti-development and anti-metastasis characteristics on cancer cells [7, 8]. Nevertheless, HDL either modified by oxidation and glycation or isolated from patients with systemic inflammation decreases its protective effects and even becomes a pro-inflammatory, pro-oxidative, and pro-apoptotic factor [9]. Consequently, the question whether HDL-C is still the "good cholesterol" becomes more bewildering to be answered.

2. HDL-C levels and CVD

For a half century, Framingham study has supported the concept that HDL-C was thought to be a "good" lipoprotein and a negative risk factor against atherosclerosis and a decreased HDL-C level emerged as an independent risk for CVD, owing to a strong inverse correlation between plasma HDL-C levels and CVD [1]. The basis of this concept mainly came from the role of HDL in RCT [10]. However, the understanding of HDL-C and its relationship to CVD has changed dramatically. Deficiency and inhibition of CEPT or mutation of scavenger receptor class B type I (SR-BI) increase plasma HDL-C levels but do not accordingly reduce CVD events in these patients, which challenge the traditional ideas.

2.1 Reverse cholesterol transport (RCT)

An excess of cholesterol production or absorption is deleterious by contributing to cholesterol accumulation in vessel wall and subsequent atherosclerosis initiation. Thus, there is a physiological need to move the excessive cholesterol from peripheral tissues; this process is called reverse cholesterol transport (RCT) [10]. RCT represents the primary mechanism by which HDL delivers cholesterol from peripheral cells to the liver. This pathway of recycling and eliminating cholesterol is the antiatherogenic basis of high HDL-C levels against CVD and also represents a rescue mechanism for atherosclerotic plaque regression.

The first step of RCT is cellular cholesterol efflux to apolipoprotein A-I (apoA-I) mediated by ATP-binding cassette transporter A1 (ABCA1). Cholesterol efflux also occurs toward mature HDL through ATP-binding cassette transporter G1 (ABCG1) and SR-BI. Cholesteryl esters (CE) of HDL can be transferred to apolipoprotein B (apoB)-containing lipoproteins through the action of CETP, with ultimate uptake by low-density lipoprotein (LDL) receptor (LDL-R) in the liver. Each step in this process may influence the plasma levels of HDL-C. Because of the failure of reducing CVD risk by elevating HDL-C, the cholesterol efflux capacity of HDL seems to be more valuable to predict CVD incidence than HDL-C levels [11].

2.2 ATP-binding cassette transporter A1/G1 (ABCA1/G1)

A major breakthrough in understanding the mechanisms of RCT came from the discovery of Tangier disease, which is characterized by low HDL-C levels and high CVD risk because of the molecular defect in ABCA1 [12]. Low HDL-C level in these patients is caused by decreased cellular cholesterol efflux owing to ABCA1 mutation as well as increased catabolism of lipid-poor apoA-I [13]. ABCA1 knockout mice have an extremely low HDL-C phenotype similar to that of Tangier disease patients [14]. Thus, ABCA1 is essential for HDL maturation. In addition, it is also worth noting that the interaction between apoA-I and ABCA1 in macrophages also displays significant anti-inflammatory effects through activating JAK2/STAT3 pathway [15]. These effects reduce the attraction of macrophages into the vessel wall and ultimately result in the decreased plaque formation. These findings implicated that ABCA1 is a direct molecular link between the cardio-protective effects of cholesterol export and the inhibition of inflammatory responses in macrophages.

In contrast to ABCA1, ABCG1 promotes cholesterol efflux from macrophages to mature HDL but not to apoA-I [16]. In addition, ABCG1 also stimulates the release of cellular phospholipids to HDL [17]. Therefore, ABCA1 and ABCG1 are assumed to act in a sequential manner, which generates nascent HDL through ABCA1 and then facilitates cholesterol efflux via ABCG1, resulting in the formation of mature HDL.

2.3 Lecithin: cholesterol acyltransferase (LCAT)

LCAT is responsible for the esterification of free cholesterol and thus for the maturation of HDL by transferring fatty acids from lecithin (phosphatidyl choline) to cholesterol [18]. Once esterified, cholesterol moves from the surface to the hydrophobic core of HDL. In the presence of LCAT, the bidirectional movement of cholesterol between cells and HDL results in cholesterol efflux. Therefore, LCAT plays a central role in the initial step of RCT.

As a result of cholesterol esterification, LCAT also maintains a gradient of free cholesterol between cell membranes and lipoproteins. The activity of LCAT is essential to maintain the normal HDL metabolism and the optimal functional properties of HDL particles. In human, LCAT deficiency is responsible for low HDL-C levels, which changes HDL distribution and composition [19]. HDL from transgenic mice overexpressing human LCAT is prior to accepting free cholesterol from fibroblast compared to control HDL. It is likely that LCAT-mediated changes in HDL composition favor cholesterol accommodation within the particles. The flux of CE to the liver is increased in human LCAT-transgenic mice as a result of increased CE content in HDL but not an increased catabolic rate of HDL [20]. Although it is clear that LCAT deficiency in human and mice is associated with reduced HDL-C levels, it is still not defined whether LCAT overexpression or deficiency is pro- or anti-atherogenesis.

2.4 Cholesterol ester transfer protein (CETP)

CETP plays a key role in the exchange of CE and triglycerides (TG) between HDL and apoB-containing lipoproteins (VLDL, IDL, and LDL). As a result of CETP activation, HDL becomes smaller and TG-enriched. It is estimated that 66% of CE in HDL returns to the liver through CETP, indicating an important role of CETP in RCT process and HDL remodeling [21]. Some studies found that deficiency of CETP in human is associated with increased plasma HDL-C levels but inversely displays a relatively increased CVD incidence [22]. Small HDL particles are not increased in CETP-deficient subjects, suggesting that ABCA1-mediated cholesterol efflux might not represent the predominant pathway of cellular cholesterol efflux. Earlier studies also demonstrated that HDL from CETP-deficient subjects is defective to mediate cholesterol efflux from cholesterol-loaded macrophages, leading to the hypothesis that enrichment of CE in HDL in homozygous subjects might not be favorable for the antiatherogenic activities of these particles [23]. However, HDL from CETP-deficient subjects has been shown to possess an increased capacity to mediate cholesterol efflux through ABCG1 [24].

Additionally, inhibition of CETP successfully elevates HDL-C levels and decreases LDL-C levels but unexpectedly does not show atheroprotections and even increases cardiovascular mortality [25–27]. Until now, almost all CETP inhibitors, including torcetrapib (Pfizer), dalcetrapib (RO4607381, Roche; JTT-705, JT), anacetrapib (MK-0859, Merck), and evacetrapib (LY2484595, Eli Lilly), were announced to be failed to reduce CVD accidence although significantly elevating plasma HDL-C levels.

2.5 Scavenger receptor class B type I (SR-BI)

As the last step in RCT, SR-BI has been shown to function as another HDL receptor that mediates selective cholesterol uptake in the liver. SR-BI knockout mice remarkably elevate HDL-C levels but paradoxically increase atherosclerosis [28]. Some studies also reported that variant of SR-BI in which leucine replaces proline 376 (P376L) abrogates its ability to uptake HDL from plasma to the liver. Consequently, these patients have a profound HDL-related phenotype and an increased CVD risk [29].

3. HDL composition

Generally, HDL particles contain apolipoproteins, enzymes, charged lipids (phospholipids and free cholesterol) on the surface, and neutral lipids (TG and CE) in the core. The compositional complexity of HDL is further verified through the quantitative and qualitative proteome and lipidome assay, which carries more than 80 different proteins, over 200 lipid species, various microRNAs, as well as other bioactive molecules [30]. This physiological heterogeneity is further increased in the inflammatory conditions (e.g., CVD, diabetes mellitus, chronic kidney disease, and rheumatic diseases). The known functions associated with these components are diverse and span physiological roles far beyond the classical roles for HDL in lipid metabolism, suggesting that novel properties of HDL may exist. Therefore, it seems not reasonable to simply make HDL-C levels reflect the compositions and functions of HDL particles and predict the risk of CVD.

3.1 Apolipoprotein A-I (apoA-I)

ApoA-I is the most abundant protein of HDL, which is synthesized in the liver and intestine and almost located in all HDL particles. Mature apoA-I is a 28-kDa protein that consists of 243 amino acids and contains 10 amphipathic helical domains. It has been found that apoA-I plays a variety of roles associated with HDL metabolism. One primary function of apoA-I is to interact with cellular surface transporters (ABCA1), mediate cholesterol efflux, and activate LCAT, which exerts the foundational effects in RCT process as described above [31, 32]. Human subjects with apoA-I deficiency and apoA-I-deficient mice fail to form mature HDL particles [33]. Liver-specific overexpression of apoA-I was found to increase apoA-I and HDL-C levels in plasma, thereby reducing atherosclerosis in hyperlipidemic mice [34, 35]. In addition, apoA-I enhances the proliferation of human endothelial progenitor cells (EPCs) and promotes angiogenesis through ATP synthase in cell surface [36]. ApoA-I restores neovascularization of the lymphatic system in tumor necrosis factor (TNF)-alpha-mediated inflammatory responses [37]. We also found that human apoA-I induces cyclooxygenase-2 (COX-2) expression and prostaglandin I-2 (PGI2) release in endothelial cells through ABCA1 [38]. ApoA-I inhibits the chemotaxis, adhesion, and activation of THP-1 monocytes induced by lipopolysaccharide (LPS) and improves HDL inflammatory index (HII) in plasma [39]. Furthermore, apoA-I displays anti-inflammatory effects in adipocytes and adipose tissues similar to their effects in other cell types [40].

3.2 Paraoxonase-1 (PON1)

PON1 is a HDL-associated lactonase, which could hydrolyze a wide variety of lactones, thiolactones, aryl esters, cyclic carbonates, and organophosphate pesticides and prevent LDL oxidation [41, 42]. Decreased PON1 activity is a risk factor for CVD development independently of HDL-C levels [43]. PON1 reduces oxidative stress, inhibits cholesterol synthesis, and promotes cholesterol efflux in macrophages [44, 45]. Low PON1 activity is associated with many inflammatory diseases, including diabetes mellitus, rheumatoid arthritis, systemic lupus erythematosus, psoriasis, and renal diseases [46]. In the presence of PON1, lipid hydroperoxide is reduced, monocyte chemotactic protein 1 (MCP-1) production is inhibited, and atherosclerotic progression is attenuated [47]. Overexpression of PON1 inhibits atherosclerosis in mice with metabolic syndrome [48]. Additionally, it has been shown that PON1 can prevent the development of diabetes mellitus in mice through its anti-oxidative properties, suggesting a possible role of PON1 in stimulating insulin biosynthesis in islet beta cells [49].

3.3 Phospholipids in HDL

Besides free cholesterol, TG, and CE, there are many kinds of phospholipids in HDL molecules, mainly including ceramide, sphingomyelin, and sphingosine-1-phosphate (S1P) [50]. These phospholipids are located 0n the surface monolayer of HDL together with free cholesterol and apolipoproteins. Ceramide mediates an inflammatory response induced by cytokines or oxidized LDL (ox-LDL), which upregulates the expression of adhesion molecules, increases the adhesion and migration of monocytes, and subsequently promotes the initiation and progression of CVD [51]. Sphingomyelin regulates cholesterol efflux from peripheral cells, which is considered to possess an inverse relationship with CVD [52].

S1P is a bioactive lipid mediator generated by the phosphorylation of sphingosine via sphingosine kinases (SphK) 1 and SphK2, which plays variously biological and pathophysiological roles through three members of G protein-coupled S1P receptors (S1P1, S1P2, and S1P3) [53]. These S1P receptors are differentially expressed, regulating proliferation, migration, adhesion, and inflammation in endothelial cells, smooth muscle cells, and macrophages, all of which play key roles in the development of CVD [54, 55]. HDL-associated S1P limits endothelial inflammation induced by TNF-alpha, including adhesion molecule abundance, monocyte-endothelial adhesion, and endothelial barrier permeability [56, 57]. S1P elevates endothelial nitric oxide synthase (eNOS) activity and promotes nitric oxide (NO) release in endothelial cells [58]. S1P induces endothelial cell migration and proliferation, prevents apoptosis and inflammation, improves vascular relaxation, and preserves endothelial barrier function [55, 59, 60]. Some studies showed that reduced HDL-S1P content contributes to HDL dysfunction in CVD patients, including induction of eNOS activation in endothelial cells and promotion of vasodilatory potential in precontracted arteries. These decreased HDL functions could be efficiently improved by loading additional S1P to HDL both *in vitro* and *in vivo* [61]. In addition, exogenously administrated S1P accelerates neovascularization and blood flow recovery in ischemic limbs, suggesting its usefulness for angiogenic therapy. Furthermore, S1P was also shown to regulate VSMC proliferation and migration and to manipulate vascular tension via G protein-coupled receptors [62].

S1P1 is mainly expressed in endothelial cells, which mediates vascular maturation and maintains vascular integrity by contributing to eNOS activation, inhibiting vascular permeability and inducing endothelial cell chemotaxis via Gi-coupled mechanisms [55]. By contrast, S1P2 is expressed in VSMCs and some types of tumor cells in high levels, which inhibits cell migration via the G(12/13)-and Rho-dependent mechanism [55]. S1P3 is also primarily expressed in endothelial cells and mediates chemotaxis and vasorelaxation through a NO-dependent manner, which plays protective roles for vascular integrity [55]. These results provide evidence for S1P receptor subtype-specific pharmacological intervention as a novel therapeutic approach to CVD [63].

3.4 MicroRNAs (miRNAs) in HDL

miRNAs are small noncoding RNAs that suppress gene expression through post-transcriptional regulation of mRNA stability. Extracellular miRNAs likely serve as the cellular messages, which are transported between cells in an endocrine form of intercellular communication via circulation. In the blood vessels, these transferring miRNAs modulate atherosclerosis and angiogenesis, and in the heart, they modulate ischemic/reperfusion (I/R) injuries, myocardial infarction, and heart failure. In plasma, they are protected from circulating ribonucleases through the association with lipoproteins [64]. Especially, HDL is reported to be the major carrier of miRNAs in plasma. Furthermore, HDL exhibits an independent miRNA profile distinct from that of plasma through the micro-transcriptome assay, which might notably influence the biological functions of HDL [65].

HDL transports endogenous miRNAs and delivers them to recipient cells with functional targeting capabilities. Cellular export of miRNAs to HDL is regulated by neutral sphingomyelinase. Injecting reconstituted HDL (rHDL) into mice retrieves distinct miRNA profiles from normal and atherogenic animal models. Furthermore, HDL-mediated delivery of miRNAs to recipient cells was demonstrated to be dependent on SR-BI. The human HDL-miRNA profiles in healthy subjects are significantly different from those of familial hypercholesterolemia subjects. Notably, HDL-miRNAs from atherosclerotic subjects induce differential gene expression [66, 67]. Collectively, these observations indicated that HDL participates in a mechanism of intercellular communication through the delivery of miRNAs.

Some studies reported that the contents of miR-486 and miR-92a in HDL are reduced in vulnerable CVD patients [68]. HDL-associated miR-223 levels are decreased after high-protein diet-induced weight loss in overweight and obese males [69]. Intestinal lymphatic HDL-associated miR-223 is reduced during insulin resistance and is restored by niacin in rats [70]. Furthermore, HDL-transferred miR-223 inhibits intercellular adhesion molecule-1 (ICAM-1) expression in endothelial cells [71].

4. HDL functions beyond RCT

HDL exerts diverse biological functions besides removing cholesterol from peripheral cells through RCT, which have attracted considerable attentions.

4.1 Endothelial cell protections of HDL

Endothelial cells play fundamental roles in regulating vascular functions [72, 73]. Many risk factors for atherosclerosis (e.g., hypercholesterolemia, hypertension, and hyperglycemia) induce the inflammation and apoptosis in endothelial cells and initiate the pathogenesis of atherosclerosis [74]. Therefore, improving endothelial dysfunction is a potential target for preventing and treating CVD. HDL could inhibit cytokine-induced expression of vascular cell adhesion molecule-1 (VCAM-1) and ICAM-1 in human umbilical vein endothelial cells (HUVECs) and reduce the adhesion of monocytes to endothelial cells [75]. Moreover, HDL induces endothelial repair by enhancing eNOS activity and increasing NO production through the SR-BI and ABCG1 pathways [76, 77]. HDL also improves vascular health by reducing apoptosis and promotes proliferation and migration in endothelial cells, which are crucial for neovascularization after vascular injuries [78].

4.2 HDL functions on vascular smooth muscle cells (VSMCs)

Many stress factors could induce VSMCs shifting from a contractile phenotype to a synthetic phenotype, and consequently the activated VSMCs proliferate and migrate from the medial layer of vessels into the intima which results in neointimal hyperplasia and artery stenosis [79, 80]. HDL counterbalances the pro-inflammatory effects of ox-LDL by inhibiting intracellular reactive oxygen species (ROS) release and subsequent nuclear factor kappa-B (NF-κB) activation in VSMCs [81]. HDL also downregulates the production of fibroblast growth factor (FGF) and represses the proliferation of VSMCs triggered by ox-LDL [82]. HDL suppresses the expression of chemokines (CCL2, CCL5, CX3CL1, CCR2, and CX3CR1) and the proliferation of VSMCs induced by TNF-alpha via the SR-BI pathway [7]. In addition, HDL-associated alpha-antitrypsin (AAT) inhibits extracellular matrix degradation, cell detachment, and apoptosis triggered by elastase in human VSMCs [83].

4.3 HDL against inflammation

HDL plays an important role against inflammatory responses [84, 85]. HDL is able to bind and neutralize LPS as well as to facilitate LPS release from the surface of macrophages, which inhibits macrophage activation and cytokine release [85–87]. HDL-bound LPS does not interact with the cellular membrane receptors in macrophages, thereby decreasing the uptake of LPS by macrophages. And apoA-I is identified as the LPS-binding molecule in HDL [88]. In rat models of LPS-mediated sepsis, infusion of rHDL significantly reduces cytokine release, organ injuries, and animal mortality [89]. In addition, elevation of plasma HDL-C levels in transgenic mice by overexpressing apoA-I protects against septic shock and death caused by LPS and severe bacterial infection [90]. Similarly, low levels of HDL-C increase the mortality in patients with sepsis/septic shock [91]. Systemic administration of rHDL blunts the deleterious effects of LPS caused by small doses of intravenous LPS injection in human volunteers, such as attenuating cytokine release, correcting procoagulant state, and downregulating CD14 expression [92, 93]. Furthermore, HDL was shown to suppress cytokine and chemokine production, downregulate co-stimulatory molecules, and inhibit antigen presentation in macrophages and monocyte-derived dendritic cells [94].

4.4 Regulation of glucose metabolism by HDL

HDL may favorably regulate glucose metabolism. HDL promotes glycogen synthesis in skeletal muscle myocytes via SR-BI and stimulates glucose uptake by adipocytes [95]. HDL and apoA-I stimulate glucose uptake by skeletal muscle myocytes via increasing adenosine monophosphate-activated protein kinase (AMPK) activity [96]. HDL also enhances insulin secretion by pancreatic beta cells, which requires ABCA1-mediated cholesterol efflux as well as SR-BI expression [97]. In patients with type 2 diabetes mellitus (T2DM), intravenously injecting rHDL increases plasma insulin levels and decreases glucose concentrations *in vivo* [98].

4.5 HDL and cancer

Epidemiological studies showed that CVD and cancer possess various similarities and possible interactions, including a number of common risk factors (e.g., smoking, obesity, and diabetes mellitus) and a shared biology [99]. Low HDL-C levels might be a prognostic factor for biliary tract cancer, prostate cancer, colon cancer, breast cancer, and gastric cancer [100, 101]. *In vitro* studies also demonstrated that native HDL could inhibit the migration and invasion of breast cancer cells [102, 103]. In addition, HDL could repress the adhesion of breast cancer cells to endothelial cells that mitigate the metastasis of breast cancer and reduce cancer growth through inhibiting tumor angiogenesis [104, 105].

5. Dysfunctional HDL in systemic inflammation

The potent atheroprotective effects of HDL originate from its unique composition and structure. If the composition or structure of HDL is altered in the setting of systemic inflammation (e.g., CVD and diabetes mellitus), it may lose its protective effects and even acquire deleterious functions, which is called dysfunctional HDL [106, 107]. Moreover, changes of HDL functions in systemic inflammation can also be the results of chemical modifications of HDL components without changing its composition. The most common modifications of HDL are oxidation and glycation of its proteins or lipids [9, 108]. Therefore, understanding the features of dysfunctional HDL might lead to a new diagnostic and therapeutic approach to CVD.

5.1 Dysfunctional HDL in CVD

In the early phase of acute myocardial infarction, the pro-inflammatory HDL particles display remarkable alterations, including increased levels of lysophosphatidylcholine (LysoPC), phosphatidic acid (PA), ceruloplasmin, and serum amyloid A (SAA); decreased amounts of apoA-I, PON1, and platelet-activating factor acetylhydrolase (PAF-AH); and reduced abilities of cholesterol efflux and anti-oxidative activity, which are implicated in the impaired functions of HDL [109, 110].

Myeloperoxidase (MPO) is released to plasma from monocytes and neutrophils in CVD, which uses hydrogen peroxide to generate hypochlorous acid (HClO) and subsequently causes oxidative modifications of lipids and proteins of lipoproteins, rendering HDL dysfunctions [111, 112]. Some studies found that MPO-dependent oxidation of HDL reduces the binding affinity of HDL to receptors and impairs its ability to stimulate cholesterol efflux from foam cells [113, 114]. Meanwhile, oxidized HDL (ox-HDL) can induce ROS production and upregulate the expression of pro-inflammatory and pro-thrombotic genes, such as TNF-alpha, matrix metalloproteinase-2/-9 (MMP-2/-9), COX-2, and plasminogen activator inhibitor-1 (PAI-1), which elevates CVD risk [115–117]. In addition, ox-HDL is dysfunctional in inducing NO production and promoting endothelial repair *in vitro* and reendothelialization of injured carotid arteries *in vivo* [118, 119]. And ox-HDL also promotes VSMC proliferation and migration by triggering intercellular ROS production [120]. Furthermore, ox-HDL has an elevated capability to induce the proliferation, migration, and invasion of breast cancer cells, thereby promoting the metastasis of breast cancer [103, 121].

5.2 Abnormal HDL in diabetes mellitus

HDL could be deficient in T2DM conditions, because of enrichment of TG, depletion of CE, and glycation of apoA-I and HDL-associated enzymes. These changes impair the structure and function of HDL, reduce receptor-mediated cholesterol efflux, and increase CVD risk [122, 123]. Glycation of HDL *in vitro* reduces its capacity to mediate cholesterol efflux from THP-1 macrophages, and incubation with glycation inhibitors (metformin and aminoguanidine) restores HDL-mediated cholesterol efflux [124, 125]. HDL from diabetic subjects reduces the abilities of anti-oxidation and anti-apoptosis as well as the capacities to mediate cholesterol efflux from THP-1 macrophages, which result from depleting HDL-associated apoA-I level and PON1 activity and elevating SAA concentration [126–128]. Both glycated HDL *in vitro* and diabetic HDL lose their protective effects on inhibition of cytokine release against LPS in macrophages [129]. Moreover, diabetic HDL is less effective to stimulate NO production, to promote proliferation and migration in endothelial cells by downregulating SR-BI, and to improve endothelium-dependent vasodilation and endothelial repairment [130, 131]. In addition, HDL from diabetic patients leads to abnormal actions on breast cancer cell adhesion to endothelial cells and extracellular matrix, thereby promoting the metastasis of breast cancer [132]. HDL from T2DM patients carries a higher level of S1P, which could be partly responsible for the abnormal functions of diabetic HDL [133]. CETP activity is elevated in diabetic patients compared to healthy subjects, resulting in changed HDL remodeling and accelerated HDL clearance [134]. ApoA-I is glycated in T2DM patients *in vivo* and by glucose or methylglyoxal *in vitro*, and such glycation may impair its anti-inflammatory effects in endothelial cells [135, 136]. Some specific lysine (K) residues of apoA-I (K12, K23, K40, K96, K106, K107, and K238) are susceptible to be glycated either *in vitro* or *in vivo*, which alter the conformation of apoA-I and consequently impair the anti-inflammatory effects of apoA-I in diabetic conditions [137].

Plasma levels of ox-HDL in T2DM patients were reported to be higher than those in healthy individuals [138]. It was found that ox-HDL is independently and positively correlated with fasting glucose levels, suggesting that high glucose levels may also contribute to HDL oxidation [139]. Glycated HDL is more susceptible to oxidation *in vitro* as shown by an increase in lipid peroxidation products and thiobarbituric acid-reactive substances (TBARS) following incubation of HDL with glucose.

6. HDL particles (HDL-P)

HDL comprises a heterogeneous group of discoid and spherical particles (7–12 nm in diameter) that differ in density, size, and electrophoretic mobility [30]. As whether plasma HDL-C levels could really reflect the whole HDL compositions and whether HDL-C is still a good predictor for CVD are questioned; quantitative measurements of HDL particles (HDL-P) might be more valuable and meaningful than HDL-C [140]. In addition, HDL is a complex carrier for many kinds of proteins, lipids, and other biochemical materials in plasma as mentioned above, which might make it to be the natural endogenous nanoparticles that deliver cargoes targeted to recipient cells.

6.1 HDL subclasses

Distinct content in proteins and lipids results in various HDL subclasses, each characterized by differences in shape, density, size, and charge. Broadly, HDL can be distinguished into two subfractions by density: HDL2 and HDL3. HDL2 is larger, less dense, and strongly associated with apoA-I, which carries the majority of cholesterol reflected in HDL-C measurements. Unlike HDL2, HDL3 carries proteins that prevent oxidative stress and receive cholesterol from RCT through ABCA1. HDL3 cholesterol is well approximated by the sum of small and medium HDL-P concentration, whereas HDL2 cholesterol correlates strongly with large HDL-P concentration. By the action of LCAT, small HDL3 is progressively transformed to CE-enriched HDL2. CETP mediates the hetero-transfer of TG and CE between

HDL2a- and TG-rich lipoproteins, resulting in the formation of HDL2b subspecies. These latter particles are then transformed back to HDL3c by hydrolysis of TG and phospholipids as a result of the combined action of phospholipid transfer protein (PLTP), hepatic lipase (HL), and CETP [141, 142].

6.2 HDL-P measurement

Experimental studies pointed out that the widely used measurements of HDL-C levels may have obvious limitations, and the quantitative evaluation of HDL-P might be a more robust biomarker for assessing HDL functions and predicting CVD risk [11, 140]. A number of epidemiological and clinical trials, including the Heart Protection Study (HPS) [143], the Justification for the Use of Statins in Prevention: An Intervention Trial Evaluating Rosuvastatin (JUPITER) [144], and the Multi-Ethnic Study of Atherosclerosis (MESA) [145], demonstrated that HDL-P is a stronger and more independent predictor of CVD risk than HDL-C. In JUPITER study, investigators evaluated the relationship of HDL-C and HDL-P in more than 10,000 subjects with CVD risk. JUPITER showed a significant inverse association between HDL-P and CVD risk. In contrast, HDL-C is not associated with CVD risk in statin-treated patients after adjustment for additional lipoprotein parameters [144]. MESA also found that HDL-P is a significant predictor of incident CVD events and carotid intima-media thickening (cIMT), even adjusting for HDL-C levels and other CVD confounders [145]. In addition, HDL-P is an independent predictor of major adverse cardiovascular events (MACE) among patients undergoing angiography [146]. Therefore, HDL-P might provide a more accurate and reliable measure of HDL than HDL-C.

6.3 HDL: nature's nanoparticles

Compared to the artificial nanocarriers (e.g., liposomes, micelles, inorganic, and polymeric nanoparticles), HDL-based drug delivery strategies have unique features that deliver drugs, peptides/proteins, nucleic acids, and imaging agents targeted to various organs more efficiently [147]. These attributes of HDL include ultrasmall size (8–12 nm), high tolerability in humans (up to 8 g of protein per infusion), long circulating half-life (12–24 h), and intrinsic targeting properties to specific recipient cells [148, 149]. A statin-loaded rHDL nanoparticle remarkably inhibits the inflammatory responses in atherosclerotic plaque [150]. Moreover, nanoparticle-labeled HDL might be used to evaluate the stability of atherosclerotic plaque through magnetic resonance imaging (MRI) after intraperitoneal application [151].

Furthermore, HDL is the natural anticancer drug delivery system for tumor imaging and treatment, which provides tumor-selective delivery of anticancer agents while reducing harmful off-target effects [152, 153]. Therefore, utilizing HDL nanoparticles would revolutionize the future strategy for the management of a broad range of cancers. Synthetic HDL nanoparticles could act synergistically and lessen the amount of mitotane/etoposide/cisplatin needed for anticancer efficacy in adrenocortical carcinoma [154]. The binding of anticancer drug valrubicin with rHDL increases the water solubility of valrubicin, which appears ideally suited for extended applications, including systemic cancer chemotherapy [155]. Artificial HDL nanoparticles using a gold nanoparticle induce B lymphoma cell apoptosis through SR-BI-mediated cholesterol starvation and selectively inhibit B-cell lymphoma growth in mice [156]. In addition, after delivering anti-angiogenic RNAi to endothelial cells, HDL strongly attenuates neovascularization *in vivo* and reduces tumor growth, which might be a potential treatment for cancer [157].

7. Apolipoprotein mimetic peptides

Synthetic peptides modeling the amphipathic helices in apoA-I and apolipoprotein E (apoE) show the similar antiatherogenic properties to native ones, including promoting cholesterol efflux, improving oxidative stress, and reducing inflammatory response.

7.1 ApoA-I mimetic peptides

Studies found that the interaction between apoA-I and ABCA1 is not sequence-specific and instead the amphipathic helices of apoA-I are identified as the key structural motifs [158]. To further understand these helices, a model of 18-amino acid peptide (18A) was developed, which is not identical in sequence to any of the individual helices of apoA-I [159]. 18A is referred to as 2F because it contains two phenylalanine (F) residues, which could solubilize phospholipids and activate LCAT. Many peptides were further designed on the basis of 2F to enhance the biological activities of the peptides [160, 161]. Among them, 4F, containing four F residues, is the most well-studied peptide, which significantly reduces atherosclerotic lesion in apoE knockout and LDL-R null mice [162, 163]. The ability of 4F to promote cholesterol efflux was also noted, although it is not as effective as lipid-free apoA-I [164]. L-4F, synthesized with natural L-amino acids, is effective but not stable when administered orally, presumably due to its susceptibility to proteolysis in the intestine [162]. This problem was circumvented by fabrication of D-4F with D-amino acids, which displays the similar biological properties to L-4F and exerts significant antiatherogenic effects upon oral administration [162]. D-4F protects endothelial cells against ox-LDL-induced injury by antagonizing the downregulation of pigment epithelium-derived factor (PEDF) [105]. We also found that D-4F alleviates ox-LDL-induced oxidative stress and promotes endothelial repair through the eNOS/HO-1 pathway [165]. Besides, D-4F accelerates vasodilatation and restrains atherosclerosis by regulating phospholipid metabolites and decreasing plasma LysoPC in LDL-R null mice [166]. Furthermore, D-4F decreases the myocardial infarction area in hyperglycemia mice through promoting NO release and decreasing ROS generation in endothelial cells [167]. Metabolomic analysis showed that D-4F alleviates ox-LDL-induced oxidative stress and abnormal glycolysis in endothelial cells [168].

In addition, 6F is also bioactive even made from L-amino acids and presented orally [169]. End-blocked 6F is more hydrophobic than 4F, more effectively activates LCAT, and is at least as effective in binding oxidized lipids [170, 171]. The 5A peptide possesses many functional attributes of native apoA-I including cholesterol efflux, inhibition of LDL oxidation, and suppression of inflammation [172]. Additionally, 5A reduces atherosclerosis and prevents the induction of asthma in mouse models [173, 174].

7.2 ApoE-mimetic peptides

ApoE is a multifunctional apolipoprotein that associates with VLDL, LDL, and subsets of HDL. It participates in the clearance of these lipoproteins from plasma, by serving as ligand for LDL-R and its family of related receptors. Like apoA-I, it is also active in RCT and has anti-inflammatory and anti-oxidative activities [175]. These properties are believed to contribute to the antiatherogenic functions of apoE. Mimetic peptides derived from apoE have been developed. AT1–5261 is an apoE-mimetic peptide containing 25 amino acids [176]. In the lipid-free state, ATI-5261 efficiently promotes ABCA1-mediated cholesterol efflux. When the peptide is complexed with phospholipids, it is still capable of promoting cholesterol efflux in a partially ABCA1-dependent fashion [176].

7.3 Dual-domain peptides

The rationale for constructing the dual-domain peptide (Ac-hE18A-NH2) is that 18A promotes the association of lipid-free apoE ligand-binding sequence with lipoproteins. The peptide Ac-hE18A-NH2 lowers plasma cholesterol levels in hyperlipidemic mice [177, 178]. *In vitro* studies, Ac-hE18A-NH2 also decreases monocyte adhesion to endothelial cells, attenuates LPS-induced inflammatory responses in HUVECs, and reduces lipid hydroperoxides in LDL [178]. Compared to 4F, Ac-hE18A-NH2 peptide was also shown to promote cholesterol efflux, improve endothelial dysfunctions, and lower plasma lipid hydroperoxides [179]. 4F binds oxidized lipid with high affinity, and Ac-hE18A-NH2 rapidly reduces plasma cholesterol levels, including lowering VLDL and LDL levels [177].

8. Other therapeutics targeted to HDL

Besides the traditional drugs (e.g., statins, niacin, and PPARs agonists), there are some emerging molecules targeted to regulating HDL metabolism [180].

8.1 RVX-208 (apabetalone)

RVX-208 is a selective antagonist of the bromodomain of bromodomain and extra-terminal (BET), which induces apoA-I mRNA and protein expression through an epigenetic mechanism in hepatocytes *in vitro*, leading to elevated levels of plasma apoA-I and HDL-C *in vivo* [181–183]. RVX-208 selectively binds to BET bromodomains, competing for a site bound by the endogenous ligand (acetylated lysine) [184]. RVX-208 also increases HDL-C levels, decreases LDL-C contents, and reduces atherosclerotic plaque formation in hyperlipidemic apoE knockout mice [185]. Thus, RVX-208 might be a promising new approach for CVD treatment.

Microarray analysis found that RVX-208 upregulates many antiatherogenic gene expression and downregulates lots of pro-atherogenic gene expression *in vivo* [186]. RVX-208 reduces the vascular inflammation *in vitro* and in CVD patients by a BET-dependent epigenetic mechanism [187]. RVX-208 remarkably represses the expression of pro-inflammatory cytokines (VCAM-1, MCP-1, and IL-6) *in vitro* and *in vivo* [185]. RVX-208 also increases 10 lipid classes in plasma HDL fractions, delays oral glucose absorption and endogenous glucose production, and reduces peripheral glucose disposal, which may protect against T2DM development [188]. RVX-208 reduces the expression of complement factors either *in vitro* or in mice and in CVD patients [189]. RVX-208 counters the trans-differentiation and calcification of VSMCs [190]. RVX-208 lowers serum alkaline phosphatase levels and improves CVD risk [191]. RVX-208 favorably modulates the vulnerability of carotid artery plaque through ultrasonic measurement, which is related to an increase of HDL-P levels [192]. These results demonstrated that the antiatherogenic functions of RVX-208 occur via a combination of lipid profile changes, anti-inflammatory activities, as well as many other protective properties.

Recently, phase II trials showed that RVX-208 reduces MACE in treated patients, over and above that of apoA-I/HDL increasing action. This MACE reducing actions of RVX-208 is largely due to its novel anti-inflammatory actions [193, 194]. Currently, a phase III trial, BETonMACE, is ongoing to look for the effects of RVX-208 in CVD patients. Therefore, RVX-208 might act in multiple ways to inhibit atherosclerosis and would be an emerging option for CVD management. However, we still need long-term phase III trial data to verify these effects on real-world CVD patients.

8.2 Liver X receptor (LXR) agonists

LXR agonists, as the key regulators of ABCA1/ABCG1 expression in macrophages, have been shown to promote cholesterol efflux in macrophages *in vitro*, raise HDL-C levels, and decrease atherosclerosis in LDL-R knockout mice [195, 196]. Studies have highlighted the primary antiatherogenic activity of LXR agonists on macrophages [197]. Unfortunately, the first generation of LXR compounds has been hampered by their capacity to promote the expression of lipogenic genes in the liver, which elevate TG levels and increase hepatic steatosis [198]. LXR activator, T091317, induces gene expression of Niemann-Pick C1/2 (NPC1/2) in macrophages, increases cholesterol content in the outer layer of macrophage membranes, and decreases atherosclerosis in mice [199]. A novel LXR agonist, ATI-111, also prevents atherosclotic plaque formation in mice [200]. LXR agonist (LXR-623) is associated with increased expression of ABCA1 and ABCG1, but adverse central nervous system-related effects are noted in more than half of patients, leading to termination of the study [201, 202]. Other agonists (AZ876 and GW3965) were also shown to reduce the progression of atherosclerotic lesions [203]. Interestingly, restricting LXR activation to the intestine might also result in an increase in intestinal HDL formation via ABCA1, without developing fatty liver [204]. An intestinal-specific LXR agonist, GW6340, promotes cholesterol efflux in macrophages and increases intestinal excretion of HDL-C [205]. Thus, LXR agonists may be a highly plausible and conceptually attractive target for the treatment of dyslipidemia and atherosclerosis, particularly if it can be accomplished with selective targeting to macrophage or the intestine.

8.3 Farnesoid X receptor (FXR) agonists

FXR is a bile acid-activated nuclear receptor that regulates cholesterol homeostasis and HDL metabolism [206]. Activation of FXR is reported to lead to both pro- and anti-atherosclerotic effects, because a major metabolic change caused by FXR agonists is a reduction of plasma HDL-C in LDL-R knockout mice [206, 207]. In addition, FXR agonists promote HDL-C excretion into feces in mice and monkeys [207]. Therefore, FXR agonists have received much attention as a potential therapeutic target, and different agonists (GW4064, 6ECDCA, FXR-450, and PX20606) have been generated as a strategy for regulating HDL metabolism [207, 208]. These observations will support further studies to investigate the potential roles of FXR activation on HDL regulation.

8.4 miRNA inhibitors

HDL is a major carrier of circulating miRNAs in plasma as mentioned above. Meanwhile, miRNAs have also emerged as the important regulators on HDL metabolism. Several studies demonstrated that miRNAs control the expression of a large number of genes associated with HDL metabolism, including ABCA1, ABCG1, and SR-BI [209, 210]. These findings strongly suggested that miRNAs regulate HDL biogenesis, cholesterol efflux, and uptake in the liver, thereby controlling the whole RCT process [211, 212].

miR-33 could repress the expression of ABCA1/ABCG1 proteins; however, knockout of miR-33 upregulates ABCA1/ABCG1 expression, promotes HDL-mediated cholesterol efflux, increases plasma HDL-C levels, and prevents the progression of atherosclerosis [213–215]. Besides raising HDL-C levels, inhibition of miR-33 also lowers VLDL-TG contents in nonhuman primates [216]. Furthermore, anti-miR-33 therapy inhibits the gene expressions that enhance mitochondrial respiration and ATP production, promotes macrophage cholesterol efflux accompanying with ABCA1 upregulation, and reduces atherosclerosis [217]. In addition, miR33 inhibition overcomes the deleterious effects of atherosclerosis plaque progression in LDL-R knockout mice and diabetic mice [218, 219].

Additionally, inhibiting miR-144 could upregulate hepatic ABCA1 expression and increase HDL-C levels through the FXR-dependent pathway [220]. However, overexpression of miR-144 in the liver reduces ABCA1 expression, attenuates cholesterol efflux in macrophages, reduces HDL-C levels, and promotes atherosclerosis development [221]. An increase in miR-145 decreases ABCA1 expression and reduces plasma HDL-C levels and glucose-stimulated insulin secretion in islets. However, inhibiting miR-145 produces the opposite effects of increasing ABCA1 expression, promoting HDL biogenesis in the liver and improving glucose-stimulated insulin secretion in islets [222]. In mice, inhibition of miR-148a increases the hepatic expression of LDL-R and ABCA1, subsequently decreases plasma LDL-C concentrations, and elevates HDL-C levels, which may decrease LDL-C/HDL-C ratio and CVD risk [223]. Furthermore, miR-185, miR-96, and miR-223 may repress selective HDL-C uptake through inhibiting hepatic SR-BI, implying a novel mode of SR-BI regulation and an important role of miRNAs in modulating cholesterol metabolism [224]. Thus, these findings strongly supported the idea of developing miRNA inhibitors for the treatment of dyslipidemia and atherosclerosis [225].

9. Conclusions

As the failure of CEPT inhibitors on reducing CVD risk, the traditional concept of HDL against CVD from Framingham study has been challenged. Besides, abnormal HDL functions in the setting of systemic diseases also make HDL more confused to be understood. Consequently, whether HDL-C is still a good predictor for CVD and whether HDL could really provide valuable protections against CVD are questioned. HDL comprises a heterogeneous group of particles composed of various of bioactive components. The compositional complexity of HDL is almost hardly to be reflected by measuring cholesterol contents loading in HDL. Thus, quantifying HDL-P numbers and evaluating HDL functions might be the more meaningful markers for CVD prediction. Meanwhile, many emerging strategies targeted to regulate HDL metabolism and increase HDL-P levels were also attempted. Expectedly, more available measurement methods and therapeutic agents about HDL would arise in the near future.

Acknowledgements

This project was supported by grant 31200884 from the National Natural Science Foundation of China; by grant 2018Y9100 from the Joint Funds for the Innovation of Science and Technology, Fujian Province; and by grant 2019HSJJ04 from high-level hospital foster grants of Fujian Provincial Hospital, Fujian Province, China.

Conflict of interest

The author declares that there are no conflicts of interest.

Notes/thanks/other declarations

I thank Dr. Yansong Guo and Dr. Na Lin for the kind help on editing and polishing this manuscript.

Appendices and nomenclature

AAT	alpha-antitrypsin
ABCA1	ATP-binding cassette transporter A1
ABCG1	ATP-binding cassette transporter G1
AMPK	adenosine monophosphate-activated protein kinase
apoA-I	apolipoprotein A-I
apoB	apolipoprotein B
BET	bromodomain and extra-terminal
CE	cholesteryl esters
CETP	cholesterol ester transfer protein
cIMT	carotid intima-media thickening
COX-2	cyclooxygenase-2
CVD	cardiovascular disease
eNOS	endothelial nitric oxide synthase
EPCs	endothelial progenitor cells
FGF	fibroblast growth factor
FXR	Farnesoid X receptor
HDL	high-density lipoprotein
HDL-C	high-density lipoprotein cholesterol
HDL-P	HDL particles
HII	HDL inflammatory index
HL	hepatic lipase
HOCl	hypochlorous acid
HUVECs	human umbilical vein endothelial cells
ICAM-1	intercellular adhesion molecule-1
I/R	ischemic/reperfusion
LCAT	lecithin cholesterol acyltransferase
LDL	low-density lipoprotein
LDL-R	low-density lipoprotein receptor
LPS	lipopolysaccharide
LXR	liver X receptor
LysoPC	lysophosphatidylcholine
MACE	major adverse cardiovascular events
miRNAs	microRNAs
MMP	metalloproteinases
MPO	myeloperoxidase
NF-κB	nuclear factor kappa-B
NO	nitric oxide
ox-HDL	oxidized HDL
ox-LDL	oxidized LDL
PA	phosphatidic acid
PAF-AH	platelet-activating factor acetylhydrolase
PAI-1	plasminogen activator inhibitor-1
PEDF	pigment epithelium-derived factor
PGI2	prostaglandin I-2

PLTP	phospholipid transfer protein
PON1	paraoxonase-1
RCT	reverse cholesterol transport
rHDL	reconstituted HDL
ROS	reactive oxygen species
S1P	sphingosine-1-phosphate
SAA	serum amyloid A
SR-BI	scavenger receptor class B type I
SM	sphingomyelin
SphK	sphingosine kinases
T2DM	type 2 diabetes mellitus
TBARS	thiobarbituric acid-reactive substances
TG	triglycerides
TNF-alpha	tumor necrosis factor-alpha
VCAM-1	vascular cell adhesion molecule-1
VSMCs	vascular smooth muscle cells

Author details

Donghui Liu
Department of Cardiology, Fujian Provincial Hospital, Fujian Provincial Key Laboratory of Cardiovascular Disease, Fujian Cardiovascular Institute, Provincial Clinical Medicine College of Fujian Medical University, Fuzhou, China

*Address all correspondence to: liudonghuis@126.com

References

[1] Gordon T, Castelli WP, Hjortland MC. High density lipoprotein as a protective factor against coronary heart disease. The Framingham study. The American Journal of Medicine. 1977;**62**:707-714. DOI: 10.1016/0002-9343(77)90874-9

[2] Gotto AM, Brinton EA. Assessing low levels of high-density lipoprotein cholesterol as a risk factor in coronary heart disease: A working group report and update. Journal of the American College of Cardiology. 2004;**43**:717-724. DOI: 10.1016/j.jacc.2003.08.061

[3] Gordon DJ, Probstfield JL, Garrison RJ. High-density lipoprotein cholesterol and cardiovascular disease. Four prospective American studies. Circulation. 1989;**79**:8-15. DOI: 10.1161/01.cir.79.1.8

[4] Barter PJ, Caulfield M, Eriksson M. Effects of torcetrapib in patients at high risk for coronary events. The New England Journal of Medicine. 2007;**357**:2109-2122. DOI: 10.1056/NEJMoa0706628

[5] Ballantyne CM, Miller M, Niesor EJ. Effect of dalcetrapib plus pravastatin on lipoprotein metabolism and high-density lipoprotein composition and function in dyslipidemic patients: Results of a phase IIb dose-ranging study. American Heart Journal. 2012;**163**: 515-521. DOI: 10.1016/j.ahj.2011.11.017

[6] Joy TR. Novel HDL-based therapeutic agents. Pharmacology & Therapeutics. 2012;**135**:18-30. DOI: 10.1016/j.pharmthera.2012.03.004

[7] van der Vorst EP, Vanags LZ, Dunn LL. High-density lipoproteins suppress chemokine expression and proliferation in human vascular smooth muscle cells. FASEB Journal. 2013;**27**:1413-1425. DOI: 10.1096/fj.12-212753

[8] Rye KA, Barter PJ. Cardioprotective functions of HDLs. Journal of Lipid Research. 2014;**55**:168-179. DOI: 10.1194/jlr.R039297

[9] Rosenson RS, Brewer HB Jr, Ansell BJ. Dysfunctional HDL and atherosclerotic cardiovascular disease. Nature Reviews. Cardiology. 2016;**13**:48-60. DOI: 10.1038/nrcardio.2015.124

[10] Miller GJ, Miller NE. Plasma-high-density-lipoprotein concentration and development of ischaemic heart-disease. Lancet. 1975;**1**:16-19. DOI: 10.1016/s0140-6736(75)92376-4

[11] Khera AV, Demler OV, Adelman SJ. Cholesterol efflux capacity, high-density lipoprotein particle number, and incident cardiovascular events: An analysis from the JUPITER trial (justification for the use of statins in prevention: An intervention trial evaluating rosuvastatin). Circulation. 2017;**135**:2494-2504. DOI: 10.1161/CIRCULATIONAHA.116.025678

[12] Brooks-Wilson A, Marcil M, Clee SM. Mutations in ABC1 in Tangier disease and familial high-density lipoprotein deficiency. Nature Genetics. 1999;**22**:336-345. DOI: 10.1038/11905

[13] Tall AR, Yvan-Charvet L, Terasaka N. HDL, ABC transporters, and cholesterol efflux: Implications for the treatment of atherosclerosis. Cell Metabolism. 2008;**7**:365-375. DOI: 10.1016/j.cmet.2008.03.001

[14] Lee JY, Parks JS. ATP-binding cassette transporter AI and its role in HDL formation. Current Opinion in Lipidology. 2005;**16**:19-25. DOI: 10.1097/00041433-200502000-00005

[15] Tang C, Liu Y, Kessler PS. The macrophage cholesterol exporter ABCA1 functions as an anti-inflammatory receptor. The Journal of

Biological Chemistry. 2009;**284**:32336-32343. DOI: 10.1074/jbc.M109.047472

[16] Wang N, Lan D, Chen W. ATP-binding cassette transporters G1 and G4 mediate cellular cholesterol efflux to high-density lipoproteins. Proceedings of the National Academy of Sciences of the United States of America. 2004;**101**:9774-9779. DOI: 10.1073/pnas.0403506101

[17] Kobayashi A, Takanezawa Y, Hirata T. Efflux of sphingomyelin, cholesterol, and phosphatidylcholine by ABCG1. Journal of Lipid Research. 2006;**47**:1791-1802. DOI: 10.1194/jlr.M500546-JLR200

[18] Czarnecka H, Yokoyama S. Regulation of cellular cholesterol efflux by lecithin: Cholesterol acyltransferase reaction through nonspecific lipid exchange. Journal of Biological Chemistry. 1996;**271**:2023-2028. DOI: 10.1074/jbc.271.4.2023

[19] Guerin M, Dachet C, Goulinet S. Familial lecithin:cholesterol acyltransferase deficiency: Molecular analysis of a compound heterozygote: LCAT (Arg147→Trp) and LCAT (Tyr171→stop). Atherosclerosis. 1997;**131**:85-95. DOI: 10.1016/s0021-9150(97)06079-6

[20] Francone OL, Haghpassand M, Bennett JA. Expression of human lecithin:cholesterol acyltransferase in transgenic mice: Effects on cholesterol efflux, esterification, and transport. Journal of Lipid Research. 1997;**38**:813-822. DOI: 10.1089/jir.1997.17.229

[21] Barter PJ, Brewer HB Jr, Chapman MJ. Cholesteryl ester transfer protein: A novel target for raising HDL and inhibiting atherosclerosis. Arteriosclerosis, Thrombosis, and Vascular Biology. 2003;**23**:160-167. DOI: 10.1161/01.atv.0000054658.91146.64

[22] Inazu A, Brown ML, Hesler CB. Increased high-density lipoprotein levels caused by a common cholesteryl-ester transfer protein gene mutation. The New England Journal of Medicine. 1990;**323**:1234-1238. DOI: 10.1056/NEJM199011013231803

[23] Ishigami M, Yamashita S, Sakai N. Large and cholesteryl ester-rich high-density lipoproteins in cholesteryl ester transfer protein (CETP) deficiency cannot protect macrophages from cholesterol accumulation induced by acetylated low-density lipoproteins. Journal of Biochemistry. 1994;**116**:257-262. DOI: 10.1093/oxfordjournals.jbchem.a124516

[24] Matsuura F, Wang N, Chen W. HDL from CETP-deficient subjects shows enhanced ability to promote cholesterol efflux from macrophages in an apoE- and ABCG1-dependent pathway. The Journal of Clinical Investigation. 2006;**116**:1435-1442. DOI: 10.1172/JCI27602

[25] de Grooth GJ, Kuivenhoven JA, Stalenhoef AF. Efficacy and safety of a novel cholesteryl ester transfer protein inhibitor, JTT-705, in humans: A randomized phase II dose-response study. Circulation. 2002;**105**:2159-2165. DOI: 10.1161/01.cir.0000015857.31889.7b

[26] Brousseau ME, Schaefer EJ, Wolfe ML. Effects of an inhibitor of cholesteryl ester transfer protein on HDL cholesterol. The New England Journal of Medicine. 2004;**350**:1505-1515. DOI: 10.1056/NEJMoa031766

[27] Schwartz GG, Olsson AG, Ballantyne CM. Rationale and design of the dal-OUTCOMES trial: Efficacy and safety of dalcetrapib in patients with recent acute coronary syndrome. American Heart Journal. 2009;**158**:896-901. DOI: 10.1016/j.ahj.2009.09.017

[28] Acton S, Rigotti A, Landschulz KT. Identification of scavenger receptor SR-BI as a high density lipoprotein

receptor. Science. 1996;**271**:518-520. DOI: 10.1126/science.271.5248.518

[29] Zanoni P, Khetarpal SA, Larach DB. Rare variant in scavenger receptor BI raises HDL cholesterol and increases risk of coronary heart disease. Science. 2016;**351**:1166-1171. DOI: 10.1126/science.aad3517

[30] Annema W, von Eckardstein A. High-density lipoproteins. Multi-functional but vulnerable protections from atherosclerosis. Circulation Journal. 2013;**77**:2432-2448. DOI: DN/JST.JSTAGE/circj/CJ-13-1025

[31] Zannis VI, Chroni A, Krieger M. Role of apoA-I, ABCA1, LCAT, and SR-BI in the biogenesis of HDL. Journal of Molecular Medicine (Berlin, Germany). 2006;**84**:276-294. DOI: 10.1007/s00109-005-0030-4

[32] Gelissen IC, Harris M, Rye KA. ABCA1 and ABCG1 synergize to mediate cholesterol export to apoA-I. Arteriosclerosis, Thrombosis, and Vascular Biology. 2006;**26**:534-540. DOI: 10.1161/01.ATV.0000200082.58536.e1

[33] Williamson R, Lee D, Hagaman J. Marked reduction of high density lipoprotein cholesterol in mice genetically modified to lack apolipoprotein A-I. Proceedings of the National Academy of Sciences of the United States of America. 1992;**89**:7134-7138. DOI: 10.1073/pnas.89.15.7134

[34] Rubin EM, Krauss RM, Spangler EA. Inhibition of early atherogenesis in transgenic mice by human apolipoprotein AI. Nature. 1991;**353**:265-267. DOI: 10.1038/353265a0

[35] Plump AS, Scott CJ, Breslow JL. Human apolipoprotein A-I gene expression increases high density lipoprotein and suppresses atherosclerosis in the apolipoprotein E-deficient mouse. Proceedings of the National Academy of Sciences of the United States of America. 1994;**91**:9607-9611. DOI: 10.1073/pnas.91.20.9607

[36] Gonzalez-Pecchi V, Valdes S, Pons V. Apolipoprotein A-I enhances proliferation of human endothelial progenitor cells and promotes angiogenesis through the cell surface ATP synthase. Microvascular Research. 2015;**98**:9-15. DOI: 10.1016/j.mvr.2014.11.003

[37] Bisoendial R, Tabet F, Tak PP. Apolipoprotein A-I limits the negative effect of tumor necrosis factor on lymphangiogenesis. Arteriosclerosis, and Thrombosis, and Vascular Biology. 2015;**35**:2443-2450. DOI: 10.1161/ATVBAHA.115.305777

[38] Liu D, Ji L, Tong X. Human apolipoprotein A-I induces cyclooxygenase-2 expression and prostaglandin I-2 release in endothelial cells through ATP-binding cassette transporter A1. American Journal of Physiology Cell Physiology. 2011;**301**:C739-C748. DOI: 10.1152/ajpcell.00055.2011

[39] Wang L, Chen WZ, Wu MP. Apolipoprotein A-I inhibits chemotaxis, adhesion, activation of THP-1 cells and improves the plasma HDL inflammatory index. Cytokine. 2010;**49**:194-200. DOI: 10.1016/j.cyto.2009.08.008

[40] Umemoto T, Han CY, Mitra P. Apolipoprotein AI and high-density lipoprotein have anti-inflammatory effects on adipocytes via cholesterol transporters: ATP-binding cassette A-1, ATP-binding cassette G-1, and scavenger receptor B-1. Circulation Research. 2013;**112**:1345-1354. DOI: 10.1161/CIRCRESAHA.111.300581

[41] Rajkovic MG, Rumora L, Barisic K. The paraoxonase 1, 2 and 3 in humans. Biochemia Medica (Zagreb). 2011;**21**:122-130. DOI: 10.11613/bm.2011.020

[42] Deakin SP, Bioletto S, Bochaton-Piallat ML. HDL-associated paraoxonase-1 can redistribute to cell membranes and influence sensitivity to oxidative stress. Free Radical Biology & Medicine. 2011;**50**:102-109. DOI: 10.1016/j.freeradbiomed.2010.09.002

[43] Mackness B, Durrington P, McElduff P.Low paraoxonase activity predicts coronary events in the Caerphilly prospective study. Circulation. 2003;**107**:2775-2779. DOI: 10.1161/01.CIR.0000070954.00271.13

[44] Aviram M, Rosenblat M. Paraoxonases 1, 2, and 3, oxidative stress, and macrophage foam cell formation during atherosclerosis development. Free Radical Biology & Medicine. 2004;**37**:1304-1316. DOI: 10.1016/j.freeradbiomed.2004.06.030

[45] Berrougui H, Loued S, Khalil A. Purified human paraoxonase-1 interacts with plasma membrane lipid rafts and mediates cholesterol efflux from macrophages. Free Radical Biology & Medicine. 2012;**52**:1372-1381. DOI: 10.1016/j.freeradbiomed.2012.01.019

[46] Goswami B, Tayal D, Gupta N. Paraoxonase: A multifaceted biomolecule. Clinica Chimica Acta. 2009;**410**:1-12. DOI: 10.1016/j.cca.2009.09.025

[47] Navab M, Imes SS, Hama SY. Monocyte transmigration induced by modification of low density lipoprotein in cocultures of human aortic wall cells is due to induction of monocyte chemotactic protein 1 synthesis and is abolished by high density lipoprotein. Journal of Clinical Investigation. 1991;**88**:2039-2046. DOI: 10.1172/JCI115532

[48] Mackness B, Quarck R, Verreth W. Human paraoxonase-1 overexpression inhibits atherosclerosis in a mouse model of metabolic syndrome. Arteriosclerosis, Thrombosis, and Vascular Biology. 2006;**26**:1545-1550. DOI: 10.1161/01.ATV.0000222924.62641.aa

[49] Rozenberg O, Shiner M, Aviram M. Paraoxonase 1 (PON1) attenuates diabetes development in mice through its antioxidative properties. Free Radical Biology & Medicine. 2008;**44**:1951-1959. DOI: 10.1016/j.freeradbiomed.2008.02.012

[50] Hammad SM, Pierce JS, Soodavar F. Blood sphingolipidomics in healthy humans: Impact of sample collection methodology. Journal of Lipid Research. 2010;**51**:3074-3087. DOI: 10.1194/jlr.D008532

[51] Chatterjee S. Sphingolipids in atherosclerosis and vascular biology. Arteriosclerosis, Thrombosis, and Vascular Biology. 1998;**18**:1523-1533. DOI: 10.1161/01.atv.18.10.1523

[52] Martinez-Beamonte R, Lou-Bonafonte JM, Martinez-Gracia MV. Sphingomyelin in high-density lipoproteins: Structural role and biological function. International Journal of Molecular Sciences. 2013;**14**:7716-7741. DOI: 10.3390/ijms14047716

[53] Hait NC, Oskeritzian CA, Paugh SW. Sphingosine kinases, sphingosine 1-phosphate, apoptosis and diseases. Biochimica et Biophysica Acta. 2006;**1758**:2016-2026. DOI: 10.1016/j.bbamem.2006.08.007

[54] Argraves KM, Argraves WS. HDL serves as a S1P signaling platform mediating a multitude of cardiovascular effects. Journal of Lipid Research. 2007;**48**:2325-2333. DOI: 10.1194/jlr.R700011-JLR200

[55] Kurano M, Yatomi Y. Sphingosine 1-phosphate and atherosclerosis. Journal of Atherosclerosis and Thrombosis. 2018;**25**:16-26. DOI: 10.5551/jat.RV17010

[56] Ruiz M, Frej C, Holmer A. High-density lipoprotein-associated

apolipoprotein M limits endothelial inflammation by delivering sphingosine-1-phosphate to the sphingosine-1-phosphate receptor 1. Arteriosclerosis, Thrombosis, and Vascular Biology. 2017;**37**:118-129. DOI: 10.1161/ATVBAHA.116.308435

[57] Lucke S, Levkau B. Endothelial functions of sphingosine-1-phosphate. Cellular Physiology and Biochemistry. 2010;**26**:87-96. DOI: 10.1159/000315109

[58] Igarashi J, Michel T. S1P and eNOS regulation. Biochimica et Biophysica Acta. 2008;**1781**:489-495. DOI: 10.1016/j.bbalip.2008.06.008

[59] Kimura T, Sato K, Kuwabara A. Sphingosine 1-phosphate may be a major component of plasma lipoproteins responsible for the cytoprotective actions in human umbilical vein endothelial cells. Journal of Biological Chemistry. 2001;**276**:31780-31785. DOI: 10.1074/jbc.M104353200

[60] Kimura T, Sato K, Malchinkhuu E. High-density lipoprotein stimulates endothelial cell migration and survival through sphingosine 1-phosphate and its receptors. Arteriosclerosis, Thrombosis, and Vascular Biology. 2003;**23**:1283-1288. DOI: 10.1161/01.ATV.0000079011.67194.5A

[61] Sattler K, Graler M, Keul P. Defects of high-density lipoproteins in coronary artery disease caused by low sphingosine-1-phosphate content: Correction by sphingosine-1-phosphate-loading. The Journal of American College of Cardiology. 2015;**66**:1470-1485. DOI: 10.1016/j.jacc.2015.07.057

[62] Xing XQ, Li YL, Zhang YX. Sphingosine kinase 1/sphingosine 1-phosphate signalling pathway as a potential therapeutic target of pulmonary hypertension. International Journal of Clinical and Experimental Medicine. 2015;**8**:11930-11935

[63] Takuwa Y, Okamoto Y, Yoshioka K. Sphingosine-1-phosphate signaling and biological activities in the cardiovascular system. Biochimica et Biophysica Acta. 2008;**1781**:483-488. DOI: 10.1016/j.bbalip.2008.04.003

[64] Das S, Halushka MK. Extracellular vesicle microRNA transfer in cardiovascular disease. Cardiovascular Pathology. 2015;**24**:199-206. DOI: 10.1016/j.carpath.2015.04.007

[65] Desgagne V, Guerin R, Guay SP. Human high-density lipoprotein microtranscriptome is unique and suggests an extended role in lipid metabolism. Epigenomics. 2019;**11**: 917-934. DOI: 10.2217/epi-2018-0161

[66] Vickers KC, Palmisano BT, Shoucri BM. MicroRNAs are transported in plasma and delivered to recipient cells by high-density lipoproteins. Nature Cell Biology. 2011;**13**:423-433. DOI: 10.1038/ncb2210

[67] Michell DL, Vickers KC. Lipoprotein carriers of microRNAs. Biochimica et Biophysica Acta. 2016;**1861**:2069-2074. DOI: 10.1016/j.bbalip.2016.01.011

[68] Niculescu LS, Simionescu N, Sanda GM. MiR-486 and miR-92a identified in circulating HDL discriminate between stable and vulnerable coronary artery disease patients. PLoS One. 2015;**10**:e0140958. DOI: 10.1371/journal.pone.0140958

[69] Tabet F, Cuesta Torres LF, Ong KL. High-density lipoprotein-associated miR-223 is altered after diet-induced weight loss in overweight and obese males. PLoS One. 2016;**11**:e0151061. DOI: 10.1371/journal.pone.0151061

[70] Mangat R, Borthwick F, Haase T. Intestinal lymphatic HDL miR-223 and ApoA-I are reduced during insulin resistance and restored with niacin. FASEB Journal. 2018;**32**:1602-1612. DOI: 10.1096/fj.201600298RR

[71] Tabet F, Vickers KC, Cuesta Torres LF. HDL-transferred microRNA-223 regulates ICAM-1 expression in endothelial cells. Nature Communications. 2014;**5**:3292. DOI: 10.1038/ncomms4292

[72] Lerman A, Zeiher AM. Endothelial function: Cardiac events. Circulation. 2005;**111**:363-368. DOI: 10.1161/01.CIR.0000153339.27064.14

[73] Gimbrone MA Jr, Garcia-Cardena G. Endothelial cell dysfunction and the pathobiology of atherosclerosis. Circulation Research. 2016;**118**:620-636. DOI: 10.1161/CIRCRESAHA.115.306301

[74] Higashi Y, Noma K, Yoshizumi M. Endothelial function and oxidative stress in cardiovascular diseases. Circulation Journal. 2009;**73**:411-418. DOI: JST.JSTAGE/circj/CJ-08-1102

[75] Cockerill GW, Rye KA, Gamble JR. High-density lipoproteins inhibit cytokine-induced expression of endothelial cell adhesion molecules. Arteriosclerosis, Thrombosis, and Vascular Biology. 1995;**15**:1987-1994. DOI: 10.1161/01.atv.15.11.1987

[76] Yuhanna IS, Zhu Y, Cox BE. High-density lipoprotein binding to scavenger receptor-BI activates endothelial nitric oxide synthase. Nature Medicine. 2001;**7**:853-857. DOI: 10.1038/89986

[77] Terasaka N, Yu S, Yvan-Charvet L. ABCG1 and HDL protect against endothelial dysfunction in mice fed a high-cholesterol diet. Journal of Clinical Investigation. 2008;**118**:3701-3713. DOI: 10.1172/JCI35470

[78] Riwanto M, Landmesser U. High density lipoproteins and endothelial functions: Mechanistic insights and alterations in cardiovascular disease. Journal of Lipid Research. 2013;**54**:3227-3243. DOI: 10.1194/jlr.R037762

[79] Johnson JL. Emerging regulators of vascular smooth muscle cell function in the development and progression of atherosclerosis. Cardiovascular Research. 2014;**103**:452-460. DOI: 10.1093/cvr/cvu171

[80] Chistiakov DA, Orekhov AN, Bobryshev YV. Vascular smooth muscle cell in atherosclerosis. Acta Physiologica (Oxford, England). 2015;**214**:33-50. DOI: 10.1111/apha.12466

[81] Robbesyn F, Garcia V, Auge N. HDL counterbalance the proinflammatory effect of oxidized LDL by inhibiting intracellular reactive oxygen species rise, proteasome activation, and subsequent NF-kappaB activation in smooth muscle cells. FASEB Journal. 2003;**17**:743-745. DOI: 10.1096/fj.02-0240fje

[82] Cucina A, Scavo MP, Muzzioli L. High density lipoproteins downregulate basic fibroblast growth factor production and release in minimally oxidated-LDL treated smooth muscle cells. Atherosclerosis. 2006;**189**:303-309. DOI: 10.1016/j.atherosclerosis.2006.01.006

[83] Ortiz-Munoz G, Houard X, Martin-Ventura JL. HDL antielastase activity prevents smooth muscle cell anoikis, a potential new antiatherogenic property. FASEB Journal. 2009;**23**:3129-3139. DOI: 10.1096/fj.08-127928

[84] Murphy AJ, Woollard KJ, Hoang A. High-density lipoprotein reduces the human monocyte inflammatory response. Arteriosclerosis, Thrombosis, and Vascular Biology. 2008;**28**:2071-2077. DOI: 10.1161/ATVBAHA.108.168690

[85] Wu A, Hinds CJ, Thiemermann C. High-density lipoproteins in sepsis and septic shock: Metabolism, actions, and therapeutic applications. Shock. 2004;**21**:210-221. DOI: 10.1097/01.shk.0000111661.09279.82

[86] van Leeuwen HJ, van Beek AP, Dallinga-Thie GM. The role of high

density lipoprotein in sepsis. The Netherland Journal of Medicine. 2001;**59**:102-110. DOI: 10.1016/S0300-2977(01)00144-9

[87] Kitchens RL, Wolfbauer G, Albers JJ. Plasma lipoproteins promote the release of bacterial lipopolysaccharide from the monocyte cell surface. Journal of Biological Chemistry. 1999;**274**:34116-34122. DOI: 10.1074/jbc.274.48.34116

[88] Grunfeld C, Feingold KR. HDL and innate immunity: A tale of two apolipoproteins. Journal of Lipid Research. 2008;**49**:1605-1606. DOI: 10.1194/jlr.E800011-JLR200

[89] McDonald MC, Dhadly P, Cockerill GW. Reconstituted high-density lipoprotein attenuates organ injury and adhesion molecule expression in a rodent model of endotoxic shock. Shock. 2003;**20**:551-557. DOI: 10.1097/01.shk.0000097249.97298.a3

[90] Levine DM, Parker TS, Donnelly TM. In vivo protection against endotoxin by plasma high density lipoprotein. Proceedings of the National Academy of Sciences of the United States of America. 1993;**90**:12040-12044. DOI: 10.1073/pnas.90.24.12040

[91] Chien JY, Jerng JS, Yu CJ. Low serum level of high-density lipoprotein cholesterol is a poor prognostic factor for severe sepsis. Critical Care Medicine. 2005;**33**:1688-1693. DOI: 10.1097/01.ccm.0000171183.79525.6b

[92] Birjmohun RS, van Leuven SI, Levels JH. High-density lipoprotein attenuates inflammation and coagulation response on endotoxin challenge in humans. Arteriosclerosis, Thrombosis, and Vascular Biology. 2007;**27**:1153-1158. DOI: 10.1161/ATVBAHA.106.136325

[93] Pajkrt D, Doran JE, Koster F. Antiinflammatory effects of reconstituted high-density lipoprotein during human endotoxemia. The Journal of Experimental Medicine. 1996;**184**:1601-1608. DOI: 10.1084/jem.184.5.1601

[94] Barter PJ, Nicholls S, Rye K-A. Antiinflammatory properties of HDL. Circulation Research. 2004;**95**:764-772. DOI: 10.1161/01.res.0000146094.59640.13

[95] Zhang Q, Zhang Y, Feng H. High density lipoprotein (HDL) promotes glucose uptake in adipocytes and glycogen synthesis in muscle cells. PLoS One. 2011;**6**:e23556. DOI: 10.1371/journal.pone.0023556

[96] Han R, Lai R, Ding Q. Apolipoprotein A-I stimulates AMP-activated protein kinase and improves glucose metabolism. Diabetologia. 2007;**50**:1960-1968. DOI: 10.1007/s00125-007-0752-7

[97] Fryirs MA, Barter PJ, Appavoo M. Effects of high-density lipoproteins on pancreatic beta-cell insulin secretion. Arteriosclerosis, Thrombosis, and Vascular Biology. 2010;**30**:1642-1648. DOI: 10.1161/ATVBAHA.110.207373

[98] Drew BG, Duffy SJ, Formosa MF. High-density lipoprotein modulates glucose metabolism in patients with type 2 diabetes mellitus. Circulation. 2009;**119**:2103-2111. DOI: 10.1161/CIRCULATIONAHA.108.843219

[99] Koene RJ, Prizment AE, Blaes A. Shared risk factors in cardiovascular disease and cancer. Circulation. 2016;**133**:1104-1114. DOI: 10.1161/CIRCULATIONAHA.115.020406

[100] Ganjali S, Ricciuti B, Pirro M. High-density lipoprotein components and functionality in cancer: State-of-the-art. Trends in Endocrinology and Metabolism. 2019;**30**:12-24. DOI: 10.1016/j.tem.2018.10.004

[101] Pirro M, Ricciuti B, Rader DJ. High density lipoprotein cholesterol and

cancer: Marker or causative? Progress in Lipid Research. 2018;**71**:54-69. DOI: 10.1016/j.plipres.2018.06.001

[102] Samadi S, Ghayour-Mobarhan M, Mohammadpour A. High-density lipoprotein functionality and breast cancer: A potential therapeutic target. Journal of Cellular Biochemistry. 2019;**120**:5756-5765. DOI: 10.1002/jcb.27862

[103] Pan B, Ren H, Ma Y. High-density lipoprotein of patients with type 2 diabetes mellitus elevates the capability of promoting migration and invasion of breast cancer cells. International Journal of Cancer. 2012;**131**:70-82. DOI: 10.1002/ijc.26341

[104] Huang X, He D, Ming J. High-density lipoprotein of patients with breast cancer complicated with type 2 diabetes mellitus promotes cancer cells adhesion to vascular endothelium via ICAM-1 and VCAM-1 upregulation. Breast Cancer Research and Treatment. 2016;**155**:441-455. DOI: 10.1007/s10549-016-3696-0

[105] Ding Y, Wang Y, Zhou J. Direct cytosolic siRNA delivery by reconstituted high density lipoprotein for target-specific therapy of tumor angiogenesis. Biomaterials. 2014;**35**:7214-7227. DOI: 10.1016/j.biomaterials.2014.05.009

[106] Ansell BJ, Fonarow GC, Fogelman AM.High-density lipoprotein: Is it always atheroprotective? Current aAherosclerosis Reports. 2006;**8**:405-411. DOI: 10.1007/s11883-006-0038-4

[107] Ansell BJ, Fonarow GC, Fogelman AM. The paradox of dysfunctional high-density lipoprotein. Current Opinion in Lipidology. 2007;**18**:427-434. DOI: 10.1097/MOL.0b013e3282364a17

[108] Smith JD. Dysfunctional HDL as a diagnostic and therapeutic target. Arteriosclerosis, Thrombosis, and Vascular Biology. 2010;**30**:151-155. DOI: 10.1161/ATVBAHA.108.179226

[109] Huang Y, DiDonato JA, Levison BS. An abundant dysfunctional apolipoprotein A_1 in human atheroma. Nature Medicine. 2014;**20**:193-203. DOI: 10.1038/nm.3459

[110] Rached F, Lhomme M, Camont L. Defective functionality of small, dense HDL3 subpopulations in ST segment elevation myocardial infarction: Relevance of enrichment in lysophosphatidylcholine, phosphatidic acid and serum amyloid A. Biochimica et Biophysica Acta. 2015;**1851**:1254-1261. DOI: 10.1016/j.bbalip.2015.05.007

[111] Zheng L, Nukuna B, Brennan ML. Apolipoprotein A-I is a selective target for myeloperoxidase-catalyzed oxidation and functional impairment in subjects with cardiovascular disease. Journal of Clinical Investigation. 2004;**114**:529-541. DOI: 10.1172/JCI21109

[112] Shao B, Oda MN, Oram JF. Myeloperoxidase: An oxidative pathway for generating dysfunctional high-density lipoprotein. Chemical Research in Toxicology. 2010;**23**:447-454. DOI: 10.1021/tx9003775

[113] Nagano Y, Arai H, Kita T. High density lipoprotein loses its effect to stimulate efflux of cholesterol from foam cells after oxidative modification. Proceedings of the National Academy of Sciences of the United States of America. 1991;**88**:6457-6461. DOI: 10.1073/pnas.88.15.6457

[114] Rifici VA, Khachadurian AK. Oxidation of high density lipoproteins: Characterization and effects on cholesterol efflux from J774 macrophages. Biochimica et Biophysica Acta. 1996;**1299**:87-94. DOI: 10.1016/0005-2760(95)00198-0

[115] Soumyarani VS, Jayakumari N. Oxidatively modified high density lipoprotein promotes inflammatory response in human monocytes-macrophages by enhanced production of ROS, TNF-alpha, MMP-9, and MMP-2. Molecular and Cellular Biochemistry. 2012;**366**:277-285. DOI: 10.1007/s11010-012-1306-y

[116] Callegari E, Norata GD, Inoue H. Oxidized-HDL3 modulates the expression of Cox-2 in human endothelial cells. International Journal of Molecular Medicine. 2006;**18**:209-213. DOI: 10.3892/ijmm.18.1.209

[117] Norata GD, Banfi C, Pirillo A. Oxidised-HDL3 induces the expression of PAI-1 in human endothelial cells. Role of p38MAPK activation and mRNA stabilization. Brtish Journal of Haematology. 2004;**127**:97-104. DOI: 10.1111/j.1365-2141.2004.05163.x

[118] Pan B, Yu B, Ren H. High-density lipoprotein nitration and chlorination catalyzed by myeloperoxidase impair its effect of promoting endothelial repair. Free Radical Biology & Medicine. 2013;**60**:272-281. DOI: 10.1016/j.freeradbiomed.2013.1002.1004

[119] Matsunaga T, Nakajima T, Sonoda M. Modulation of reactive oxygen species in endothelial cells by peroxynitrite-treated lipoproteins. Journal of Biochemistry. 2001;**130**: 285-293. DOI: 10.1093/oxfordjournals.jbchem.a002984

[120] Wang Y, Ji L, Jiang R. Oxidized high-density lipoprotein induces the proliferation and migration of vascular smooth muscle cells by promoting the production of ROS. Journal of Atherosclerosis and Thrombosis. 2014;**21**:204-216. DOI: DN/JST.JSTAGE/jat/19448

[121] Pan B, Ren H, Lv X. Hypochlorite-induced oxidative stress elevates the capability of HDL in promoting breast cancer metastasis. Journal of Translational Medicine. 2012;**10**:65. DOI: 10.1186/1479-5876-10-65

[122] Kontush A, Chapman MJ. Why is HDL functionally deficient in type 2 diabetes? Current Diabetes Reports. 2008;**8**:51-59. DOI: 10.1007/s11892-008-0010-5

[123] Mooradian AD. Dyslipidemia in type 2 diabetes mellitus. Nature Clinical Practice. Endocrinology & Metabolism. 2009;**5**:150-159. DOI: 10.1038/ncpendmet1066

[124] Matsuki K, Tamasawa N, Yamashita M. Metformin restores impaired HDL-mediated cholesterol efflux due to glycation. Atherosclerosis. 2009;**206**:434-438. DOI: 10.1016/j.atherosclerosis.2009.03.003

[125] Hoang A, Murphy AJ, Coughlan MT. Advanced glycation of apolipoprotein A-I impairs its anti-atherogenic properties. Diabetologia. 2007;**50**:1770-1779. DOI: 10.1007/s00125-007-0718-9

[126] Nobecourt E, Jacqueminet S, Hansel B. Defective antioxidative activity of small dense HDL3 particles in type 2 diabetes: Relationship to elevated oxidative stress and hyperglycaemia. Diabetologia. 2005;**48**:529-538. DOI: 10.1007/s00125-004-1655-5

[127] Murakami H, Tanabe J, Tamasawa N. Reduction of paraoxonase-1 activity may contribute the qualitative impairment of HDL particles in patients with type 2 diabetes. Diabetes Research and Clinical Practice. 2013;**99**:30-38. DOI: 10.1016/j.diabres.2012.10.022

[128] Tsun JG, Shiu SW, Wong Y. Impact of serum amyloid A on cellular cholesterol efflux to serum in type 2 diabetes mellitus. Atherosclerosis. 2013;**231**:405-410. DOI: 10.1016/j.atherosclerosis.2013.10.008

[129] Wang J, Liu K, Shen L. Small interfering RNA to c-myc inhibits vein graft restenosis in a rat vein graft model. The Journal of Surgical Research. 2011;**169**:e85-e91. DOI: 10.1016/j.jss.2011.03.060

[130] Sorrentino SA, Besler C, Rohrer L. Endothelial-vasoprotective effects of high-density lipoprotein are impaired in patients with type 2 diabetes mellitus but are improved after extended-release niacin therapy. Circulation. 2010;**121**:110-122. DOI: 10.1161/CIRCULATIONAHA.108.836346

[131] Pan B, Ma Y, Ren H. Diabetic HDL is dysfunctional in stimulating endothelial cell migration and proliferation due to down regulation of SR-BI expression. PLoS One. 2012;7:e48530. DOI: 10.1371/journal.pone.0048530

[132] Pan B, Ren H, He Y. HDL of patients with type 2 diabetes mellitus elevates the capability of promoting breast cancer metastasis. Clinical Cancer Research. 2012;**18**:1246-1256. DOI: 10.1158/1078-0432.CCR-11-0817

[133] Tong X, Peng H, Liu D. High-density lipoprotein of patients with type 2 diabetes mellitus upregulates cyclooxgenase-2 expression and prostacyclin I-2 release in endothelial cells: Relationship with HDL-associated sphingosine-1-phosphate. Cardiovascular Diabetology. 2013;**12**:27. DOI: 10.1186/1475-2840-12-27

[134] Srivastava RAK. Dysfunctional HDL in diabetes mellitus and its role in the pathogenesis of cardiovascular disease. Molecular and Cellular Biochemistry. 2018;**440**:167-187. DOI: 10.1007/s11010-017-3165-z

[135] Nobecourt E, Davies MJ, Brown BE.The impact of glycation on apolipoprotein A-I structure and its ability to activate lecithin:cholesterol acyltransferase. Diabetologia. 2007;**50**:643-653. DOI: 10.1007/s00125-006-0574-z

[136] Nobecourt E, Tabet F, Lambert G. Nonenzymatic glycation impairs the antiinflammatory properties of apolipoprotein A-I. Arteriosclerosis, Thrombosis, and Vascular Biology. 2010;**30**:766-772. DOI: 10.1161/ATVBAHA.109.201715

[137] Liu D, Ji L, Zhao M. Lysine glycation of apolipoprotein A-I impairs its anti-inflammatory function in type 2 diabetes mellitus. Journal of Molecular and Cellular Cardiology. 2018;**122**:47-57. DOI: 10.1016/j.yjmcc.2018.08.001

[138] Ueda M, Hayase Y, Mashiba S. Establishment and evaluation of 2 monoclonal antibodies against oxidized apolipoprotein A-I (apoA-I) and its application to determine blood oxidized apoA-I levels. Clinical Chimica Acta. 2007;**378**:105-111. DOI: 10.1016/j.cca.2006.11.002

[139] Kotani K, Sakane N, Ueda M. Oxidized high-density lipoprotein is associated with increased plasma glucose in non-diabetic dyslipidemic subjects. Clinical Chimica Acta. 2012;**414**:125-129. DOI: 10.1016/j.cca.2012.08.021

[140] Perez-Mendez O, Pacheco HG, Martinez-Sanchez C. HDL-cholesterol in coronary artery disease risk: Function or structure? Clinical Chimica Acta. 2014;**429**:111-122. DOI: 10.1016/j.cca.2013.12.001

[141] Kontush A. HDL particle number and size as predictors of cardiovascular disease. Frontiers in Pharmacology. 2015;**6**:218. DOI: 10.3389/fphar.2015.00218

[142] Superko HR, Pendyala L, Williams PT. High-density lipoprotein subclasses and their relationship to cardiovascular disease. The Journal of Clinical Lipidology. 2012;**6**:496-523. DOI: 10.1016/j.jacl.2012.03.001

[143] Parish S, Offer A, Clarke R. Lipids and lipoproteins and risk of different vascular events in the MRC/BHF heart protection study. Circulation. 2012;**125**:2469-2478. DOI: 10.1161/CIRCULATIONAHA.111.073684

[144] Mora S, Glynn RJ, Ridker PM. High-density lipoprotein cholesterol, size, particle number, and residual vascular risk after potent statin therapy. Circulation. 2013;**128**:1189-1197. DOI: 10.1161/CIRCULATIONAHA.113.002671

[145] Mackey RH, Greenland P, Goff DC. High-density lipoprotein cholesterol and particle concentrations, carotid atherosclerosis, and coronary events: MESA (multi-ethnic study of atherosclerosis). The Journal of American College of Cardiology. 2012;**60**:508-516. DOI: 10.1016/j.jacc.2012.03.060

[146] May HT, Anderson JL, Winegar DA. Utility of high density lipoprotein particle concentration in predicting future major adverse cardiovascular events among patients undergoing angiography. Clinical Biochemistry. 2016;**49**:1122-1126. DOI: 10.1016/j.clinbiochem.2016.09.004

[147] Mutharasan RK, Foit L, Thaxton CS. High-density lipoproteins for therapeutic delivery systems. Journal of Materials Chemistry B. 2016;**4**:188-197. DOI: 10.1039/C5TB01332A

[148] Kuai R, Li D, Chen YE. High-density lipoproteins: Nature's multifunctional nanoparticles. ACS Nano. 2016;**10**:3015-3041. DOI: 10.1021/acsnano.5b07522

[149] Varshosaz J, Vakilzadeh H, Ghassami E. Use of lipoprotein like nanoparticles used in drug and gene delivery. Current Pharmaceutical Design. 2016;**22**:3466-3485. DOI: CPD-EPUB-73561

[150] Duivenvoorden R, Tang J, Cormode DP. A statin-loaded reconstituted high-density lipoprotein nanoparticle inhibits atherosclerotic plaque inflammation. Nature Communications. 2014;**5**:3065. DOI: 10.1038/ncomms4065

[151] Jung C, Kaul MG, Bruns OT. Intraperitoneal injection improves the uptake of nanoparticle-labeled high-density lipoprotein to atherosclerotic plaques compared with intravenous injection: A multimodal imaging study in ApoE knockout mice. Circulation. Cardiovascular Imaging. 2014;**7**:303-311. DOI: 10.1161/CIRCIMAGING.113.000607

[152] McMahon KM, Foit L, Angeloni NL. Synthetic high-density lipoprotein-like nanoparticles as cancer therapy. Cancer Treatment and Research. 2015;**166**:129-150. DOI: 10.1007/978-3-319-16555-4_6

[153] Raut S, Mooberry L, Sabnis N. Reconstituted HDL: Drug delivery platform for overcoming biological barriers to cancer therapy. Frontiers in Pharmacology. 2018;**9**:1154. DOI: 10.3389/fphar.2018.01154

[154] Subramanian C, Kuai R, Zhu Q. Synthetic high-density lipoprotein nanoparticles: A novel therapeutic strategy for adrenocortical carcinomas. Surgery. 2016;**159**:284-295. DOI: 10.1016/j.surg.2015.08.023

[155] Shah S, Chib R, Raut S. Photophysical characterization of anticancer drug valrubicin in rHDL nanoparticles and its use as an imaging agent. Journal of Photochemistry and Photobiology. B Biology. 2016;**155**:60-65. DOI: 10.1016/j.jphotobiol.2015.12.007

[156] Yang S, Damiano MG, Zhang H. Biomimetic, synthetic HDL nanostructures for lymphoma. Proceedings of the National Academy of Sciences of the United States of America. 2013;**110**:2511-2516. DOI: 10.1073/pnas.1213657110

[157] Tripathy S, Vinokour E, McMahon KM. High density lipoprotein nanoparticles deliver RNAi to endothelial cells to inhibit angiogenesis. Particle & Particle Systems Characterization. 2014;**31**:1141-1150. DOI: 10.1002/ppsc.201400036

[158] Remaley AT, Thomas F, Stonik JA. Synthetic amphipathic helical peptides promote lipid efflux from cells by an ABCA1-dependent and an ABCA1-independent pathway. Journal of Lipid Research. 2003;**44**:828-836. DOI: 10.1194/jlr.M200475-JLR200

[159] Anantharamaiah GM, Jones JL, Brouillette CG. Studies of synthetic peptide analogs of the amphipathic helix. Structure of complexes with dimyristoyl phosphatidylcholine. The Journal of Biological Chemistry. 1985;**260**:10248-10255

[160] Datta G, Chaddha M, Hama S. Effects of increasing hydrophobicity on the physical-chemical and biological properties of a class A amphipathic helical peptide. Journal of Lipid Research. 2001;**42**:1096-1104

[161] Navab M, Anantharamaiah GM, Reddy ST. Apolipoprotein A-I mimetic peptides. Arteriosclerosis, Thrombosis, and Vascular Biology. 2005;**25**:1325-1331. DOI: 10.1161/01.ATV.0000165694.39518.95

[162] Navab M, Anantharamaiah GM, Hama S. Oral administration of an Apo A-I mimetic peptide synthesized from D-amino acids dramatically reduces atherosclerosis in mice independent of plasma cholesterol. Circulation. 2002;**105**:290-292. DOI: 10.1161/hc0302.103711

[163] Navab M, Ruchala P, Waring AJ. A novel method for oral delivery of apolipoprotein mimetic peptides synthesized from all L-amino acids. Journal of Lipid Research. 2009;**50**:1538-1547. DOI: 10.1194/jlr.M800539-JLR200

[164] Tang C, Vaughan AM, Anantharamaiah GM. Janus kinase 2 modulates the lipid-removing but not protein-stabilizing interactions of amphipathic helices with ABCA1. Journal of Lipid Research. 2006;**47**:107-114. DOI: 10.1194/jlr.M500240-JLR200

[165] Liu D, Ding Z, Wu M. The apolipoprotein A-I mimetic peptide, D-4F, alleviates ox-LDL-induced oxidative stress and promotes endothelial repair through the eNOS/HO-1 pathway. Journal of Molecular and Cellular Cardiology. 2017;**105**:77-88. DOI: 10.1016/j.yjmcc.2017.01.017

[166] Ou Z-J, Li L, Liao X-L. Apolipoprotein AI mimetic peptide inhibits atherosclerosis by altering plasma metabolites in hypercholesterolemia. American Journal of Physiology, Endocrinology and Metabolism. 2012;**303**:E683-E694. DOI: 10.1152/ajpendo.00136.2012

[167] Baotic I, Ge Z-D, Sedlic F. Apolipoprotein A-1 mimetic D-4F enhances isoflurane-induced eNOS signaling and cardioprotection during acute hyperglycemia. American Journal of Physiology, Heart and Circulatory Physiology. 2013;**305**:H219-H227. DOI: 10.1152/ajpheart.00850.2012

[168] Xu W, Qian M, Huang C. Comparison of mechanisms of endothelial cell protections between high-density lipoprotein and apolipoprotein A-I mimetic peptide. Frontiers in Pharmacology. 2019;**10**:817. DOI: 10.3389/fphar.2019.00817

[169] Chattopadhyay A, Navab M, Hough G. A novel approach to oral apoA-I mimetic therapy. Journal of Lipid Research. 2013;**54**:995-1010. DOI: 10.1194/jlr.M033555

[170] Anantharamaiah GM, Mishra VK, Garber DW. Structural requirements for antioxidative and anti-inflammatory properties of apolipoprotein A-I

mimetic peptides. Journal of Lipid Research. 2007;**48**:1915-1923. DOI: 10.1194/jlr.R700010-JLR200

[171] Van Lenten BJ, Wagner AC, Jung CL. Anti-inflammatory apoA-I-mimetic peptides bind oxidized lipids with much higher affinity than human apoA-I. Journal of Lipid Research. 2008;**49**:2302-2311. DOI: 10.1194/jlr.M800075-JLR200

[172] D'Souza W, Stonik JA, Murphy A. Structure/function relationships of apolipoprotein A-I mimetic peptides: Implications for antiatherogenic activities of high-density lipoprotein. Circulation Research. 2010;**107**:217-227. DOI: 10.1161/CIRCRESAHA.110.216507

[173] Amar MJ, D'Souza W, Turner S. 5A apolipoprotein mimetic peptide promotes cholesterol efflux and reduces atherosclerosis in mice. The Journal of Pharmacology and Experimental Therapeutics. 2010;**334**:634-641. DOI: 10.1124/jpet.110.167890

[174] Yao X, Dai C, Fredriksson K. 5A, an apolipoprotein A-I mimetic peptide, attenuates the induction of house dust mite-induced asthma. Journal of Immunology (Baltimore, Md.:1950). 2011;**186**:576-583. DOI: 10.4049/jimmunol.1001534

[175] Getz GS, Reardon CA. Apoprotein E as a lipid transport and signaling protein in the blood, liver, and artery wall. Journal of Lipid Research. 2009;**50**(Suppl):S156-S161. DOI: 10.1194/jlr.R800058-JLR200

[176] Bielicki JK, Zhang H, Cortez Y. A new HDL mimetic peptide that stimulates cellular cholesterol efflux with high efficiency greatly reduces atherosclerosis in mice. Journal of Lipid Research. 2010;**51**:1496-1503. DOI: 10.1194/jlr.M003665

[177] Garber DW, Handattu S, Aslan I. Effect of an arginine-rich amphipathic helical peptide on plasma cholesterol in dyslipidemic mice. Atherosclerosis. 2003;**168**:229-237. DOI: 10.1016/s0021-9150(03)00101-1

[178] Nayyar G, Handattu SP, Monroe CE. Two adjacent domains (141-150 and 151-160) of apoE covalently linked to a class A amphipathic helical peptide exhibit opposite atherogenic effects. Atherosclerosis. 2010;**213**:449-457. DOI: 10.1016/j.atherosclerosis.2010.09.030

[179] Nayyar G, Garber DW, Palgunachari MN. Apolipoprotein E mimetic is more effective than apolipoprotein A-I mimetic in reducing lesion formation in older female apo E null mice. Atherosclerosis. 2012;**224**:326-331. DOI: 10.1016/j.atherosclerosis.2012.05.040

[180] Barylski M, Toth PP, Nikolic D. Emerging therapies for raising high-density lipoprotein cholesterol (HDL-C) and augmenting HDL particle functionality. Best Practice & Research Clinical Endocrinology & Metabolism. 2014;**28**:453-461. DOI: 10.1016/j.beem.2013.11.001

[181] Bailey D, Jahagirdar R, Gordon A. RVX-208: A small molecule that increases apolipoprotein A-I and high-density lipoprotein cholesterol in vitro and in vivo. Journal of the American College of Cardiology. 2010;**55**:2580-2589. DOI: 10.1016/j.jacc.2010.02.035

[182] Nicholls SJ, Gordon A, Johansson J. Efficacy and safety of a novel oral inducer of apolipoprotein a-I synthesis in statin-treated patients with stable coronary artery disease a randomized controlled trial. Journal of the American College of Cardiology. 2011;**57**:1111-1119. DOI: 10.1016/j.jacc.2010.11.015

[183] Picaud S, Wells C, Felletar I. RVX-208, an inhibitor of BET transcriptional regulators with selectivity for the second bromodomain. Proceedings

of the National Academy of Sciences of the United States of America. 2013;**110**:19754-19759. DOI: 10.1073/pnas.1310658110

[184] McLure KG, Gesner EM, Tsujikawa L. RVX-208, an inducer of ApoA-I in humans, is a BET bromodomain antagonist. PLoS One. 2013;**8**:e83190. DOI: 10.1371/journal.pone.0083190

[185] Jahagirdar R, Zhang H, Azhar S. A novel BET bromodomain inhibitor, RVX-208, shows reduction of atherosclerosis in hyperlipidemic ApoE deficient mice. Atherosclerosis. 2014;**236**:91-100. DOI: 10.1016/j.atherosclerosis.2014.06.008

[186] Gilham D, Wasiak S, Tsujikawa LM. RVX-208, a BET-inhibitor for treating atherosclerotic cardiovascular disease, raises ApoA-I/HDL and represses pathways that contribute to cardiovascular disease. Atherosclerosis. 2016;**247**:48-57. DOI: 10.1016/j.atherosclerosis.2016.01.036

[187] Tsujikawa LM, Fu L, Das S. Apabetalone (RVX-208) reduces vascular inflammation in vitro and in CVD patients by a BET-dependent epigenetic mechanism. Clinical Epigenetics. 2019;**11**:102. DOI: 10.1186/s13148-019-0696-z

[188] Siebel AL, Trinh SK, Formosa MF. Effects of the BET-inhibitor, RVX-208 on the HDL lipidome and glucose metabolism in individuals with prediabetes: A randomized controlled trial. Metabolism. 2016;**65**:904-914. DOI: 10.1016/j.metabol.2016.03.002

[189] Wasiak S, Gilham D, Tsujikawa LM. Downregulation of the complement cascade in vitro, in mice and in patients with cardiovascular disease by the BET protein inhibitor apabetalone (RVX-208). Journal of Ccardiovascular Translational Research. 2017;**10**:337-347. DOI: 10.1007/s12265-017-9755-z

[190] Gilham D, Tsujikawa LM, Sarsons CD. Apabetalone downregulates factors and pathways associated with vascular calcification. Atherosclerosis. 2019;**280**:75-84. DOI: 10.1016/j.atherosclerosis.2018.11.002

[191] Haarhaus M, Ray KK, Nicholls SJ. Apabetalone lowers serum alkaline phosphatase and improves cardiovascular risk in patients with cardiovascular disease. Atherosclerosis. 2019;**290**:59-65. DOI: 10.1016/j.atherosclerosis.2019.09.002

[192] Shishikura D, Kataoka Y, Honda S. The effect of bromodomain and extra-terminal inhibitor apabetalone on attenuated coronary atherosclerotic plaque: Insights from the ASSURE trial. American Journal of Cardiovascular Drugs: Drugs, Devices, and Other Interventions. 2019;**19**:49-57. DOI: 10.1007/s40256-018-0298-8

[193] Ghosh GC, Bhadra R, Ghosh RK. RVX 208: A novel BET protein inhibitor, role as an inducer of apo A-I/HDL and beyond. Cardiovascular Therapeutics. 2017;**35**:e12265. DOI: 10.1111/1755-5922.12265

[194] Nicholls SJ, Ray KK, Johansson JO. Selective BET protein inhibition with apabetalone and cardiovascular events: A pooled analysis of trials in patients with coronary artery disease. American Journal of Cardiovascular Drugs: Drugs, Devices, and Other Interventions. 2018;**18**:109-115. DOI: 10.1007/s40256-017-0250-3

[195] Naik SU, Wang X, Da Silva JS. Pharmacological activation of liver X receptors promotes reverse cholesterol transport in vivo. Circulation. 2006;**113**:90-97. DOI: 10.1161/CIRCULATIONAHA.105.560177

[196] Terasaka N, Hiroshima A, Koieyama T. T-0901317, a synthetic liver X receptor ligand, inhibits development of atherosclerosis in

LDL receptor-deficient mice. FEBS Letters. 2003;**536**:6-11. DOI: 10.1016/s0014-5793(02)03578-0

[197] Levin N, Bischoff ED, Daige CL. Macrophage liver X receptor is required for antiatherogenic activity of LXR agonists. Arteriosclerosis, Thrombosis, and Vascular Biology. 2005;**25**:135-142. DOI: 10.1161/01.ATV.0000150044.84012.68

[198] Oosterveer MH, Grefhorst A, Groen AK. The liver X receptor: Control of cellular lipid homeostasis and beyond implications for drug design. Progress in Lipid Research. 2010;**49**:343-352. DOI: 10.1016/j.plipres.2010.03.002

[199] Rigamonti E, Helin L, Lestavel S. Liver X receptor activation controls intracellular cholesterol trafficking and esterification in human macrophages. Circulation Research. 2005;**97**:682-689. DOI: 10.1161/01.RES.0000184678.43488.9f

[200] Peng D, Hiipakka RA, Xie JT. A novel potent synthetic steroidal liver X receptor agonist lowers plasma cholesterol and triglycerides and reduces atherosclerosis in LDLR(−/−) mice. British Journal of Pharmacology. 2011;**162**:1792-1804. DOI: 10.1111/j.1476-5381.2011.01202.x

[201] Quinet EM, Basso MD, Halpern AR. LXR ligand lowers LDL cholesterol in primates, is lipid neutral in hamster, and reduces atherosclerosis in mouse. Journal of Lipid Research. 2009;**50**:2358-2370. DOI: 10.1194/jlr.M900037-JLR200

[202] Katz A, Udata C, Ott E. Safety, pharmacokinetics, and pharmacodynamics of single doses of LXR-623, a novel liver X-receptor agonist, in healthy participants. Journal of Clinical Pharmacology. 2009;**49**:643-649. DOI: 10.1177/0091270009335768

[203] Bradley MN, Hong C, Chen M. Ligand activation of LXR beta reverses atherosclerosis and cellular cholesterol overload in mice lacking LXR alpha and apoE. The Journal of Clinical Investigation. 2007;**117**:2337-2346. DOI: 10.1172/JCI31909

[204] Lo Sasso G, Murzilli S, Salvatore L. Intestinal specific LXR activation stimulates reverse cholesterol transport and protects from atherosclerosis. Cell Metabolism. 2010;**12**:187-193. DOI: 10.1016/j.cmet.2010.07.002

[205] Yasuda T, Grillot D, Billheimer JT. Tissue-specific liver X receptor activation promotes macrophage reverse cholesterol transport in vivo. Arteriosclerosis, Thrombosis, and Vascular Biology. 2010;**30**:781-786. DOI: 10.1161/ATVBAHA.109.195693

[206] Mencarelli A, Fiorucci S. FXR an emerging therapeutic target for the treatment of atherosclerosis. Journal of Cellular and Molecular Medicine. 2010;**14**:79-92. DOI: 10.1111/j.1582-4934.2009.00997.x

[207] Hambruch E, Miyazaki-Anzai S, Hahn U. Synthetic farnesoid X receptor agonists induce high-density lipoprotein-mediated transhepatic cholesterol efflux in mice and monkeys and prevent atherosclerosis in cholesteryl ester transfer protein transgenic low-density lipoprotein receptor (−/−) mice. The Journal of Pharmacology and Experimental Therapeutics. 2012;**343**:556-567. DOI: 10.1124/jpet.112.196519

[208] Fiorucci S, Cipriani S, Baldelli F. Bile acid-activated receptors in the treatment of dyslipidemia and related disorders. Progress in Lipid Research. 2010;**49**:171-185. DOI: 10.1016/j.plipres.2009.11.001

[209] Canfran-Duque A, Ramirez CM, Goedeke L. microRNAs and HDL life cycle. Cardiovascular Research.

2014;**103**:414-422. DOI: 10.1093/cvr/cvu140

[210] Baldan A, de Aguiar Vallim TQ. miRNAs and high-density lipoprotein metabolism. Biochimica et Biophysica Acta. 2016;**1861**:2053-2061. DOI: 10.1016/j.bbalip.2016.01.021

[211] Canfran-Duque A, Lin CS, Goedeke L. Micro-RNAs and high-density lipoprotein metabolism. Arteriosclerosis, Thrombosis, and Vascular Biology. 2016;**36**:1076-1084. DOI: 10.1161/ATVBAHA.116.307028

[212] Rayner KJ, Moore KJ. MicroRNA control of high-density lipoprotein metabolism and function. Circulation Research. 2014;**114**:183-192. DOI: 10.1161/CIRCRESAHA.114.300645

[213] Rayner KJ, Suarez Y, Davalos A. MiR-33 contributes to the regulation of cholesterol homeostasis. Science. 2010;**328**:1570-1573. DOI: 10.1126/science.1189862

[214] Horie T, Baba O, Kuwabara Y. MicroRNA-33 deficiency reduces the progression of atherosclerotic plaque in ApoE−/− mice. Journal of American Heart Association. 2012;**1**:e003376. DOI: 10.1161/JAHA.112.003376

[215] Fernandez-Hernando C, Moore KJ. MicroRNA modulation of cholesterol homeostasis. Arteriosclerosis, Thrombosis, and Vascular Biology. 2011;**31**:2378-2382. DOI: 10.1161/ATVBAHA.111.226688

[216] Rayner KJ, Esau CC, Hussain FN. Inhibition of miR-33a/b in non-human primates raises plasma HDL and lowers VLDL triglycerides. Nature. 2011;**478**:404-407. DOI: 10.1038/nature10486

[217] Karunakaran D, Thrush AB, Nguyen MA. Macrophage mitochondrial energy status regulates cholesterol efflux and is enhanced by anti-miR33 in atherosclerosis. Circulation Research. 2015;**117**:266-278. DOI: 10.1161/CIRCRESAHA.117.305624

[218] Rotllan N, Ramirez CM, Aryal B. Therapeutic silencing of microRNA-33 inhibits the progression of atherosclerosis in Ldlr−/− mice—Brief report. Arteriosclerosis, Thrombosis, and Vascular Biology. 2013;**33**:1973-1977. DOI: 10.1161/ATVBAHA.113.301732

[219] Distel E, Barrett TJ, Chung K. miR33 inhibition overcomes deleterious effects of diabetes mellitus on atherosclerosis plaque regression in mice. Circulation Research. 2014;**115**:759-769. DOI: 10.1161/CIRCRESAHA.115.304164

[220] de Aguiar Vallim TQ, Tarling EJ, Kim T. MicroRNA-144 regulates hepatic ATP binding cassette transporter A1 and plasma high-density lipoprotein after activation of the nuclear receptor farnesoid X receptor. Circulation Research. 2013;**112**:1602-1612. DOI: 10.1161/CIRCRESAHA.112.300648

[221] Ramirez CM, Rotllan N, Vlassov AV. Control of cholesterol metabolism and plasma high-density lipoprotein levels by microRNA-144. Circulation Research. 2013;**112**:1592-1601. DOI: 10.1161/CIRCRESAHA.112.300626

[222] Kang MH, Zhang LH, Wijesekara N. Regulation of ABCA1 protein expression and function in hepatic and pancreatic islet cells by miR-145. Arteriosclerosis, Thrombosis, and Vascular Biology. 2013;**33**:2724-2732. DOI: 10.1161/ATVBAHA.113.302004

[223] Goedeke L, Rotllan N, Canfran-Duque A. MicroRNA-148a regulates LDL receptor and ABCA1 expression to control circulating lipoprotein levels. Nature Medicine. 2015;**21**:1280-1289. DOI: 10.1038/nm.3949

[224] Wang L, Jia XJ, Jiang HJ. MicroRNAs 185, 96, and 223 repress selective high-density lipoprotein cholesterol uptake through posttranscriptional inhibition. Molecular and Cellular Biology. 2013;**33**:1956-1964. DOI: 10.1128/MCB.01580-12

[225] Aryal B, Singh AK, Rotllan N. MicroRNAs and lipid metabolism. Current Opinion in Lipidology. 2017;**28**:273-280. DOI: 10.1097/MOL.0000000000000420

Chapter 3

Monoglycerides as an Antifungal Agent

Febri Odel Nitbani and Jumina Jumina

Abstract

Monoglyceride is a part of a lipid group compound. As a derivative of triglycerides, monoglycerides could be produced from renewable resources like fat or vegetable oils. Structurally, monoglyceride has lipophilic and hydrophilic properties in its molecule. Lipophilic properties could be donated by an acyl group from fatty acid and hydrophilic properties from two hydroxyl residues. Therefore, it was referred to as an organic amphiphilic compound. Monoglycerides have potency as antifungal agents. Based on its chemical structure, monoglyceride allows to bind to lipid bilayer and other components on the cell membrane of fungal microorganism and damage it. In this chapter, we will describe the structure and classification, physical and chemical properties, as well as reaction path synthesis of monoglyceride from vegetable oils and mechanism of action of monoglyceride as antifungal agents.

Keywords: monoglyceride, amphiphilic, antifungal, vegetable oil

1. Introduction

Monoglyceride or monoacylglycerol is a part of lipid compounds. Lipid compounds can be derived from two sources, vegetable oils and animal fats. The main component in vegetable oil or animal fats is triglycerides or triacylglycerides that can be converted to monoglycerides. In this case, triglycerides from vegetable oils or animal fats are losing two acyl groups to produce monoglycerides via the formation of diglycerides.

Based on their structure, monoglycerides are monoester compounds from glycerol and fatty acid. The position of the acyl group can determine the types of monoglycerides, whether it is 1-monoglyceride or 2-monoglycerides. A type of fatty acid chain also defines the type of monoglycerides. For example, lauric acid (saturated fatty acid) can be derived into monolaurin, while oleic acid (unsaturated fatty acid) into monoolein.

Monoglycerides have two hydroxyl groups (OH) and anion ester from fatty acids that bound to the propane chain. These structures make monoglyceride an amphiphilic compound with a lipophilic and hydrophilic part in the molecules. Amphiphilic properties of monoglyceride are used as an emulsifier and antibacterial, antifungal, antiviral, antioxidant, and anti-arteriosclerosis agent [1, 2]. Monoglycerides are also widely used in food, cosmetics, pharmaceutical, detergents, and plasticizer industries [3].

Amphiphilic properties of monoglycerides are similar to the amphiphilic properties of polyene that have been proven to have antifungal properties.

An amphiphilic drug such as a polyene compound could interact with the cell membrane of fungi via chemical bonding that led to the damage of the cell [4]. Monoglycerides such as monolaurin were reported to be active against *Candida albicans* [5]. Monocaprin and monomiristine were also proven to have antifungal activity because of their amphiphilic properties.

In this chapter, chemical structure and classification of monoglycerides, as well as their physical and chemical properties, are described. Furthermore, the reaction pathways to produce monoglycerides from vegetable oils and the mechanism of action of monoglycerides as antifungal agents are presented.

2. Chemical structure and classification of monoglycerides

Monoglycerides are lipid compounds that have a wide range of applications in human life. As a part of lipid biomolecule, monoglycerides are primarily composed of carbon (C), hydrogen (H), and oxygen (O) atoms. Monoglycerides can be referred to as the derivatives of triglycerides or tri-acylglycerol or triester glycerol. The lysis process of triglycerides into monoglycerides can be carried out via chemical reaction in methanol or ethanol and catalyzed by base catalysts or lipase enzymes. This reaction is referred to as a transesterification reaction because it converts an ester into another ester compound.

Monoglycerides can be afforded as the intermediate from the transesterification reaction of triglycerides of vegetable oils with short-chain alcohol. This reaction produced methyl esters or ethyl esters and glycerol as the by-products. In other words, monoglycerides are intermediates in the production of biodiesel from vegetable oil through the lysis of triglyceride using methanol in the presence of base catalysts. Limiting the amount of methanol and the reaction time is expected to produce a large amount of monoglyceride, which is referred to as a partial transesterification reaction [6].

The position of the acyl group attached to the glycerol skeleton can define the type of monoglycerides. Based on the acyl group position, monoglycerides can be classified into two types, i.e., 1-monoglycerides (α-monoglycerides) and 2-monoglycerides (β-monoglycerides) as seen in **Figure 1**.

1-Monoglycerides or α-monoglycerides can be identified by the acyl groups attached to the C_1 or C_3 atom of glycerol while 2-monoglycerides (β-monoglycerides) on the C_2 atom. The type of acyl group can be differentiated by the type of the alkyl (—R) groups bound with fatty acid (**Figure 1**). If the —R groups are derived from saturated fatty acid, then it will produce saturated monoglycerides. Monolaurin, monocaprin, monomiristine, monopalmitin, and monostearin are saturated monoglycerides in the form of solid [7–10]. On the other hand, if the –R groups are from unsaturated fatty acid, it could generate unsaturated

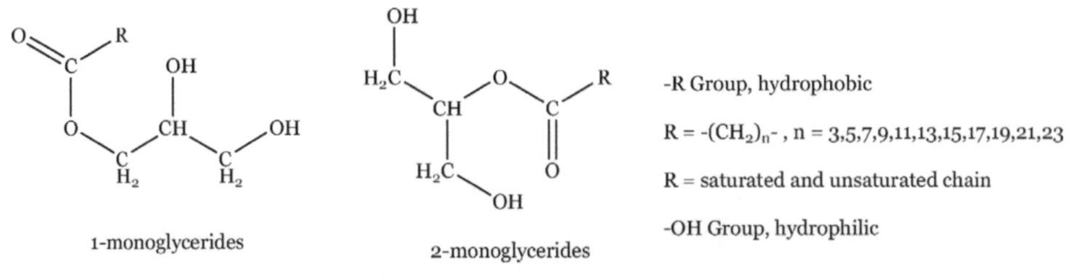

Figure 1.
The structure of 1- and 2-monoglycerides.

monoglycerides such as monoolein and monolinolein that are tend to be in the form of thick liquid [11].

Based on the length of the acyl groups on the glycerol, monoglycerides can be categorized into medium- and long-chain monoglycerides. Monoglycerides with the carbon chain length ranging from 8 to 14 atoms in the acyl groups are considered as medium-chain monoglycerides, for example, monocaprylin, monolaurin, monocaprin, and monomiristine [3]. Meanwhile, monoglycerides with a carbon chain length greater than 14 atoms are classified as long-chain monoglycerides. Monopalmitin, monostearin, and monoolein are the examples of the long-chain monoglycerides.

Previously, it was mentioned that the type of fatty acid has an essential role in determining the type of monoglyceride. Therefore, knowledge about fatty acids is important to discuss. Fatty acids can be divided into saturated and unsaturated fatty acids. Some essential saturated fatty acids are lauric acid and stearic acid, while the unsaturated fatty acids are oleic and linoleic acid. There are also two types of polyunsaturated fatty acids (PUFAs), namely, eicosapentaenoic acid (EPA) and docosahexaenoic acid (DHA) [1]. Fatty acids can also be classified based on the chain lengths, namely, medium- and long-chain fatty acids. Caprylic acid, capric acid, lauric acid, and myristic acid are some examples of medium-chain fatty acids. On the other hand, palmitic acid, stearic acid, and oleic acid are included as long-chain fatty acids. The chemical structure of some fatty acids from vegetable oils sample is presented in **Table 1**.

Vegetable oil is one of the raw material sources for producing monoglycerides because it contains triglycerides. Triglycerides from each type of vegetable oil are different, and it depends on the type and composition of fatty acids. Certain fatty acids that construct a triglyceride in vegetable oil tend to present as a major component besides other minor fatty acids. The major fatty acid in coconut oil (*Cocos nucifera L.*) is lauric acid (54%) [12], in castor oil (*Ricinus communis L.*) is ricinoleic acid (93%) [13], in olive oil is oleic acid [14], in sunflower oil is oleic and linoleic acid [15], and in palm oil is palmitic acids [16].

Vegetable oils with certain high levels of fatty acids can be used as raw materials for the synthesis of monoglycerides through various chemical reaction approaches such as partial alkaline-catalyzed transesterification, glycerolysis, and alcoholysis. An alternative reaction approach is by firstly isolating the fatty acid methyl esters or ethyl esters before carrying out the synthesis of monoglyceride. The other reaction pathways in the synthesis of monoglycerides from vegetable oils are esterification reactions of fatty acid and glycerol, transesterification of the fatty acid methyl ester with glycerol, and transesterification of fatty acid methyl ester with protected glycerol reactions followed by deprotection reactions using an acid resin. Hence, in the production of monoglycerides, vegetable oil samples are the primary source because it is rich in triglycerides and fatty acids.

3. Physicochemical properties and applications of monoglycerides

Monoglycerides consist of a lipophilic and hydrophilic part in the molecules (**Figure 1**). Acyl groups from the fatty acid contribute to the lipophilic properties, while the two hydroxyl groups are responsible for the hydrophilic properties of monoglyceride. Because of its unique structure, monoglycerides are also known as an amphiphilic compound, which are widely used as surfactants. Surfactants are an active compound with lipophilic tail and hydrophilic head in a molecule that has a function to decrease the surface tension of molecules. Hydrophilic properties of —OH groups enable monoglycerides to interact with a water molecule, while the lipophilic from acyl groups make it possible to interact with oil or lipid.

Fatty acid structure	Name
Saturated fatty acid	
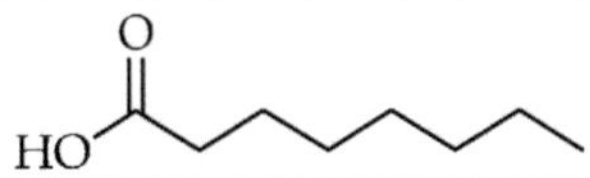	Caprylic acid
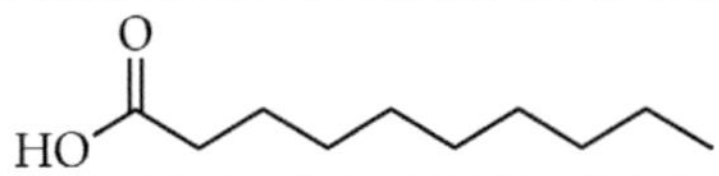	Capric acid
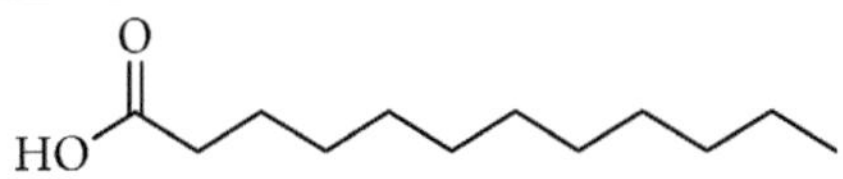	Lauric acid
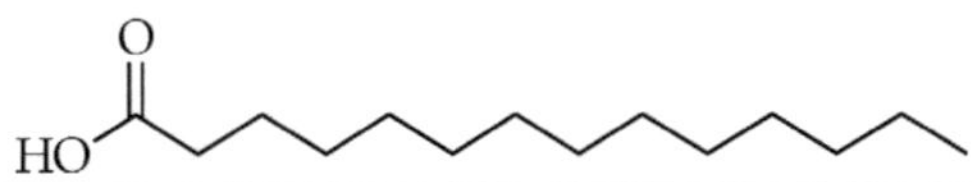	Myristic acid
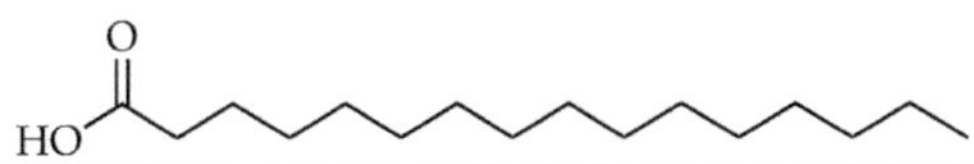	Palmitic acid
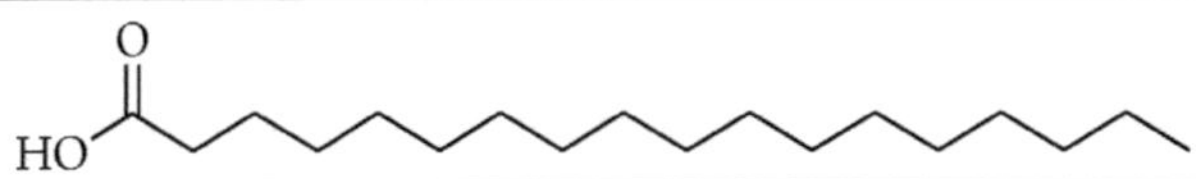	Stearic acid
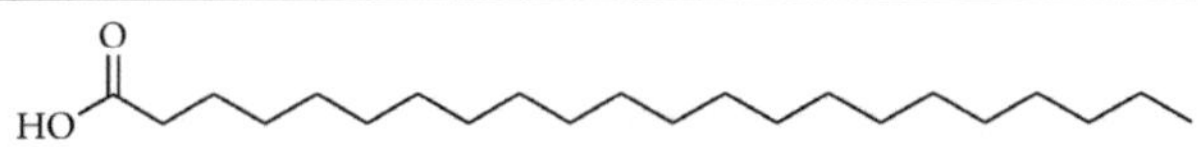	Behenic acid
Unsaturated fatty acid	
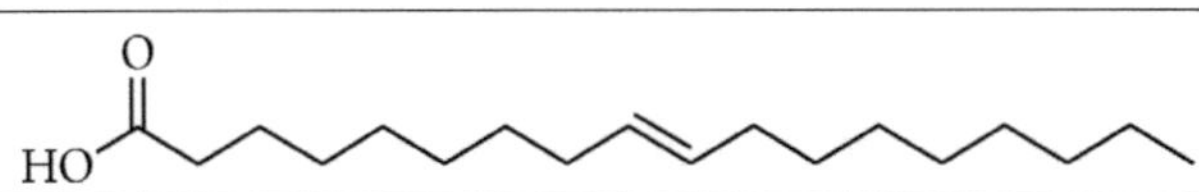	Oleic acid
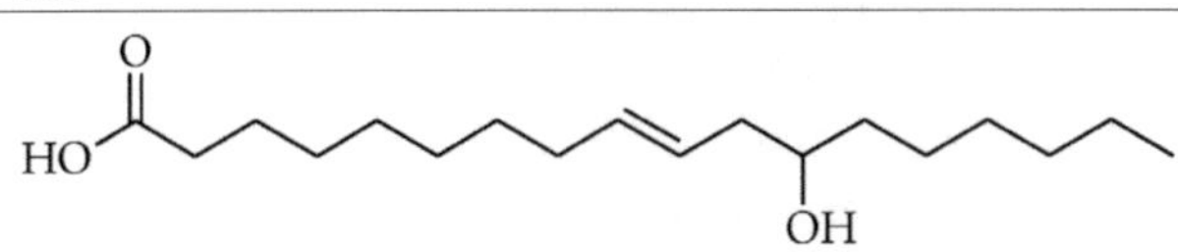	Ricinoleic acid
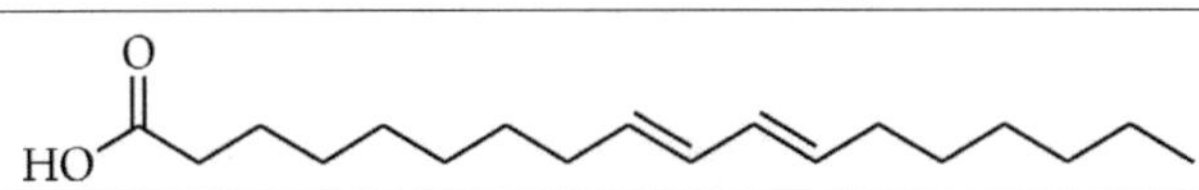	Linoleic acid
Polyunsaturated fatty acid (PUFA)	
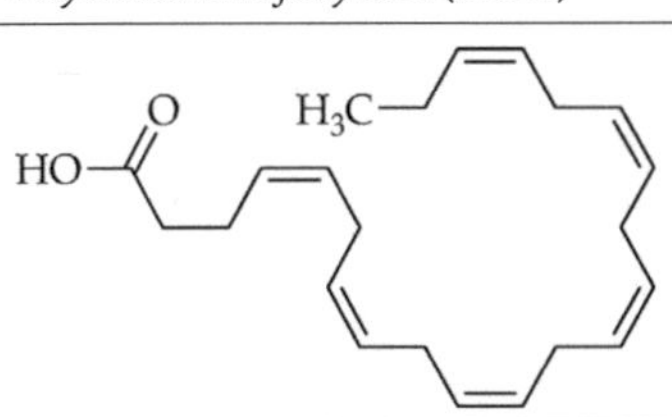	Docosahexaenoic acid
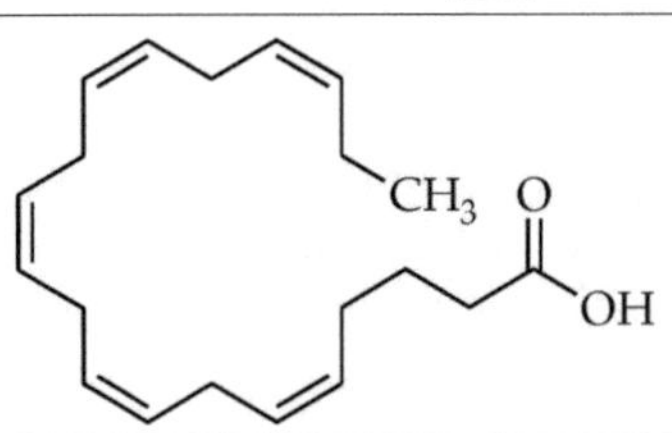	Eicosapentaenoic acid

Table 1.
Fatty acids from vegetable oils.

Monoglycerides are classified as nonionic surfactants because the acyl groups do not have any charges. Based on its properties, monoglycerides are suitable to be used as an emulsifier to mix oils and water. Monoglycerides from vegetable oils are expected to be nontoxic emulsifier. Therefore, monoglycerides have broad applications in human life. Some studies supported the claim that monoglycerides have crucial and excellent applications as a safe emulsifier. This fact makes monoglycerides to be widely used in the food industry, detergents, plasticizers, cosmetics, and pharmaceutical formulations [3]. Naik et al. [15] have reported that almost 75% of emulsifiers in the food industry are using monoglycerides, making them the main food emulsifier. About 85,000,000 kg monoglycerides are purchased in the United States each year [17]. The long-chain monoglycerides (monopalmitin and monostearate) have an excellent emulsifier properties compared to medium-chain monoglycerides (monolaurin and monocaprin). Moreover, the saturated monoglycerides have better emulsifying ability than unsaturated monoglycerides [3].

Some studies have reported that monoglycerides are promising antibacterial, antifungal, and antiviral agents [2, 9, 11, 18]. Medium-chain monoglycerides are revealed to have better antimicrobial activity than long-chain monoglycerides. The high antimicrobial activity of monoglycerides is contributed by its unique chemical structures and excellent amphiphilic properties. The amphiphilic nature of monoglycerides enables the formation of effective interactions with various chemical compositions that construct the membrane cell of pathogenic bacteria, fungi, or viruses, paralyzing these microorganisms.

Monoglycerides from polyunsaturated fatty acids, such as EPA (eicosapentaenoic acid) and DHA (docosahexaenoic acid), play an important function for human health. Wang et al. [1] have reported that monoolein has antioxidant and anti-atherosclerotic activity. Naik et al. [15] have described the antibacterial properties of monoglycerides, allowing them to be used as drug coating agents. Various applications of monoglycerides that have been presented previously prove that the nature and chemical structure of monoglycerides have an important role in determining their wide range of applications in human life.

The compound of 1-monoglycerides can exist in four different polymorphic forms at different temperatures [3]. In the polymorphic form α, 1-monoglyceride is unstable but has active chemical properties and excellent emulsion ability. The transition of 1-monoglycerides from the polymorphic form α to sub polymorphic α is observed in a lower temperature condition (35–50°C). If 1-monoglycerides are stored for a long time at room temperature, it will change to a more stable polymorphic form β. The 1-monoglyceride compound can also crystallize rapidly in certain solvents and form a stable β'-form. For example, 1-monocaprin, 1-monolaurin, and 1-monomiristine will crystallize quickly to form white solids from n-hexane solvent [6]. Unsaturated 1-monoglyceride compounds such as 1-monoolein also exist in the β-polymorphic form [19]. The symmetrical molecular structure of 2-monoglycerides causes this compound to always be in the form of polymorphic β [19]. This fact can be seen in 2-monolaurin that can crystallize rapidly in n-hexane solvents to form white solids and is very stable at room temperature [18].

The saturated monoglycerides can be found in three physical appearances, such as thick liquid (monocaprylin), fatty solid (monocaprin, monolaurin, monomiristine, monopalmitin), and waxy solid (monostearate and monobehenate). The longer carbon chain length of the saturated monoglyceride leads to an increase in the melting point. Monocaprylin has a melting point of 40–42°C. As the Carbon chain length increases from monocaprylin to monobehenate, its melting point increases to 65–77°C. Unsaturated monoglycerides such as monoolein and monolinolein are present in liquid form with melting points below 35°C. Regarding the solubility, almost all of the saturated monoglycerides with the acyl carbon chain length of C_{10}–C_{18} are soluble in ethanol but not soluble in water. Only monocaprylin (C_8) is slightly soluble in water but completely soluble in ethanol. That means the hydrophilic properties of monoglycerides are inversely proportional to the carbon chain length of the acyl groups [3].

4. Preparation of monoglycerides as antifungal agents

Almost all monoglycerides are conventionally made from vegetable oils or animal fat through a lysis reaction approach using alcohol from glycerol. The reaction to break down triglycerides in vegetable oils or animal fats to monoglycerides using glycerol is called glycerolysis reaction. The glycerolysis reaction is a transesterification reaction that needs base catalysts, such as inorganic bases NaOH, KOH, and $Ca(OH)_2$, and sodium alkoxide [20, 21]. The glycerolysis reaction of vegetable oils to produce monoglycerides always takes place at a high temperature around 220–260°C and under inert N_2 gas conditions [17]. However, there are some weaknesses from the production of monoglyceride through the glycerolysis reaction of vegetable oils, i.e.:

a. Monoglycerides are obtained in a dark color and burnt smell.

b. It requires high energy consumption.

c. High reaction temperatures are not suitable for the production of heat-sensitive monoglycerides such as monoglycerides from EPA and DHA.

d. It requires purification of monoglyceride products with molecular distillation.

The development of alternative reaction pathways for the synthesis of monoglyceride is essential, considering the high demand for monoglycerides as safe emulsifiers, antimicrobial agents (antibacterial, antifungal, antiviral), and other important applications. In this chapter, some reaction routes of monoglycerides are summarized, and their antifungal activities are also discussed.

4.1 Monolaurin

Monolaurin is a saturated monoglyceride from lauric acid (C12: 0) with 12 carbon atoms in the acyl group (**Table 1**). Lauric acid is a saturated fatty acid that can be found in coconut oil. Nitbani et al. [12] have reported that the percentage composition of lauric acid in coconut oil was 54% as methyl laurate by GC–MS analysis. Generally, derivatization of lauric acid and fatty acids are needed because their boiling points are too high to be detected by gas chromatography unless converted into their esters such as methyl laurate. From this brief explanation, it can be concluded that monolaurin can be developed or derived from the raw ingredients of coconut oil, lauric acid, or methyl laurate. Monolaurin can be classified into 1-monolaurin and 2-monolaurin based on the position of the acyl group, as can be seen in **Figure 2**.

Monolaurin is a well-known monoglyceride for its killing activities of various species of fungi, especially *Candida albicans* [22]. Monolaurin has also been reported to be very active against *Penicillium* spp., *Aspergillus* spp. [23], and *Fusarium* spp. [24], the dominant food spoilage organisms. Those fungi species contribute to the decrease of the economic value of some foods such as meat, fruits, and cereals. Burhannuddin et al. [25] have reported that monoglyceride from virgin coconut oil (VCO), with monolaurin as the dominant compound, showed good

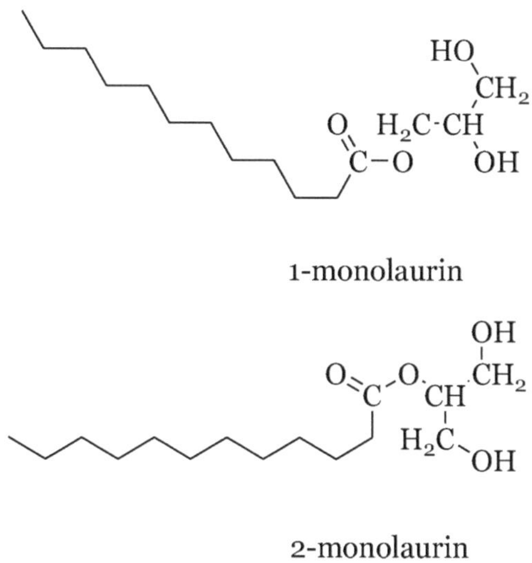

Figure 2.
The structure of monolaurin.

antifungal activity against *C. albicans* fungus. The ability of monolaurin to inhibit the growth of *C. albicans* was also reported by Chinatangkul et al. [26]. Monolaurin was also able to be used against common foodborne pathogens, including *C. albicans* with a minimum inhibitory concentration (MIC) value of 0.64 mg/mL and a minimum bactericidal concentration (MBC) value of 5.0 mg/mL [27].

Various literature presented the synthesis routes of 1-monolaurin through three reaction pathways that are summarized below. The reaction routes are esterification reaction of lauric acid and glycerol using either homogeneous or heterogeneous acid catalysts or a lipase enzyme catalyst and transesterification reaction of methyl laurate with glycerol.

4.1.1 Esterification reaction of lauric acid and glycerol catalyzed by acid catalysts

Monolaurin can be synthesized via esterification of lauric acid and glycerol in the presence of acid catalysts such as sulfuric acid (H_2SO_4) and *p*-toluenesulfonic acid (*p*TSA). The disadvantage of homogeneous catalyst is that it cannot be reused and needs to be neutralized. Utilization of heterogeneous acid catalysts in the esterification reaction of lauric acid and glycerol is expected to fulfill the drawbacks of homogeneous catalyst. The advantage of heterogeneous acid catalysts is that it can be easily separated from the reaction mixture and is reusable.

Table 2 summarizes the conditions of the esterification reaction of lauric acid and glycerol in the presence of homogeneous and heterogeneous acid catalysts from several publications. From these data, it can be concluded that the *p*TSA (homogeneous acid catalyst) shows excellent catalytic activity in the synthesis of monolaurin with a moderate yield (43.54%) and high purity (100%). Among the heterogeneous acid catalysts, sulfated zirconia-loaded SBA-15 displays an excellent catalytic ability in the esterification reaction to produce monolaurin with high yield (79.1%) and selectivity (83.42%).

4.1.2 Esterification reaction of lauric acid and glycerol catalyzed by lipase enzymes

Various types of enzyme catalysts, especially lipase enzymes, have been applied to synthesize monolaurin via the esterification reaction of lauric acid and glycerol.

Type of catalyst	Reaction condition	Yield	References
H_2SO_4	Temperature 130°C, 6 h, molar ratio of 1:1 (glycerol/lauric acid), 5% H_2SO_4 (w/w from lauric acid), solvent-free	31.05% yield of monolaurin, 91% purity	[28]
pTSA	Temperature 60°C, 6 h, molar ratio of 8:1 (glycerol/lauric acid), 2.5% pTSA (w/w from lauric acid), solvent-free	43.54% yield of monolaurin, 100% purity	[7]
Silica gel-coated with propyl sulfonic acids	Temperature 112°C, 8 h, molar ratio of 1:1 (glycerol/lauric acid), 5 wt% catalyst (relative to glycerol), solvent-free	51% yield of monoglyceride	[29]
Sulfated zirconia-loaded SBA-15	Temperature 160°C, 6 h, molar ratio of 4:1 (glycerol/lauric acid), sulfated zirconia-loaded SBA-15 catalyst with 16 wt.% zirconium oxychloride loading, solvent-free	79.1% yield and 83.42% selectivity of monolaurin, lauric acid conversion 94.9%	[30]
Zeolite Y CBV 712	Temperature 120–140°C, 7 h, molar ratio of 8:1 (glycerol/lauric acid), 15 wt% zeolite Y with dealumination, solvent-free	Reaction conversion: 97.8%, selectivity of 65%, and 59.5% yield of monolaurin	[31]

Table 2.
Reaction conditions in the synthesis of monolaurin from lauric acid and glycerol catalyzed by an acid catalyst.

Table 3 presents some publications related to the synthesis condition of monolaurin. The Novozym 435 lipase enzyme produced by Novozym Inc. has the highest catalytic activity to catalyze the esterification reaction of glycerol and lauric acid to produce monolaurin with high yield and selectivity (up to 100%).

4.1.3 Transesterification reaction of methyl laurate and glycerol

Monolaurin can be produced from methyl laurate and glycerol via transesterification reaction. The methyl laurate can be obtained from the transesterification reaction of coconut oil [12]. Glycerol itself is a by-product of the biodiesel production process. A mass yield of glycerol produced from the biodiesel production from vegetable oils is around 10% [37–39].

The reaction conditions and the results from the reaction of glycerol and methyl laurate to produce monolaurin can be seen in **Table 4**. Based on the data, 1-monolaurin compounds can be produced in high yield (almost reached 83%) in a binary solvent of *tert*-butanol/*iso*propanol (20:80; wt/wt). The solvent plays an important role in creating effective collisions between methyl laurate molecule and glycerol so that the activation energy is reached and the product of monolaurin is obtained. The presence of the lipase enzyme also determines the success of the formation of monolaurin products via a transesterification reaction of methyl laurate and glycerol.

Compound 2-monolaurin can be synthesized through the ethanolysis reaction of coconut oil using the Lipozyme TL IM catalyst [42]. Coconut oil is rich in lauric acid so that it can be used as a raw material for synthesis 2-monolaurin. In this case, ethanol can break down the triglycerides from coconut oil into monoglycerides, catalyzed by the sn-1,3-specific lipase enzyme. Because it uses an sn-1,3-specific enzyme, the acyl groups released are in positions 1 and 3 of triglycerides. This reaction is referred to as the alcoholysis reaction of coconut oils. Nitbani et al. [18] have successfully synthesized 2-monolaurin with high purity through the

Type of catalyst	Reaction condition	Yield	References
Lipozyme IM-20	Temperature 55°C, 6 h, molar ratio of 1:1 (glycerol/lauric acid), 3% (w/w) Lipozyme IM-20, solvent-free	45.5% yield of monolaurin, 26.8% of dilaurin	[32]
The partially purified lipase from *Rhizopus* sp.	Temperature 50°C, 72 h, stirring at 200 rpm, the molar ratio of 1:1 (glycerol/lauric acid), 100% (w/w) molecular sieve, 2 mg of partially purified lipase from *Rhizopus* sp., solvent-free	17.52% yield of monolaurin	[33]
CALB lipase (*Candida antarctica B*) immobilized on polysiloxane-polyvinyl alcohol particles (POS-PVA)	Temperature 45–60°C, 6 h, stirring at 200 rpm, molar ratio of 3:1 (glycerol/lauric acid), 0.5 g CALB L immobilized on POS-PVA (from the total weight of reactant), solvent-free	36% yield of monolaurin	[34]
Lipase G (*Penicillium camembertii lipase*) immobilized on epoxy SiO_2-PVA composite	Temperature 60°C, 6 h, stirring at 200 rpm, molar ratio of 8:1 (glycerol/lauric acid), 5% (w/w) lipase G immobilized on SiO_2-PVA loading, solvent-free	59.45% yield of monoglycerides, 62.91% selectivity	[17]
Lipozyme RM IM (*Rhizomucor mayhem lipase*)	Temperature 60°C, 3 h, molar ratio of 4:1 (glycerol/lauric acid), 4% (w/w) Lipozyme RM IM, solvent-free	50% yield of monolaurin, 34.6% yield of dilaurin	[35]
Novozym 435	Temperature 60°C, 8 h, stirring at 200 rpm, molar ratio of 4:1 (glycerol/lauric acid), Novozym 435 (60mg per mmol of carboxylic acid), nonaqueous reaction media: ionic liquid $[C_{12}mim][BF_4]$	100% selectivity and 100% yield of monolaurin	[36]

Table 3.
Reaction conditions in the synthesis of monolaurin from lauric acid and glycerol catalyzed by lipase enzymes.

Type of catalyst	Reaction condition	Yield	References
Novozym 435	Temperature 50°C, 24 h, stirring at 250rpm, molar ratio of 1:1 (glycerol/ methyl laurate), 5% (w/w) Novozym-435, solvent-free	47.6% monolaurin	[40]
Lipozyme 435 (immobilized-*Candida Antarctic lipase* on a macroporous acrylic polymer resin)	Temperature 50°C, 1.5 h, molar ratio of 1:6 (glycerol/ methyl laurate), 5% (w/w) Novozym 435, solvent: 15% (wt.) a binary solvent system (*tert*-butanol/*iso*propanol, 20:80, wt./wt.), a continuous flow system at a flow rate of 0.1 mL/min	82.5 ± 2.5 (wt. %) yield of monolaurin	[41]

Table 4.
Reaction conditions in the synthesis of monolaurin from methyl laurate catalyzed by lipase enzymes.

alcoholysis reaction of coconut oil using the Lipozyme TL IM enzyme. The 2-monolaurin compound was obtained in a yield of 30.1% and purity of 100% after purification using TLC preparation with a mixture of chloroform/acetone/methanol (9.5:0.45:0.05) as the eluent solvent. The Lipozyme TL IM enzyme is an sn-1,3-

specific lipase enzyme that has excellent catalytic activity for lipid substrate-related reactions such as hydrolysis, alcoholysis (transesterification), esterification, and acidolysis [42].

4.2 Monomyristin

Monomyristin compound is a medium-chain saturated monoglyceride with a number of carbon atoms in the acyl chain as 14 (C14). Altieri et al. [24] reported that both myristic acid and monomyristin compounds could inhibit the growth of fungi *F. oxysporum* DSMZ 2018 and *F. avenaceum* DSMZ 62161, although they were still weaker than lauric acid and monolaurin. Jumina et al. [9] also reported that 1-monomyristin has high activity against *C. albicans* compared with 2-monomyristin (not active). Therefore, the synthesis route of 1-monomyristin is important to be discussed. The chemical structure of 1-monomyristin is shown in **Figure 3**.

The 1-monomyristin compound can be obtained from the reaction of myristic acid and glycerol in the presence of lipase enzyme catalyst [17]. Another reaction pathway is through the transesterification of ethyl myristate with protected glycerol (1,2-O-isopropylidene glycerol) followed by a deprotection reaction using Amberlyst-15 to produce 1-monomyristin. Some works related to the synthesis of 1-monomyristin are presented in **Table 5**.

From **Table 5**, it can be noticed that the Novozym 435 enzyme has displayed extraordinary catalytic activity in the esterification reaction of myristic acid and glycerol to produce 1-monomyristin with 100% selectivity and yield. It should be noted that this reaction takes place in an ionic liquid [C12mim][BF4]. The [C12mim][BF4] is an example of a temperature switchable ionic liquid/solid phase used for the selective synthesis of monoglycerides [36]. Monomyristin can also be produced through two stages of reaction of ethyl myristate with 1,2-O-isopropylidene glycerol (mol ratio 1:8) using K_2CO_3 as a catalyst, which produces an intermediate isopropylidene glycerol myristate as a yellowish liquid with a yield of 32.12% and purity of 95.55%. In the second step, the isopropylidene glycerol myristate compound is deprotected with Amberlyst-15 (mol ratio 1:10) for 30 hours at room temperature to produce 1-monomyristin in the form of white solid in 100% of yield.

4.3 Monocaprin

Monocaprin is a monoglyceride from capric acid comprising 10 carbon atoms (C10) (**Figure 4**).

Monocaprin is produced from vegetable oils that are relatively safe or nontoxic to the body. Monocaprin is known as a safe food additive, so it is widely used as an emulsifier in the food industry. Solutions containing monocaprin can also act as

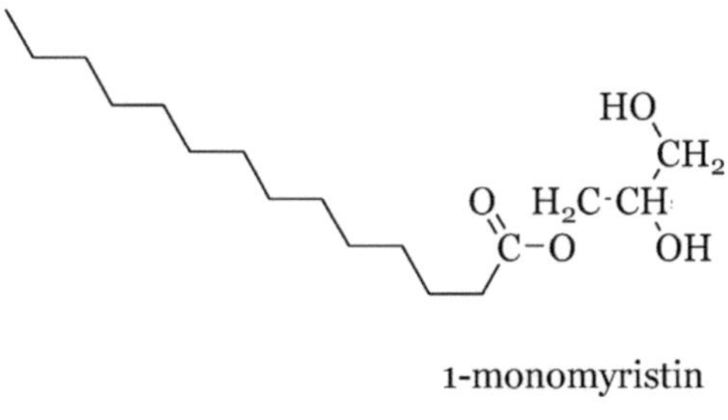

Figure 3.
The structure of 1-monomyristin.

Type of catalyst	Reaction condition	Yield	References
Lipase G (*Penicillium camembertii* lipase) immobilized on epoxy SiO_2-PVA composite	Temperature 60°C, 6 h, stirring at 200 rpm, molar ratio 8:1 (glycerol/ myristic acid), 5% Lipase G immobilized on SiO_2-PVA loading (w/w), solvent-free	47.92% MAG, 79.92% selectivity	[17]
Novozym 435	Temperature 60°C, 8 h, stir at 200 rpm, molar ratio 4:1 (glycerol/ myristic acid); Novozym 435 (60mg per mmol of carboxylic acid), nonaqueous reaction media: ionic liquid [C_{12}mim][BF_4]	100% selectivity and 100% yield of monomyristin	[36]
Potassium carbonate (K_2CO_3)	a. Step 1: reaction of ethyl myristate and 1,2-acetonide glycerol: temperature 140°C, 30 h, molar ratio 1:8 (ethyl caprate/1,2-acetonide glycerol), 5% K_2CO_3 (w/w), isolation product of isopropylidene glycerol myristate using diethyl ether b. Step 2: reaction of isopropylidene glycerol myristate and Amberlyst-15: room temperature, 30 h, ethanol as solvent, weight ratio 1:10 (Amberlyst-15/ isopropylidene glycerol myristate), isolation of 1-monomyristin by filtration and evaporation	100% yield of 1-monomyristin	[9]

Table 5.
Reaction conditions in the synthesis of 1-monomyristin.

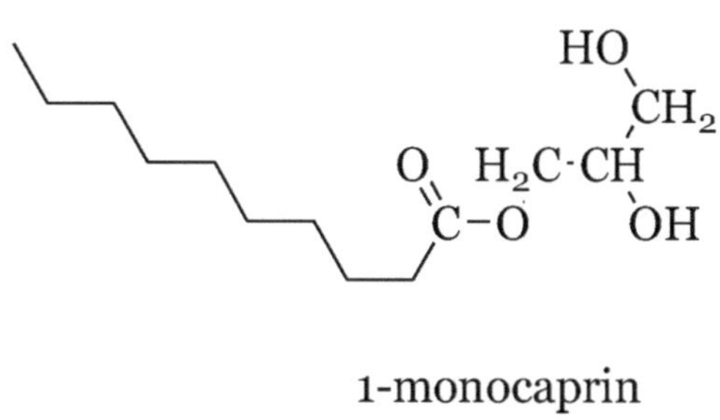

Figure 4.
The structure of 1-monocaprin.

microbicidal agents against *Candida albicans* [43]. Lipid compounds, including monoglycerides, are also known as antifungal agents [44]. Monocaprin has been reported to have antifungal activity against three food spoilage fungi *Saccharomyces cerevisiae*, *Aspergillus niger*, and *Penicillium citrinum* with minimum inhibitory concentration values of 0.31, 0.63, and 0.63 mg/mL, respectively. The minimum fungicidal concentrations were also reported to be 1.25, 2.50, and 2.50 mg/mL, respectively [45].

The production of monocaprin also has similarities with monomiristine and monolaurin, which involves esterification reactions of capric acid and glycerol. Another reaction pathway for the synthesis of monocaprin is by utilizing ethyl caprate and 1,2-acetonide glycerol (a protected glycerol compound). A summary of

Type of catalyst	Reaction condition	Yield	References
Candida antarctica (CAL)	Temperature 60°C, 6 h, water (12% in glycerol (w/w)), molar ratio of 1:1 (glycerol/capric acid), lipase dosage (100 U/g) of capric acid, solvent-free, batch reactor	Capric acid conversion as high as 96.9%	[46]
Porcine liver carboxylesterase (PLE)	Temperature 60 °C; 4 h; pH = 7; reverse micelles: isooctane (reaction medium) and bis(2-ethylhexyl) sodium sulfosuccinate (AOT, anionicsurfactant); R-value ([water]/[surfactant]): 0.1; G/F-value ([glycerol]/[fatty acid]) 3.0	The degree of esterification at equilibrium state 62.7%	[47]
Sodium carbonate (Na_2CO_3)	a. *Step 1: reaction of ethyl caprate and 1,2-acetonide glycerol* Temperature 110°C; 24 h; molar ratio of ethyl caprate/1,2-acetonide glycerol, 1:8; 5% Na_2CO_3 (w/w); isolation product (1,2-asetonide-3-capryl glycerol) using n-hexane b. *Step 2: reaction of 1,2-acetonide-3-capryl glycerol and Amberlyst-15* Room temperature, 24 h, ethanol as a solvent, ratio weight 1:7 (Amberlyst-15/ 1,2-acetonide-3-capryl glycerol); isolation of crude 1-monocaprin with recrystallization in n-hexane, purification of 1-monocaprin using preparative thin-layer chromatography	a. Yield, 88.12%; purity, 87% (1,2-acetonide-3-capryl glycerol) b. Yield, 78.37%; purity, 100% (1-monocaprin)	[6]

Table 6.
Reaction conditions in the synthesis of 1-monocaprin.

the reaction conditions on the synthesis of monocaprin can be seen in **Table 6**. It is noticed that monocaprin can be produced with high yield and purity via two reaction steps. The first stage involves the transesterification reaction of ethyl caprate with 1,2-acetonide glycerol to produce intermediate 1,2-acetonide-3-capryl glycerol with a yield of 88.12%. The second step is deprotection reaction of 1,2-acetonide-3-capryl glycerol using a heterogeneous catalyst (Amberlyst-15) to produce 1-monocaprin with 100% purity and a yield of 78.37% after purification. A lipase enzyme *Candida antarctica* (CAL) is also reported to be used as the catalyst in the esterification of capric acid and glycerol, with conversion rate reaching 96.9%. Production of monocaprin in a reverse micelle system using isooctane (reaction medium) and bis(2-ethylhexyl) sodium sulfosuccinate (AOT, anionic surfactant) has marked the conversion rate of esterification reaction of capric acid and glycerol at equilibrium condition at 62.7%, which is achieved using *Porcine liver carboxylesterase* (PLE) as a catalyst.

5. Mechanism of monoglycerides as antifungal agents

Several publications related to the antifungal activity of monoglycerides such as monolaurin, monomiristine, and monocaprin have been explained in the previous section. The studies showed that monoglycerides with good amphiphilic properties were not only able to inhibit the fungal growth but also kill it, especially *C. albicans* (22) (9) and some food spoilage fungi species such as *Saccharomyces cerevisiae*, *Aspergillus niger*, and *Penicillium citrinum* [45]. Publication data supported a hypothesis that the saturated medium-chain monoglycerides such as monolaurin, monomiristine, and monocaprin have good antifungal activity. Based on their structure, each monoglyceride has an acyl group derived from lauric acid (C12: 0), myristic acid (C14: 0), and capric acid (C10: 0).

It is noted that the chemical structure of monoglyceride plays an important role in its antifungal activity by affecting the interaction with fungal organisms. Monoglycerides with lipophilic acyl groups and two hydrophilic hydroxyls (-OH) groups (**Figure 1**) are beneficial to interact with various chemical components that build fungal organisms.

Prasad et al. [4] have reported that the current development of antifungal agents is aimed at interactions with the fungal cell wall. There are two main targets of antifungal agents, firstly, by targeting the interactions with chemical components of fungal cell walls such as mannans, glucans, and chitins. The second target is aiming at the interactions with several enzymes that responsible for bioactivity and the biosynthesis pathway of ergosterol. Ergosterol is one of the main sterol components that build fungal cell membranes and regulate the fluidity, permeability, and structure of the membranes [48, 49].

One of the antifungal drugs that have been developed and are quite useful in invading fungal infections is polyene. There are three main types of fungal drugs from polyene compounds, i.e., Amphotericin B, Nystatin, and Natamycin [48]. These three fungal drugs are known as macrolides, which structurally are cyclic amphiphilic organic molecules. The amphiphilic aspects of macrolides are contributed by the unsaturated alkyl chain (around 14 C) as the lipophilic part that attached to the macrolactone ring and some hydroxyl (-OH) groups as the hydrophilic part.

The amphiphilic properties of some macrolide compounds (Amphotericin B, Nystatin, and Natamycin) from the polyene group are the main factors in their mechanism of action as the antifungal agents. Their amphiphilic structure allows these compounds to bind chemically to the components of lipid membrane, especially ergosterol, through van der Waals interactions. Interactions between these molecules will trigger the formation of pores on the cell membrane. Moreover, in the end, the pores will destabilize the cell membrane, damaging the balance of ions in the cell membrane and further resulting in the cell death [4, 48, 50].

Other interaction models can be predicted based on the chemical structure of macrolides as well as the chemical components of the fungal cell wall, such as mannans, glucans, and chitins. The hydrophilic part (hydroxyl groups) in macrolide compounds is can possibly interact with polar groups found in mannans, glucans, and chitins through hydrogen bonding. This interaction is also predicted to contribute to the fungal cell wall damage, so the cell lysis can occur resulting in cell death.

Monoglyceride compounds such as monolaurin, monomiristine, and monocaprin that have been proven to be antifungal agents are assumed to follow the inhibitory mechanism of macrolides (polyene) compounds. This assumption is very rational, considering that monoglycerides are having excellent amphiphilic properties (**Figures 1** and **2**). The acyl group of the lipophilic part of monoglycerides is expected to interact via van der Waals interaction with ergosterol in fungal cell walls and causes lysis and cell death. Meanwhile, the two hydroxyl groups in monoglycerides are responsible for the antifungal activity by forming hydrogen bonding with other polar components in the fungal cell wall (glucans, chitins) and assisting the cell membrane lysis process.

Comparing the number of hydroxyl groups in monoglycerides, macrolides compounds have more hydroxyl groups. Therefore, the hydrogen bonding interactions of monoglycerides with polar components in the fungal cell wall is expected to be less effective than in macrolides. Based on the prediction of the mechanism of action and in vitro data of monoglycerides as antifungal agents, it can be concluded that monoglycerides, especially monolaurin, monomiristine, and monocaprin, have the potential to be developed as antifungal drugs. Thus, there will be new candidates for antifungal drugs from monoglyceride-based lipid.

Author details

Febri Odel Nitbani[1*] and Jumina Jumina[2]

1 Faculty of Science and Engineering, Department of Chemistry, Universitas Nusa Cendana, Kupang, Indonesia

2 Faculty of Mathematics and Natural Sciences, Department of Chemistry, Universitas Gadjah Mada, Yogyakarta, Indonesia

*Address all correspondence to: febri_nitbani@yahoo.com

References

[1] Wang X, Jin Q, Wang T, Huang J, Wang X. An improved method for the synthesis of 1-monoolein. Journal of Molecular Catalysis B: Enzymatic. 2013; **97**:130-136

[2] Nitbani FO, Jumina SD, Sholikhah E. Reaction path synthesis of monoacylglycerol from fat and oils. International Journal of Pharmaceutical Sciences and Research. 2015;**35**(1): 126-136

[3] Miao S, Lin D. Monoglycerides: Categories, Structures, Properties, Preparations, and Applications in the Food Industry. Elsevier: Encyclopedia of Food Chemistry; 2019

[4] Prasad R, Shah AH, Rawal MK. Antifungals: Mechanism of action and drug resistance. In: Ramos J, Sychrová H, Kschischo M, editors. Yeast Membrane Transport. Cham: Springer International Publishing; 2016

[5] Seleem D, Chen E, Benso B, Pardi V, Murata RM. *In vitro* evaluation of antifungal activity of monolaurin against *Candida albicans* biofilms. PeerJ. 2016;**4**:1-17

[6] Nitbani FO. Synthesis of monoacylglycerol as antibacterial and immunostimulant agents from coconut oil (*Cocos nucifera L.*) [Ph.D Thesis]. Universitas Gadjah Mada; 2017

[7] Nitbani FO, Jumina, Siswanta D, Sholikhah EN, Fitriastuti D. Synthesis and antibacterial activity 1-monolaurin. Oriental Journal of Chemistry. 2018; **34**(2):863-867

[8] Nitbani FO, Siswanta D, Sholikhah EN. Synthesis and antibacterial activity test of 1-monocaprin. International Journal of Pharmaceutical Sciences and Research. 2016;**39**(16):74-80

[9] Jumina, Nurmala A, Fitria A, Pranowo D, Sholikhah E, Kurniawan Y, et al. Monomyristin and Monopalmitin derivatives: Synthesis and evaluation as potential antibacterial and antifungal agents. Molecules. 2018;**23**(12):3141

[10] Yu CC, Lee Y-S, Cheon BS, Lee SH. Synthesis of glycerol monostearate with high purity. Bulletin of the Korean Chemical Society. 2003;**24**(8):1229-1231

[11] Jumina J, Lavendi W, Singgih T, Triono S, Kurniawan YS, Koketsu M. Preparation of Monoacylglycerol derivatives from Indonesian edible oil and their antimicrobial assay against *Staphylococcus aureus* and *Escherichia coli*. Scientific Reports. 2019;**9**(1):1-8

[12] Nitbani FO, Jumina, Siswanta D, Solikhah EN. Isolation and antibacterial activity test of Lauric acid from crude coconut oil (*Cocos nucifera L.*). Procedia Chemistry. 2016;**18**:132-140

[13] Ola PD, Tambaru D. Sintesis Biodiesel dan Minyak Epoksi dari Minyak Jarak (*Ricinnus communis L.*). Cendana, Kupang, Indonesia: Laporan Penelitian, Universitas Nusa; 2009

[14] Orsavova J, Misurcova L, Ambrozova JV, Vicha R, Mlcek J. Fatty acids composition of vegetable oils and its contribution to dietary energy intake and dependence of cardiovascular mortality on dietary intake of fatty acids. International Journal of Molecular Sciences. 2015;**16**(6):12871-12890

[15] Naik MK, Naik SN, Mohanty S. Enzymatic glycerolysis for conversion of sunflower oil to food based emulsifiers. Catalysis Today. 2014;**237**: 145-149

[16] Mba OI, Dumont M-J, Ngadi M. Palm oil: Processing, characterization and utilization in the food industry – A review. Food Bioscience. 2015;**10**: 26-41

[17] Freitas L, Paula A, Santos J, Zanin G, Castro H. Enzymatic synthesis of monoglycerides by esterification reaction using Penicillium camembertii lipase immobilized on epoxy SiO2-PVA composite. Journal of Molecular Catalysis B: Enzymatic. 2010;**65**:87-90

[18] Nitbani FO, Jumina, Siswanta D, Sholikhah EN, Fitriastuti D. Synthesis and antibacterial activity of 2-monolaurin. Oriental Journal of Chemistry. 2016;**32**(6):3113-3120

[19] Vereecken J, Meeussen W, Foubert I, Lesaffer A, Wouters J, Dewettinck K. Comparing the crystallization and polymorphic behaviour of saturated and unsaturated monoglycerides. Food Research International. 2009;**42**(10): 1415-1425

[20] Corma A, Iborra S, Miquel S, Primo J. Catalysts for the production of fine chemicals: Production of food emulsifiers, monoglycerides, by glycerolysis of fats with solid base catalysts. Journal of Catalysis. 1998; **173**(2):315-321

[21] Tüter M, Aksoy HA. Enzymatic glycerolysis of palm and palm kernel oils. Chemical Engineering Communications. 2005;**192**(1):14-17

[22] Seleem D, Freitas-Blanco VS, Noguti J, Zancope BR, Pardi V, Murata RM. In vivo antifungal activity of Monolaurin against Candida albicans biofilms. Biological & Pharmaceutical Bulletin. 2018;**41**(8):1299-1302

[23] Luo C, Zeng Z, Gong D, Zhao C, Liang Q, Zeng C. Evaluation of monolaurin from camphor tree seeds for controlling food spoilage fungi. Food Control. 2014;**46**:488-494

[24] Altieri C, Bevilacqua A, Cardillo D, Sinigaglia M. Antifungal activity of fatty acids and their monoglycerides against *Fusarium* spp. in a laboratory medium. International Journal of Food Science and Technology. 2009;**44**(2):242-245

[25] Burhannuddin, Karta IW, Tresnanda B, IGND P, IPA D, IDA P, et al. Daya Hambat Virgin Coconut Oil Terhadap Pertumbuhan Jamur *Candida albicans* Isolat Vagina. Sainteknol: Jurnal Sains dan Teknologi. 2017;**6**(2):209-219

[26] Chinatangkul N, Limmatvapirat C, Nunthanid J, Luangtana-Anan M, Sriamornsak P, Limmatvapirat S. Design and characterization of monolaurin loaded electrospun shellac nanofibers with antimicrobial activity. Asian Journal of Pharmaceutical Sciences. 2018;**13**(5):459-471

[27] Zhang S, Xiong J, Lou W, Ning Z, Zhang D, Yang J. Antimicrobial activity and action mechanism of triglycerol monolaurate on common foodborne pathogens. Food Control. 2019;**98**:113-119

[28] Widiyarti G, Hanafi M, Soewarso WP. Study on the synthesis of monolaurin as antibacterial agent against *Staphylococcus aureus*. Indian Journal of Chemistry. 2010;**9**(1):99-106

[29] Bossaert WD, De Vos DE, Van Rhijn WM, Bullen J, Grobet PJ, Jacobs PA. Mesoporous sulfonic acids as selective heterogeneous catalysts for the synthesis of monoglycerides. Journal of Catalysis. 1999;**182**(1):156-164

[30] Abdullah AZ, Gholami Z, Ayoub M, Gholami F. Selective monolaurin synthesis through esterification of glycerol using sulfated zirconia-loaded SBA-15 catalyst. Chemical Engineering Communications. 2016;**203**(4):496-504

[31] Setianto WB, Wibowo TY, Yohanes H, Illaningtyas F, Anggoro DD. Synthesis of glycerol mono-laurate from lauric acid and glycerol for food antibacterial additive. IOP Conference Series: Earth and Environmental Science. 2017;**65**:012046

[32] Pereira CCB, Langone MAP. Enzymatic synthesis of monolaurin. Applied Biochemistry and Biotechnology. 2004;**113**:434-445

[33] Koblitz MGB, Pastore GM. Contribution of response surface design to the synthesis of monoacylglycerols catalyzed by Rhizopus sp. lipase. Journal of Food Science. 2005;**70**(8):c503-c505

[34] Freitas L, Perez VH, Santos JC, Castro HFD. Enzymatic synthesis of glyceride esters in solvent-free system: Influence of the molar ratio, lipase source and functional activating agent of the support. Journal of the Brazilian Chemical Society. 2007;**18**(7):1360-1366

[35] Mustafa A, Karmali A, Abdelmoez W. Optimisation and economic assessment of lipase-catalysed production of monoesters using Rhizomucor miehei lipase in a solvent-free system. Journal of Cleaner Production. 2016;**137**:953-964

[36] Lozano P, Gomez C, Nieto S, Sanchez-Gomez G, García-Verdugo E, Luis SV. Highly selective biocatalytic synthesis of monoacylglycerides in sponge-like ionic liquids. Green Chemistry. 2017;**19**(2):390-396

[37] Tan HW, Abdul Aziz AR, Aroua MK. Glycerol production and its applications as a raw material: A review. Renewable and Sustainable Energy Reviews. 2013;**27**:118-127

[38] Nitbani FO. Sampah Biodiesel Bernilai Emas. Yogyakarta, Indonesia: Deepublish; 2018

[39] Adeniyi AG, Ighalo JO. A review of steam reforming of glycerol. Chemical Papers. 2019;**73**(11):2619-2635

[40] Jamlus NNA, Salimon J, Derawi D. Gliserolisis berenzim oleh metil laurat menggunakan Lipase b Candida antarctica. The Malaysian Journal of Analytical Sciences. 2016;**20**(6): 1365-1372

[41] Chen F, Zhang G, Liu C, Zhang J, Zhao F, Xu B. Highly selective synthesis of monolaurin via enzymatic transesterification under batch and continuous flow conditions. Journal of Oleo Science. 2019;**68**(11):1125-1132

[42] Fernandez-Lafuente R. Lipase from *Thermomyces lanuginosus*: Uses and prospects as an industrial biocatalyst. Journal of Molecular Catalysis B: Enzymatic. 2010;**62**:197-212

[43] Ósk Thorgeirsdóttir T, Kristmundsdóttir T, Thormar H, Axelsdóttir I, Peter HW. Antimicrobial activity of monocaprin: A monoglyceride with potential use as a denture disinfectant. Acta Odontologica Scandinavica. 2006;**64**(1):21-26

[44] Thormar H, Hilmarsson H. The role of microbicidal lipids in host defense against pathogens and their potential as therapeutic agents. Chemistry and Physics of Lipids. 2007;**150**(1):1-11

[45] Ma M, Wen X, Xie Y, Guo Z, Zhao R, Yu P, et al. Antifungal activity and mechanism of monocaprin against food spoilage fungi. Food Control. 2018; **84**:561-568

[46] Xia YM, Fang Y, Zhang KC, Shi GY, Brown JJ. Enzymatic synthesis of partial glycerol caprate in solvent-free media. Journal of Molecular Catalysis B: Enzymatic. 2003;**23**(1):3-8

[47] Park KM, Lee JH, Hong SC, Kwon CW, Jo M, Choi SJ, et al. Selective production of 1-monocaprin by porcine liver carboxylesterase-catalyzed esterification: Its enzyme kinetics and catalytic performance. Enzyme and Microbial Technology. 2016;**82**:51-57

[48] Vandeputte P, Ferrari S, Coste AT. Antifungal resistance and new strategies to control fungal infections. International Journal of Medical Microbiology. 2012;**2012**:1-26

[49] Rodrigues ML. The multifunctional fungal ergosterol. MBio. 2018;**9**(5):1-5

[50] Sanglard D, Coste A, Ferrari S. Antifungal drug resistance mechanisms in fungal pathogens from the perspective of transcriptional gene regulation. FEMS Yeast Research. 2009; **9**(7):1029-1050

Chapter 4

Multi-Scale Mathematical Modeling of Prion Aggregate Dynamics and Phenotypes in Yeast Colonies

Mikahl Banwarth-Kuhn and Suzanne Sindi

Abstract

Prion diseases are a multi-scale biological phenomenon that requires understanding intracellular processes as well as how cells interact with each other and their environment. In mammals, prion diseases are progressive, untreatable, and fatal. Yeast prion phenotypes are harmless and reversible, which suggests a deep understanding of the reversal of prion phenotypes in yeast may be informative to mammalian diseases. In yeast, the loss of some prion phenotypes appears to be stochastic and spatially dependent, suggesting a cell-based model of yeast prion dynamics would be a powerful tool for comparisons with experimental results and hypothesis generation. In this work, we consider the components necessary to develop such a model that depicts both the biochemical-, intracellular-, and colony-level scales in yeast prion phenotypes. We first review the literature of mathematical models of the intracellular processes of prion disease. We then review common approaches to cell-based modeling of multicellular systems and how they have led to biological insights in other systems. This chapter ends with a discussion of future studies aimed at motivating how these two types of models can be coupled to produce multi-scale models of prion phenotypes.

Keywords: agent-based models, cell-based model, prion disease dynamics, protein aggregation, multi-scale models

1. Introduction

Prion proteins are most commonly associated with fatal progressive neurodegenerative diseases in humans and other mammals [1–4]. Intriguingly, all mammalian prion diseases are the result of a single host expressed protein, PrP. In mammalian prion disease, a misfolded form of PrP appears, and rather than being cleared by protein quality control mechanisms, this misfolded form persists and associates into aggregates. These aggregates then act as templates that convert normally folded protein to the misfolded form and may break into multiple aggregates further amplifying the conversion of normal protein [5, 6]. The initiating event of prion disease, the initial appearance of an infectious protein agent, can occur spontaneously, as in sporadic Creutzfeldt-Jakob disease [7]; because of a genetic mutation, as in fatal familial insomnia [8]; and through interaction with

IntechOpen

another infected host from the same (as in scrapie [9], chronic wasting disease [10], or Kuru [11]) or different species, as in when Bovine spongiform encephalopathy ("mad cow" disease) is acquired by humans [12] or scrapie is spread from sheep to goats [13]. Further, some prion diseases can occur through multiple modes, such as Creutzfeldt-Jakob which has been associated with each of these modes of transmission [14]. More generally, the amyloid structure of prion aggregates is common to non-prion proteins associated with neurodegenerative disorders such as Alzheimer's, Parkinson's, and Huntington's disease, and evidence continues to suggest commonalities between such disorders and prions [15, 16], including their transmissibility [17]. At present, prion disease, and other amyloid disease such as Alzheimer's [18], has no effective methods for treatment or early detection. However, one potential source to gain insight into these pathogenic phenotypes is to study instances in biology where amyloids confer beneficial consequences, such as growth advantages in yeast and long-term memory storage in mammals [19], and therefore the amyloid dynamics can be studied absent of the disease state [16, 20].

One promising candidate for probing the mechanisms behind prion-associated phenotypes is the yeast *S. cerevisiae* [21–23]. In yeast, prion proteins were first identified in the 1980s when interrogating the genetic basis for non-Mendelian phenotypes [21, 23]. Today, nearly a dozen harmless phenotypes in yeast are known to propagate through misfolded protein aggregates [6]. Unlike their mammalian counterparts, yeast prion phenotypes have been associated with many different proteins including Sup35 ($[PSI^+]$ phenotype), Rnq1 ([*PIN*] phenotype), and Ure2 ([*URE*3] phenotype). As for mammalian prions, each of these phenotypes has multiple variants (or strains) which themselves are associated with different properties such as infectivity and phenotype presentation. However, unlike mammalian prion disease, which can take months of years to present, yeast prion phenotypes manifest quickly and, most remarkably, can be made to appear or disappear through mild experimental manipulations which do not impact the host cell [6, 16, 24, 25]. The underlying aggregate dynamics are the same regardless of organism. As shown in **Figure 1**, the establishment of prion phenotypes in yeast (and of course mammals) is an inherent multi-scale process. First, there is the biochemistry of the interactions between normal and misfolded protein. Then, there is the level of an individual cell where these protein aggregates interact not only with the cellular environment including molecular chaperones and protein degradation factors.

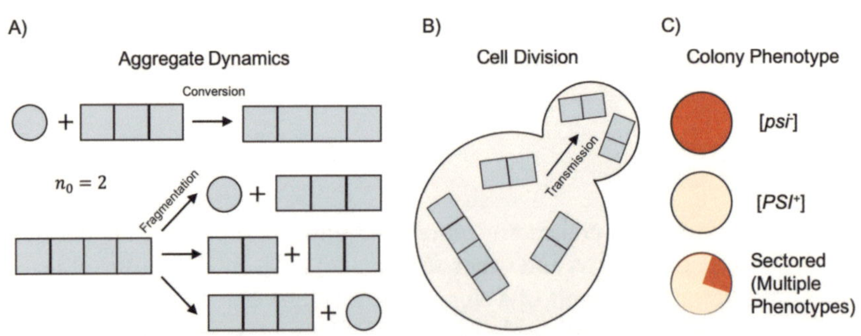

Figure 1.
Multi-scale dynamics of yeast prion Phenotypes. (A) Prion aggregates are thought to be ordered linear structures (squares) capable of converting *normally folded protein (circles) to the prion form and* fragmenting *into multiple aggregates. Prion aggregates are thought to have a minimal stable size* n_0 *below which the aggregate structure will not be stable. (B) Aggregate dynamics occur within an individual yeast cell (large circle) but are spread to daughter cells through transmission through the bud neck. (C) Phenotypes in yeast are observed at the level of a colony which consists of thousands of individual cells.* $[PSI^+]$ *phenotypes are associated with a color red ([psi−] or no prion), pink* ($[PSI^+]$)*, or sectored (mixture of phenotypes).*

Then, these protein aggregates must be transmitted to other cells. Note that in yeast, this transmission occurs between mother and daughter cells during division [26, 27], while in mammals, this transmission typically occurs between cells in a tissue [28, 29]. Finally, prion phenotypes are determined by examining populations of cells. In yeast, prion phenotypes are assayed at the level of a yeast colony, and mammalian prion phenotypes are typically distinguished by their patterns of neurological decay [14].

Although there is growing appreciation for the multi-scale nature of mammalian and yeast prion phenotypes, the vast majority of mathematical models developed for studying such phenotypes typically focus on only one scale. In particular, the majority of mathematical model have focused on the isolated prion aggregate dynamics. While these models have been particularly useful in comparisons to in vitro prion aggregation, they are limited in their ability to insight to the in vivo systems. In a population of living cells, different cell behaviors such as growth, diffusion, and division are known to impact the abundances and concentrations of reactants and could have a large impact on protein aggregation or more specifically propagation of prion aggregates. Moreover, in the case of yeast, since the proteins are only transmitted during cell division [6] and yeast have an asymmetric cell cycle [30], a colony of yeast with a prion phenotype will exhibit considerable heterogeneity in prion aggregates and potentially other relevant cellular constitutes [27].

One particular yeast prion phenotype that is intimately linked to population heterogeneity and yet to be mathematically characterized is that of colony sectoring for the $[PSI^+]$ yeast prion phenotype (**Figure 1(C)**). The $[PSI^+]$ phenotype is associated with the color of a colony: a red colony indicates a $[psi-]$ (prion-free) phenotype and $[PSI^+]$ colony is in a shade of white or pink. The actual color displayed by the colony is associated with a loss of function of the normal Sup35 protein as in when in a prion aggregate the Sup35 monomers have greatly reduced normal function [31]. A sectored colony represents a case when cells in the colony display both $[psi-]$ and $[PSI^+]$ phenotypes. Such a colony is thought to appear when the founding cell contains prion aggregates, but at some point through division of cells and growth of the colony, some cells lose the ability to propagate the prion phenotype. Because spontaneous appearance of Sup35 prion aggregates is a rare event, any subsequent daughter cells will also lack prion aggregates. These sub-colonies associated with the loss of the prion phenotypes are thought to correspond to the red sectors. Since the number and pattern of red sectors is fairly similar under the same condition, this suggests an intimate connection between the lineage of ancestry of the cells which were the first to lose the ability to propagate the prion phenotype. Indeed a previous study that did not consider the spatial distribution of individual cells suggested that loss of prion aggregates for a particular variant of $[PSI^+]$ occurred in a lineage-dependent fashion [27]. What is the mechanistic basis for this partial loss? Can these mechanisms be exploited in the context of mammalian prion diseases? Answering these questions requires developing mathematical frameworks that accurately represent all scales of the yeast prion system.

Beyond mammalian prion disease, yeast prions offer a particularly unique opportunity to investigate and model the emergence and role of heterogeneity in cellular populations. In the past, our knowledge of individual cell behaviors was based on ensemble measurements that do not account for effects of cellular heterogeneity within a population. However, emerging technologies such as single-cell gene expression analysis, single-cell genome sequencing, and single-cell imaging technologies [32] provide an unprecedented opportunity to quantitatively measure heterogeneity at the single-cell level. Molecular and live imaging experiments investigating growth and development of multicellular systems provide rich data sets that can be used for the first time to understand the different signaling cascades and components that are necessary for cell behaviors to arise. This new level of insight makes it possible to reveal a more accurate picture of cellular behavior and highlights the importance of understanding cellular variation in a wide range of biological contexts. Developing models that describe heterogeneity throughout an entire population provides a quantitative way to understand the dynamical behavior of heterogeneous cell population characteristics/biological importance of heterogeneity.

In this chapter, we consider the different tools available for developing a mathematical framework for distinct scales of prion dynamics. We first review mathematical models for prion disease dynamics and discuss the few models which have considered multiple scales of prion aggregation. We then discuss cell-based models as a promising approach for coupling intracellular and intercellular scales. We discuss several common classes of cell-based models and emphasize the contributions they have made to the understanding of biological systems. We conclude with a discussion how one could build a multi-scale model for prion disease dynamics in yeast that couples current models for aggregation of proteins with a spatial model of yeast growth and proliferation on the scale of an entire population.

2. Intracellular dynamics

As we discussed in Section 1, prion dynamics is an inherently multi-scale process. However, the bulk of mathematical modeling on prion aggregation has considered only the microscopic scale, intracellular dynamics. Because the formation of the initial stable aggregate is a rare occurrence, most of these models have focused on the dynamics of the prion aggregates themselves in a well-mixed compartment leading to models that reflect what occurs microscopically within a single cell or are taken to be the average behavior of a population or tissue. In addition, these models assume either discrete or continuous aggregate sizes employing ordinary differential equations (ODEs) or partial differential equations (PDEs), respectively. This literature has emphasized establishing results on the global stability of prion aggregates as well as their asymptotic size distribution. More recently, mathematical approaches have been developed to consider the spread of prions between cells in a tissue or even organisms, but only under severely simplified molecular scales. At present, no single multi-scale framework has yet been formulated for prion dynamics leaving many unexplained phenotypes, such as yeast sectored colony phenotypes (see **Figure 1(C)**).

In this section we review the development of mathematical models in the study of prion dynamics. (Readers interested in a more detailed view on mathematical models of prion disease can consult [33].) Although many of these models share commonality with more general aggregation processes in neurodegenerative disease, we focus only on prion aggregation. (Readers interested in mathematical models of more general neurodegenerative phenomena can consult [34].)

2.1 Intracellular models of prion phenotypes

The prion hypothesis, the idea that proteins themselves could encode information about their structure without a DNA intermediary, was first suggested by experimental studies in 1966–1967 (see [35–37]). Remarkably, the first mathematical formulation of prion aggregation was published in 1967 when Griffiths [38] offered a simple mathematical autocatalytic process that was consistent with a

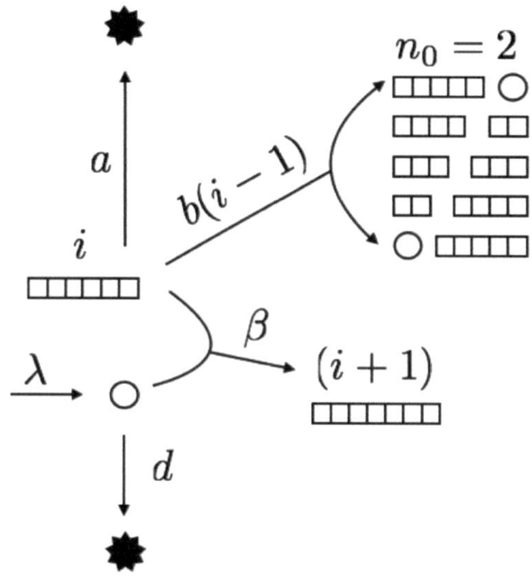

Figure 2.
Nucleated polymerization model. The NPM (see [39, 40]) is the canonical model for prion aggregate dynamics. In this formulation normally folded protein (circles) can be created with rate λ, decayed with rate d, converted to the misfolded form, and added onto the end of an existing aggregate (linear complex of squares) with rate β. Aggregates have their own degradation rate, a, that is independent of aggregate size, and number of aggregates increases through fragmentation which occurs at a rate b times the number of fragmentation sites in an aggregate. (Note that in this diagram, we allow normal protein to be recovered if an aggregate is fragmented below the minimal nucleus size $n_0 = 2$ depicting the aggregate dynamics evolving according to Eqs. (7)–(9).)

protein-based infectious agent of scrapie. Since then, a large number of mathematical models have emerged which track the size density of prion aggregates under different biochemical processes but typically include conversion of normal protein to prion aggregates and fragmentation of aggregates. Most are based whole or in part on the notion of the nucleated polymerization model (NPM).

Nowak and colleagues developed the NPM in a series of papers [39, 40] based on a model previously introduced by Eigen [41]. In this mathematical formulation, the state of the system at time t is the concentration of proteins in the normal conformation, $x(t)$, and prion aggregates of every discrete size i, $y_i(t)$. Protein in the normal conformation is assumed to be created at rate λ and decays at rate d, while aggregates of all sizes decay at rate a (see **Figure 2**) Conversion occurs through contact between aggregates and protein in the normal conformation at a rate depending on the size of the aggregate, β_i. Finally, the total number of aggregates increases through fragmentation. In their most general formulation, Nowak et al. specify the rate that aggregates of size j fragment to create an aggregate of size i as $b_{j,i}$ and that during fragmentation no mass is lost, i.e., an aggregate of size j is always fragmented into two aggregates of size i and $(j-i)$. Translating these biochemical kinetic assumptions into a set of differential equations results in the following infinite system of ordinary differential equations:

$$\frac{dx}{dt} = \lambda - dx(t) - \sum_{i=1}^{\infty} \beta_i x(t) y_i(t) \tag{1}$$

$$\frac{dy_i}{dt} = \beta_{i-1} x(t) y_{i-1}(t) - \beta_i x(t) y_i(t) - a y_i(t) + \sum_{j=i+1}^{\infty} \left(b_{j,i} + b_{j,i-j}\right) y_j(t) - \sum_{j=1}^{i-1} b_{i,j} y_i(t) \tag{2}$$

for $i = 1, 2, \ldots,$ since there could possibly be infinitely many aggregate sizes. They then considered two simplifying assumptions that allowed a moment closure. First, the conversion rate is assumed to be independent of aggregate size. This occurs when assuming that the templated conversion of normal protein to the misfolded state can only occur when interacting with either end of a linear aggregate. Second, the fragmentation of an aggregate is assumed to be linearly proportional to its size, and the fragmentation kernel is assumed to be uniform (i.e., fragmentation is equally likely to occur at any monomer junction in a linear aggregate). Under these two simplifications, the infinite system of differential equations has the following three-dimensional moment closure:

$$\frac{dx}{dt} = \lambda - dx(t) - \beta x(t)Y(t) \tag{3}$$

$$\frac{dY}{dt} = bZ(t) - (a+b)Y(t) \tag{4}$$

$$\frac{dZ}{dt} = \beta x(t)Y(t) - aZ(t) \tag{5}$$

where $Y(t) = \sum_{i=1}^{\infty} y_i(t)$ represents the total number of aggregates and $Z(t) = \sum_{i=1}^{\infty} iy_i(t)$ represents the total amount of prion protein. We note that mathematically $Y(t)$ and $Z(t)$ correspond to the zeroth and first moments of the distribution of aggregate sizes and, as such, the time evolution under these kinetic simplifications is determined by the zeroth and first moments.

Nowak and colleagues demonstrated this model (Eqs. (3)–(5)) was mathematically equivalent to previously developed viral models studied in mathematical epidemiology and derived a basic reproductive number for their model. The basic reproductive number, or R_0, is a common parameter in epidemiology and specifies the number of secondary infections (in this case infectious aggregate) created by an infectious aggregate in an initially susceptible population. In the case that $R_0 > 1$, we expect exponential growth of aggregates and therefore a stable nontrivial prion aggregate distribution. If $R_0 < 1$, we expect the prion aggregates to exponentially decay in number and the system to return to the aggregate-free state.

In their final formalization, Nowak and colleagues [39] considered the nucleated polymerization assumption of prion aggregates. Namely, it is believed that aggregates below a critical size are not thermodynamically stable. (This is thought to be part of the reason spontaneous appearance of prion disease is so rare because it is a nucleation-limited process [42, 43].) Since Nowak and colleagues did not consider spontaneous nucleation, the nucleus size enters only when aggregate fragmentation would create an aggregate of size below the minimum stable nucleation size n_0. In this case, the monomers in that aggregate return to the pool of normally folded protein. Under the previous simplifications on kinetic rates, this changes the resulting moment closure of the infinite system of ODEs as follows:

$$Y(t) = \sum_{i=n_0}^{\infty} y_i(t) \quad \text{and} \quad Z(t) = \sum_{i=n_0}^{\infty} iy_i(t) \tag{6}$$

$$\frac{dx}{dt} = \lambda - dx(t) - \beta x(t)Y(t) + b(n_0)(n_0 - 1)Y(t) \tag{7}$$

$$\frac{dY}{dt} = bZ(t) - (a + b(2n_0 - 1))Y(t) \tag{8}$$

$$\frac{dZ}{dt} = \beta x(t)Y(t) - aZ(t) - b(n_0)(n_0 - 1)Y(t). \tag{9}$$

Eqs. (7)–(9) (or appropriately modified Eqs. (1) and (2)) are referred to as the nucleated polymerization model (NPM). The dynamics of the NPM are similar to those presented in Nowak's first model; however, the minimal nucleus size modifies the resulting equations slightly. First, the quantities $Y(t)$ and $Z(t)$ now represent the aggregates above this critical minimal size, n_0.

The NPM, and its variants, has been extensively studied. In their paper, Masel, Jensen, and Nowak conducted an detailed analysis of the NPM [39] including extensive linking of experimental observations on the time to appearance of prion disease symptoms with the kinetic parameters of the NPM and determining a viable range of minimal nucleus sizes n_0. Overall, there was remarkable consistency between parameters predicted from different experimental data sets analyzed providing support at the time for this mathematical formulation. In addition, Masel et al. [39] (and then Greer and colleagues with a generalization [44]) demonstrated that the dynamics of aggregates under the NPM are consistent with the long incubation time observed for prion phenotypes. If prion disease begins with the introduction of a small amount of prion protein (in the form of aggregates), those aggregates will first have to increase in size until there are enough fragmentation sites to permit aggregate amplification through fragmentation.

Mathematicians continued to formalize the NPM through the twenty-first century. Prüss and colleagues [45] demonstrated that the prion phenotypes were globally asymptotically stable, and not merely locally stable, through deriving a Lyapunov function. Engler et al. [46] analyzed the well-posedness of a generalization of the NPM where aggregate sizes were continuous, instead of discrete. As such, rather than an infinite system of ordinary differential equations, the system consisted of a single ODE for protein in the normal configuration and a PDE specifying the distribution of aggregate sizes. While this formulation departs from the physically discrete nature of aggregates, in the limit of large aggregate sizes, these formalisms are provably equivalent [47], and the use of PDEs permits a wider array of mathematical techniques. Most notably, the continuous relaxation on aggregate sizes has permitted determination of the explicit asymptotic density [44, 46]. In comparison, the asymptotic density for the aggregate model with discrete aggregate sizes, while first approximated in 2003 by Pöschel et al. [48], was derived only recently by Davis and Sindi and required special functions [49]. Mathematical models of prion aggregate dynamics have been formulated under many more general kinetic assumptions such as nonlinear conversion rates, aggregate joining, and general fragmentation kernels (see [50–53]). More recently, models have been developed which consider prion aggregates along with other intracellular species (such as molecular chaperones) which impact their biochemical dynamics [54, 55], and others have considered interactions between prions and other protein aggregation diseases such as Alzheimer's [56].

2.2 Multicellular models

While some of the models discussed in the previous section were not explicit about their biological context, in nearly all cases the mathematical models of prion aggregation processes neglected the multicellular scale of the establishment of prion phenotypes. However, there is an emerging literature on multicellular models of prion disease dynamics. In all cases, the intracellular scale is either simplified or ignored.

2.2.1 Multicellular yeast colonies

Several models have emerged to consider the special case of heterogeneity of prion aggregates in yeast colonies. Such models have benefitted from the extensive work on yeast cell division which has facilitated the cellular scale of prion dynamics. We discuss two approaches taken to model this heterogeneity.

First, a number of models have emerged that allow for direct comparison between yeast "aggregate counting assays" [26, 57–60]. In such assays, an original cell is sampled, and and it is exposed to GdnHCl which is thought to severely reduce the fragmentation rate without altering the ability of cells to divide. Because the number of aggregates remains (relatively) constant while the number of cells increases exponentially, eventually each cell in the population is assumed to have at most one aggregate. When these resulting cells are restored to normal growth conditions (i.e., aggregate fragmentation is normal) and allowed to find their own independent colony, the number of [PSI^+] colonies observed is taken as reflective of the number of aggregates in the original cell. (For the purpose of narrative simplicity, we have greatly simplified both the experimental assay and its conclusions assay. Technically this assay counts the number of "propagons" in a cell, and here we assume that propagon is synonymous with prion aggregate although the true case is more complicated [6, 60].) Because the number of aggregates does not increase in most aggregate counting assays, this allows the intracellular processes to be greatly simplified. That is, the number of aggregates changes only through cell division when aggregates are distributed perhaps according to experimentally observed biases in cell division.

One recent study took a different approach by considering the related aggregate recovery assay using a generation and aggregate structured population model [61]. (Once again, the experimental process is too complicated to fully describe, so we offer a simplified description.) Aggregate recovery assays monitor the number of aggregates in a typical cell in time. This is a two-stage experiment where all cells are driven to have (ideally) at most one aggregate in a single cell. Then the cells are released, and aggregates will thus amplify within each cell, and cells will continue to grow and divide as normal. In this assay, cells from the recovering population are sampled in time, and the number of aggregates counted according to the assay in the previous paragraph. Both for mathematical simplicity and because this experiment only monitors the number of aggregates and not their size of concentration of protein in the normal conformation, Banks et al. [61] considered a simplified model of intracellular dynamics where the number of aggregates in a single cell increased according to either an exponential or logistic growth equation. They developed a structured population model that separately tracked the number of aggregates in cells having undergone 0, 1, 2, etc. cell divisions since the start of the experiment. Although this model greatly simplified the intracellular aggregate dynamics, researchers were able to recapitulate know biological relationships between prion strains.

At present, only two studies have considered both a detailed view of the intercellular aggregate dynamics and multicellular yeast population [27, 62]. Because these models were analytically intractable, both researchers developed a multi-scale stochastic simulation framework. Each cell in the population was modeled as a well-mixed compartment with discrete numbers of proteins in the normal conformation and in each possible aggregate size. The intracellular aggregate dynamics were forward simulated according to the Gillespie algorithm for sampling from the chemical master equation. During cell division aggregates were distributed between cells, and the lineage relationships between all cells were tracked. Tanaka et al. [62] used this stochastic multi-scale framework as a way to validate conclusions that drew from a system of ODEs they developed which considered population averages. Derdowski et al. [27] considered explicit comparisons with single-cell experiments in yeast. They used their model to conclude that in vivo fragmentation occurs in a rate-limiting fashion and that aggregate transmission was biased such that larger aggregates are retained by mother cells. However, neither study considered the spatial growth of cells in a colony and, as such, cannot be adapted to the question of yeast colony sectoring.

2.2.2 Mammalian tissue and organismal-level models

In mammals, mathematical models have been developed to study the spread of prion (and more generally protein aggregation) in the brain and between organisms. In the study of Alzheimer's, researchers have leveraged knowledge about the connectivity between regions of the brain to develop a network diffusion model of disease progression [63, 64]. Remarkably, they discovered that characteristic patterns of atrophy in patients can be explained by eigenelements of their network model. However, as in the models of aggregate counting assays in yeast, they assumed a highly simplified model of protein dynamics. That is, they did not consider molecular scale processes of aggregation but only the diffusion of an agent through the brain and associated atrophy itself with higher concentrations of the agent. However, because the early phases of mammalian prion and neurodegenerative are poorly characterized in vivo, it is possible that their models are consistent with late stages of neurodegenerative disease.

Finally, because prion disease spreads between animals in the same population, epidemic models have been employed to study the population-level dynamics of these diseases. In particular, researchers have studied the spread of scrapie in flocks of sheep [65–67] and chronic wasting disease in elk [68, 69]. Although these models have been informative of the specific mechanisms which prion disease is likely to spread in these populations (mother to offspring, indirect environmental transmission), they again consider only the organismal scale and not the molecular mechanisms responsible for within the host aggregation dynamics.

Cell-based models, a type of agent-based simulation framework, offer the ability to consider complex intracellular dynamics, distinguish between individual cells, and track the spatial distribution of cells within a colony. In the next section, we discuss common cell-based model formulations with a focus on the mathematical foundation for these models as well as their contributions to particular biological domains.

3. Cell-based modeling approaches

As described in Sections 1 and 2, prion diseases offer a particularly intriguing biological phenomenon for mathematical and computational analysis because such diseases cover many different systems and spatial scales. At the level of a population, prion diseases can be studied as a classical epidemic model where infections are spread among a susceptible population. Alternatively, prion diseases can be studied on a microscopic scale as a genetic disease whose phenotype is caused by prion aggregates that cause a gain of function mutation in certain genes. Current models of protein aggregation in yeast have successfully provided further insight into important mechanisms driving prion disease dynamics (i.e., conversion and fragmentation), but there is a need to develop models that consider the underlying microscopic processes of protein aggregation together with macroscopic properties of the environment in which they are taking place. This requires modeling frameworks that consider the impact of processes taking place on many different spatio-temporal scales. One such class of models that has been developed primarily for studying biological processes from many different scales is cell-based models.

In this section we review cell-based modeling approaches and their various applications. Cell-based models are mathematical models that represent individual cells as discrete entities and produce simulations to predict large-scale, collective behavior of a population or group of cells from the behavior and interactions of individual cells. The inputs to a cell-based model are experimentally observed cell behaviors including how individual cells respond to both intracellular and extracellular cues. Cell behaviors are encoded in a set of biologically relevant rules for cells to follow during simulations. The outputs of a cell-based model are measurable population-level and cell-level characteristics that follow nontrivially from cell–cell coordination and individual cell response to the changing microenvironment, including changes caused by the cells themselves.

Within this class of models a further distinction is made between lattice-based or cellular automaton (CA) models in which the particles live on the coordinates of a lattice and off-lattice methods that use real numbers to describe the coordinates of each particle. Biologically motivated CA models describe cell-cell and cell-environment interactions through phenomenological local rules, which allow for efficient simulation of many different biological systems ranging from bacteria, excitable media, and chemotactic aggregation to chicken embryonic tissues and tumors. (For reviews see [70–77].) CA models are well suited for studying the dynamics of systems with a large number of cells, i.e., growing monolayers or tumors, because their computation time is efficient and they allow for systematic sensitivity analysis of different parameters [70, 73, 75, 76, 78, 79]. In addition, CA models provide an easy starting point from which to derive continuum equations for cell densities which are particularly helpful in studying the mechanisms driving pattern formation and the growth process of multicellular systems with billions of cells [73, 78, 80, 81]. However, CA models are rule-based, and model parameters do not represent a direct physical description of biophysical properties of cells [73, 78]. For this reason, we focus our discussion in this chapter on lattice-free modeling approaches.

In lattice-free modeling approaches, individual cells are represented using either a single particle or agent or a group of particles or agents that are free to move to any location in the computational domain [75, 76, 78, 82–86]. In simulations, each biological cell is modeled as a discrete entity and endowed with well-defined individual characteristics such as intracellular reaction kinetics or detailed biophysical properties. As stated by T. Newman, cell shape and cell response to local mechanical forces is one of the most challenging biological characteristics to build into a cell model [78]. For this reason, lattice-free modeling approaches were developed as a way to explicitly model cell shape as well as investigate how individual cells respond to local mechanical forces and interactions with neighboring cells.

In lattice-free modeling approaches, interactions between agents are described using forces or potential functions, and the position of an individual cell is determined by solving an equation of motion for each agent that belongs to that cell. Moreover, it is standard to make the following two simplifying approximations for the equation of motion of each agent: (1) motion may be described by considering each vertex to be embedded in a viscous medium that applies a drag force on it with mobility coefficient η, and (2) inertia is vanishing [83, 87–90]. This leads to first-order dynamics with the evolution of the position $\mathbf{x}_i$ of agent i to be determined by

$$\eta \frac{d\mathbf{x}_i}{dt} = \mathbf{F}_i \qquad (10)$$

where $\mathbf{F}_i(t)$ denotes the total force acting on vertex i at time t. The mobility coefficient η determines the timescale over which mechanical relaxation occurs and is often referred to as the damping coefficient.

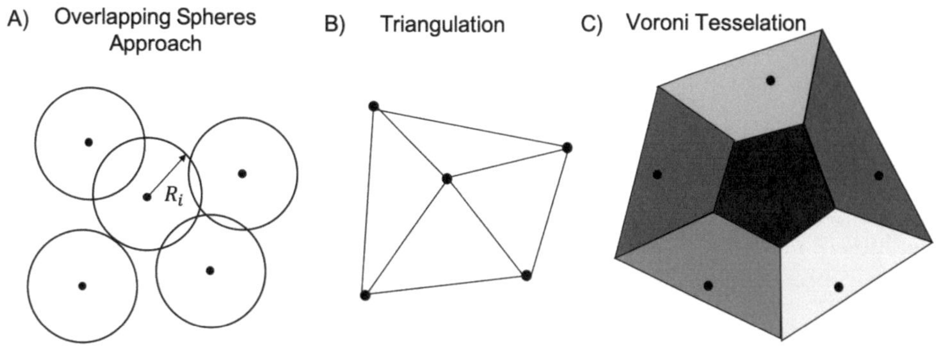

Figure 3.
Determining cell connectivity in center-based models. (A) In the overlapping spheres approach, each cell has an intrinsic radius, R_i, and two cells interact if their corresponding spheres overlap, i.e., if they are within a specified fixed distance from each other. (B) Triangulation involves computing a triangulation of the simulation domain using cell centers as nodes. Edges of the resulting mesh determine which cells are connected. (C) The Voronoi tessellation determines a region for each cell defined as a set of points in the plane that are closer to that cell center than any other.

Lattice-free modeling approaches vary in complexity from those that track the center of each cell as a single point [91–93] to those that represent individual cells as collections of membrane and cytoplasm nodes to allow for more biologically relevant, emerging cell shapes [89, 94, 95]. Although a main advantage of CA methods has been their computational efficiency, as computational power continues to increase, lattice-free modeling approaches offer equally powerful tools for realistically modeling biological cell behaviors such as shape change, polarized cell growth, division, differentiation, and intercellular signaling dynamics in great detail without adding computational complexity. Lattice-free models have been used to investigate the impact of individual cell behaviors and make predictions about mechanisms controlling many different biological and pathological processes such as determining shape of a tissue, size of a cell colony, or biological function of an organ [75, 76, 78, 82, 84, 86, 96, 97]. In this section, we review three main categories of lattice-free modeling approaches, namely, center-based models, vertex models, and subcellular element models, as well as provide examples of how lattice-free modeling approaches have been used to study various biological problems.

3.1 Center-based models

Center-based modeling (CBM) approaches, or cell center models, track the center of each cell as a single point and can be classified by two main components, a definition of cell connectivity (**Figure 3**) and a definition of how cells interact or a force law (**Figure 4**) [76, 78, 83, 91]. There are several commonly used methods for determining cell connectivity, including the overlapping spheres approach, triangulation, or Voronoi tessellation as described in **Figure 3**.

In center-based modeling approaches, the cell-cell interaction force is usually written as a function of the location of each cell centers and almost always acts in the direction of the vector connecting the two interacting cells [76, 78, 83, 91]. Let r_i be the position of the cell center of cell i, and let $\mathbf{F}_{ij}$ denote the force on cell i due to cell j. If we assume that the force $\mathbf{F}_{ij}$ acts in the direction of the vector connecting the cells, then the force on cell i due to cell j is written as

$$\mathbf{F}_{ij} = F_{ij} \frac{r_i - r_j}{\|r_i - r_j\|} \tag{11}$$

where F_{ij} is the signed magnitude of $\mathbf{F}_{ij}$. The total force on cell i is then

$$\mathbf{F}_i = \sum_j \mathbf{F}_{ij} \tag{12}$$

where the sum is over all cells j connected to cell i. As mentioned above, this force is usually taken to be balanced by a viscous drag as the cells move, so that the equation of motion for r_i, the position of the cell center of cell i, is:

$$\gamma \frac{dr_i}{dt} = \mathbf{F}_i \tag{13}$$

where γ is the viscosity coefficient which could, for example, represent adhesion between a cell and the underlying substrate. In simulations there are several choices for the numerical method used to update the position of each cell center, but the most simple is forward Euler discretization. Thus, the position of a cell center at time $t + \Delta t$ is given by

$$r_i(t + \Delta t) = r_i(t) + \frac{\Delta t}{\gamma} \mathbf{F}_i. \tag{14}$$

The cell–cell interaction force, $\mathbf{F}_{ij}$, is usually a function of the overlap, δ_{ij}, between two interacting cells which is given by

$$\delta_{ij} = R_i + R_j - \|r_i - r_j\| \tag{15}$$

where R_i and R_j are the radii of cell i and j, respectively, and r_i and r_j are the locations of the centers of cell i and cell j, respectively. Different definitions of the contact area between cells are possible to further extend the model [76, 78, 83, 91]. There exists a wide range of force laws that have been used in the literature, ranging from simple linear laws to more complex nonlinear models that can incorporate nonhomogeneous properties such as cell-cell adhesion, cell-substrate adhesion, and

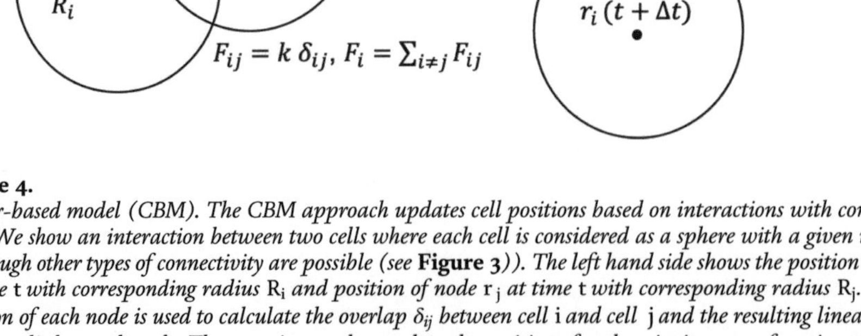

Figure 4.
Center-based model (CBM). The CBM approach updates cell positions based on interactions with connected cells. We show an interaction between two cells where each cell is considered as a sphere with a given radius (although other types of connectivity are possible (see **Figure 3***)). The left hand side shows the position of node* r_i *at time* t *with corresponding radius* R_i *and position of node* r_j *at time* t *with corresponding radius* R_j. *The position of each node is used to calculate the overlap* δ_{ij} *between cell* i *and cell* j *and the resulting linear force* F_{ij} *to be applied to each node. The equation used to update the position of node* r_i *is given as a function of the force* F_{ij}, *time step* Δt, *and damping coefficient* γ. *The right hand side shows the new positions of node* r_i *and* r_j *at time* $t + \Delta t$ *after each position has been updated. (See Section 3.1 for more information).*

cell bond breaking [76, 78, 83, 91]. For simplicity, we consider the linear law given by

$$F_{ij} = k_1\delta_{ij} \tag{16}$$

where k_1 is a stiffness parameter. Since the linear law implies there is a large attraction between distant cells, it is reasonable to define a cutoff range for interacting cells such that

$$F_{ij} = \begin{cases} k_1\delta_{ij} & \text{for} \quad \delta_{ij} \geq \delta_{min} \\ 0 & \text{for} \quad \delta_{ij} < \delta_{min}. \end{cases} \tag{17}$$

Some other choices for force laws include the Johnson-Kendall-Roberts (JKR) model [83, 91], the Hertz model [98, 99], and harmonic-like interaction [100–102]. The JKR model describes the interaction between two isotropic homogeneous spheres that are strongly adhesive, i.e., as soon as the two cells come into contact range, they immediately form a contact area of finite size by active deformation of the cell membrane. However, a unique property of the JKR model is that when two cells are pulled apart, they continue to interact for distances greater than the standard contact range but less than a specified distance (i.e., $\delta_{min} > \delta_{ij} > \delta_{ij}^c$). This phenomenon is referred to as hysteresis behavior. Similar to the JKR model, the Hertz model approximates each cell as a homogeneous sphere, but in the Hertz model, hysteresis behavior is not included. In the Hertz model, the potential interaction is represented as the sum of a repulsive and an attractive force (i.e., $F_{ij} = F_{ij}^a ttr + F_{ij}^r ep$). On the other hand, harmonic-like force laws provide a simple model for the short-range cell-cell interaction due to cell-cell adhesion and elastic deformation that approximate two adhesively interacting cells by cuboidal objects with a non-deformable core, linked by linear springs. Each of these definitions for cell-cell interaction force offers their own set of advantages and disadvantages.

3.2 Application of center-based models in biology

Center-based models have been used to analyze multicellular processes in tumors [91, 99, 100], intestinal crypts and epithelial tissues [85, 103], cell migration in extracellular matrix [85, 100], and tissue regeneration, growth, and organization [91, 99, 100, 103, 104]. In addition, several center-based models have been developed for studying macroscopic properties of yeast colonies [92, 93, 105]. (See Section 4 for more details.) Although the mechanical information that can be extracted from these models is rather approximate, various biophysical aspects in these problems have explained important mechanisms for proper cell organization and speed of growth in tissues [78, 97, 106].

3.3 Vertex models

Two-dimensional vertex models are a well-known class of lattice-free, cell-based models that provide a more detailed description of cell shape [84, 87, 107, 108]. In vertex models, each cell membrane is represented as a polygon composed of vertices and edges that are shared between adjacent cells (**Figure 5A**). A set of rules or an equation of motion defines how each vertex moves, and collective movement of vertices leads to changes in cell shape over time (**Figure 5A**). Most often, the equation of motion is a force or potential function based on the current

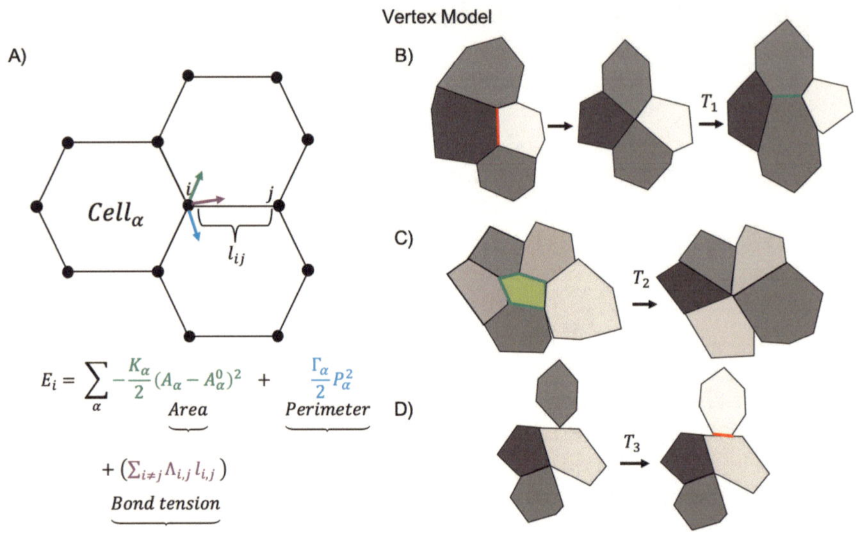

Figure 5.
Vertex model. Vertex models depict each cell membrane as a polygon of vertices and edges that are shared between adjacent cells. At each time step, an equation of motion determines how each vertex moves, and collective movement of vertices over time results in changes in cell shape. (A) Example forces acting on vertex i due to interactions with shared vertices in cell α according to Eq. (18). (B–D) Cells in the model form and break bonds due to neighbor rearrangements. Here we describe three basic cell neighbor rearrangements accounted for by most vertex models [87, 107]. (B) T1 transitions occur when two vertices sharing a short edge (defined by a minimum distance much smaller than the average length of an edge) merge into a single vertex which is then reassigned to two new vertices, thus changing the local network topology. (C) T2 transitions occur when a given cell shrinks to an approximately zero area and is removed to represent apoptosis. (D) T3 transitions occur when the intersection of a vertex with an edge is avoided by replacing the approaching vertex with two new vertices. For more detail see Section 3.3.

configuration of vertices, i.e., location and connection between pairs of vertices, as well as other geometrical features such as area or perimeter of cells [84, 87, 107, 108]. Many vertex models also incorporate rules to govern changes in connections between vertices to allow for rearrangements in cell neighbor relationships [107]. The precise equations of motion used differ between models and may be adapted to suit each particular biological problem. Below we review some of the common forms of the equation of motion and typical rules for rearrangements of vertex connectivity.

In most cases, the equation of motion for vertices is deterministic since a reasonable assumption for tightly packed cell aggregates is that stochastic motion of cells is mitigated by strong interactions between cells [84, 87, 107, 108]. One difference among vertex models in the literature lies in the definition of the force. The choice of form for the function $\mathbf{F}_i$ reflects which forces are thought to dominate cell mechanics for the system being studied. Some commonly modeled forces include tension or elastic forces, due to the combined action of a cell's actomyosin cortex [107] and adherens junctions [107], or pressure, due to hydrostatic pressure [87]. Note that the forces acting on each vertex may either be given explicitly, or else an energy function may be specified, whose gradient is assumed to exert a force on each vertex.

For a concrete example, we will consider an equation of motion built from an energy function given by Farhadifar et al. which has recently seen extensive use in modeling wing disk epithelia [87]. This energy function encodes constraints associated with the limited ability of cells to undergo elastic deformations, volume changes, and other movements due to adhesion to other cells. The energy function is defined by

$$E = \sum_{\alpha=1}^{N} \left(\frac{K_\alpha}{2} \left(A_\alpha - A_\alpha^{(0)} \right)^2 + \frac{\Gamma_\alpha}{2} P_\alpha^2 \right) + \sum_{i,\, j} \Lambda_{ij} l_{ij} \tag{18}$$

where the area elastic modulus K_α, bond tension parameter Λ_{ij}, perimeter coefficient Γ_α, and preferred cell area $A_\alpha^{(0)}$ are all model parameters whose values must be chosen in a biologically relevant manner.

In vertex models that use this energy function, the choice of preferred cell area $A_\alpha^{(0)}$ depends on how cell growth is incorporated into the model. For a given energy function U, the force on each vertex is determined by the negative derivative of the energy with respect to the coordinates of that vertex $F_i = -\nabla_i U$, where ∇_i denotes the gradient operator evaluated at $\mathbf{x}_i$. Computing the gradient of Eq. (18) and exploiting the fact that the movement of vertex i affects only the energy of the cells associated with it, the force F_i may be written explicitly as

$$F_i = \sum_{\alpha \in S_i} \left(-K_\alpha \left(A_\alpha - A_\alpha^{(0)} \right) \nabla_i A_\alpha + \Gamma_\alpha P_\alpha \nabla_i P_\alpha \right) - \sum_{j} \Lambda_{ij} \nabla_i l_{ij} \tag{19}$$

where now the first sum runs over cells associated with vertex i and the second sum runs over vertices sharing an edge with it.

Many of the earliest vertex models simulated cross-sections of epithelial tissues and so did not account for cell neighbor rearrangements [87]. However, later models were formulated to simulate top-down dynamics of epithelial cells in a plane which required modeling cell neighbor rearrangements [107]. Namely, to accurately describe planar epithelial cell dynamics, cells in the model must be allowed to form and break bonds by changing connectivity among vertices as described in **Figure 5**.

3.4 Application of vertex models in biology

Vertex models were initially developed to study the packing of bubbles in foams [109]. Similar to foams, cells in epithelial tissues are tightly packed and mechanically coupled to their neighbors by adhesion molecules along their common interfaces [84, 87, 107, 108]. In addition, cells exert forces onto each other and their environment. Since vertex models are well suited for modeling tightly packed cell ensembles with negligible intercellular space, the vertex modeling framework was later adapted to study two-dimensional packing and rearrangement of apical cell surfaces in planar epithelia [84, 87, 107, 108]. In these studies, epithelial cells are described by a planar two-dimensional network of vertices that defines the apical cell surfaces as polygons with straight interfaces between neighboring cells. The underlying assumption of two-dimensional apical vertex models is that the main forces acting to deform the cells are generated along the apical cell surfaces or can be effectively absorbed in the apical representation. The work of Odell et al. [110] first demonstrated the use of vertex models to study how spatial patterning in either active forces or passive mechanical properties may lead to tissue deformation. Subsequent studies by several groups have examined a variety of alternative patterns of force and material properties that can give rise to similar tissue deformations [84, 87, 88, 107, 108, 111].

An important behavior of biological cells that does not affect the dynamics of foams but plays a large role in the structure of biological tissues is the ability of cells to grow, divide, and die. Model components representing cell division and death require the modification of tissue connectivity which can add significant computational complexity, but these cell behaviors are essential mechanisms of many of the biological process being studied. Thus, one biological question that has been addressed using vertex models extended to include cell death and division is the control of packing geometries in an epithelial sheet [84, 87, 88, 107, 108, 111]. A highly cited example is the work of Farhadifar et al. [87] who performed a systematic analysis of the equilibrium cell packing geometries and their dependence on cell mechanical and proliferative parameters using the *Drosophila* wing disk as a model system. By comparing simulations with experimental results, the authors arrived at a set of parameter values for which their model accounts for vertex movements based on biologically relevant cell area variations, cell rearrangements due to laser ablation, and epithelial packing geometries seen in vivo. This work demonstrates how vertex models may be parametrized and tested using experimental data.

In many biological systems, patterns of mechanical stress and resulting macroscopic tissue shape changes may happen concurrently and can feed-back into each other. Vertex models can be easily modified to incorporate both mechanical and chemical feedbacks. For example, Odell et al. [110] included a simplified feedback in their earliest model by assuming that contraction of individual cells was activated by high levels of stretching due to tissue deformation. Although vertex models have been successful in testing many of the mechanisms that govern the macroscopic behavior of tightly packed epithelial tissues, this class of model generally ignores contributions from important components such as cell-matrix interactions [112], actomyosin contractility and other biophysical properties of the cell membrane [113], and active remodeling of cytoskeletal components. Thus, there have been several notable extensions of the vertex model to address these important factors [110, 114, 115]. Although vertex models provide a promising modeling framework for further investigation of different biological systems, in the next section, we review a different modeling framework that was developed to take into account more detailed representations of cell deformation and provide an extension to the types and complexity of mechanical stress patterning and feedback considered.

3.5 Subcellular element model

The subcellular element (SCE) model provides a framework for simulating lattice-free multicellular structures in which the shape of each cell dynamically emerges from interactions with the local environment due to model assumptions about the underlying mechanical properties of the system [82, 86, 89, 90]. In the model, each cell is composed of a large, and possibly varying, number of small nodes called subcellular elements (**Figure 6**). Each subcellular element of a cell is modeled as a single point at its center of mass, which changes position over time subject to three processes: (i) weak random fluctuations, (ii) elastic interaction with elements of the same cell, and (iii) elastic interaction with elements of other cells.

In the SCE modeling approach, the membrane and cytoplasm of each cell can be represented together as one set of homogeneous elements/nodes or separately using two different sets of elements/nodes (**Figure 6**). Collective interactions between pairs of nodes from the same cell represent bulk cytoplasmic contents, and collective interactions between pairs of adjacent nodes from neighboring cells represent volume exclusion of cells as well as adhesive properties. In models that consider the membrane and cytoplasm separately, collective interactions between pairs of cytoplasm nodes and membrane nodes represent the cytoplasmic pressure of the cell applied to the cell membrane. Biomechanical and adhesive properties of cells are modeled through viscoelastic interactions between elements represented by phenomenological potential functions such as linear springs or Morse potential functions, which are used to simulate close-range repulsion (modeling volume exclusion

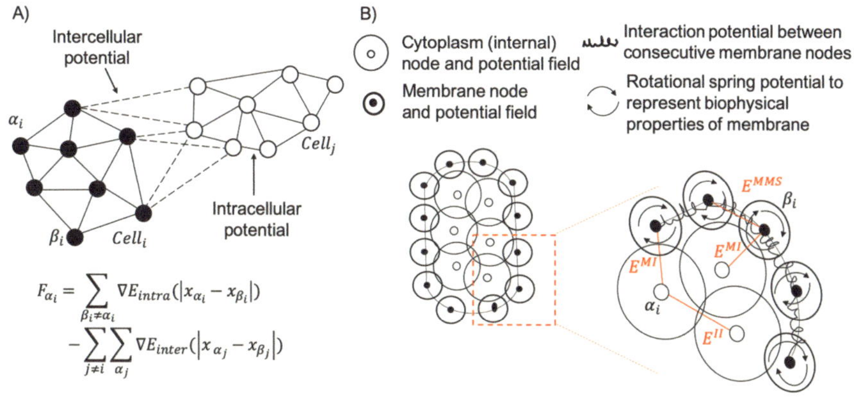

Figure 6.
Subcellular element model (SCE). The SCE model uses sets of elements/nodes to represent individual cells. At each time step, an equation of motion is used to update the location of each node based on resulting forces from interactions between nodes due to cell growth, deformation, etc. (A) Illustration of a SCE model where membrane and cytoplasm of individual cells are represented by one set of homogeneous elements/nodes. Elements/nodes of the same cell are held together via intracellular interactions (solid lines) modeled using short-range potential functions (∇E_{intra}). Elements/nodes of two neighboring cells interact via intercellular interactions (dashed line) modeled using potential functions that weakly bind adjacent elements/nodes (∇E_{inter}). The force applied to node α_i is calculated based on interactions with all other nodes $\beta_i \in cell_i$ and adjacent nodes α_j in neighboring cell, $cell_j$. (B) SCE model where the membrane (filled nodes) and cytoplasm (clear nodes) of one cell are represented separately using two different sets of elements/nodes. The force applied to cytoplasm node α_i is calculated based on intracellular interactions which include interactions with all other cytoplasm nodes in the same cell (E^{II}) and interactions with all other membrane nodes of the same cells (E^{MI}) via short-range potential functions. The force applied to membrane node β_i is calculated based on both intracellular interactions and intercellular interactions. Intracellular interactions for membrane node β_j include interactions with all cytoplasm nodes of the same cell (E^{MI}) and interactions with adjacent membrane nodes of the same cell (E^{MMS}), and intercellular interactions for membrane node β_j include interactions with adjacent membrane nodes from neighboring cells. For more details about these interactions, refer to Section 3.5.

of neighboring segments of cytoskeleton) and medium-range attraction between elements of the same or different cells (modeling the adhesive forces between segments of cytoskeleton) [116].

The potential functions described above are used in model equations to calculate the displacement of each element/node at each time step based on their interactions with neighboring elements/nodes resulting in the deformation of cells within the tissue (**Figure 6**). As mentioned previously, nodes are assumed to be in an overdamped regime so that inertia forces acting on the nodes are neglected [83, 87–90]. This leads to the following two equations of motion describing the movement of element/node α_i in cell i:

$$F_{\alpha_i} = -\left(\sum_{\alpha_i \neq \beta_i} \nabla E_{\text{intra}} - \sum_{i \neq j} \sum_{\alpha_i} \nabla E_{\text{inter}}\right) \tag{20}$$

where η is the damping coefficient, x_i is the position of node i, N is the total number of nodes, and ∇E is the potential function used to describe the interaction of node i with node j. This equation is discretized in time using the forward Euler method, and the position of node x_i is incremented at discrete times as follows:

$$x_{\alpha_i}(t + \Delta t) = x_{\alpha_i}(t) - F_{\alpha_i} \frac{\Delta t}{\eta} \tag{21}$$

where Δt is the time step size.

One of the important features of the SCE modeling approach is the ability to change parameters of the potential functions that are used to describe interactions between elements to calibrate model representations of biomechanical properties of a particular type of a cell directly using experimental data. More specifically, the SCE model can be used to perform in silico bulk rheology experiments on a single cell in order to scale the parameters such that the passive biomechanical properties of each cell are independent of the number of elements used to represent each cell [117]. As a result, SCE simulation output captures the underlying biomechanical properties of the real biological system being studied and remains relevant regardless of the choice of the number of elements used in the model.

As indicated in Fletcher et al. [84], computational experiments follow a creep-stress protocol in which a constant extensile force is applied to the end of an SCE cell whose opposite end is fixed. Before the extensile force is released, the strain is measured as the extension of the cell in the direction of the force relative to its initial linear size. In silico estimates of the viscoelastic properties of cells modeled using the SCE approach have been shown in many biological applications to agree with in vitro rheology measurements [117, 118]. This indicates that the simple phenomenological dynamics of the SCE modeling approach are enough to capture low to intermediate responses of cytoskeletal networks over short timescales (~10s) [118]. Over longer timescales (~100 s), cells respond actively to external stresses by undergoing cytoskeletal remodeling, and this phenomenon can be incorporated into the SCE modeling approach by inserting and removing subcellular elements of a cell in regions under high or low stress [86].

The generalized Morse potential functions implemented in SCE modeling approach are commonly used in physics and chemistry to model intermolecular interactions [119] and in biology to represent volume exclusion of neighboring regions of the cytoskeleton [94, 95, 120–125]. While it is difficult to associate specific potential functions directly with specific cytoskeletal components of cells, computational studies of bulk properties at the tissue level have suggested that the precise functional form of the potential used in modeling has a small impact on overall system dynamics [97, 117].

3.6 Application of SCE models in biology

The subcellular element (SCE) modeling approach has been successful in modeling mechanical properties of individual cells as well as their components and determining individual cell impact on the emerging properties of growing multicellular tissue as well as describing cellular interactions with mediums such as the extracellular matrix and fluids [76, 84, 86, 89, 90, 95, 96, 111, 117, 120, 122, 124–127]. The general modeling framework was initially developed by Newman et al. [89] for simulating the detailed dynamics of cell shapes as an emergent response to mechanical stimuli. Recent applications of the SCE modeling approach show that it is flexible enough to model additional diverse biological processes such as intracellular signaling [122], cell differentiation [94], and motion of cells in fluid [124].

To date, the SCE has not been widely used to study biological processes outside the area of epithelial morphogenesis. Christley et al. [122] developed a model of epidermal growth on a basal membrane that incorporates cell growth through the addition on new elements and division by redistributing a cell's group of elements between two new daughter cells. This mode was a novel implementation because it was coupled to a subcellular gene network representing intercellular Notch signaling. The SCE modeling framework has also been coupled to a fluid flow model to simulate the attachment of platelets to blood vessel walls [124]. Using the SCE framework for modeling individual platelets in simulations provides a detailed mechanical model of individual platelet behavior that allowed the authors to examine the relationship between platelet stiffness and movement in the fluid. This facilitated the generation of new hypothesis about mechanisms platelets use to adhere to injured sites on the blood vessel walls. In the context of developmental biology, the primary application of the SCE to date has been a computational study of primitive streak formation by Sandersius et al. [86]. In addition, the SCE model has been applied to model development in both animal [95] and plant [94] systems.

4. Conclusion

In this chapter, we described the medical and scientific importance of studying misfolded protein diseases as well as provided a broad description of several classes of mathematical models that can be used to further investigate the underlying mechanisms governing protein aggregation and propagation in a multicellular system. In Section 2, we gave an overview of different modeling frameworks that have been developed and validated for studying protein aggregation in both yeast and mammalian systems. In Section 3, we described in detail various cell-based modeling frameworks that have been successfully used to gain insight about the impact of individual cell behaviors on macroscale properties of tumors, developmental tissues, and yeast colonies. However, our main goal in this chapter was to familiarize the reader with each class of model in order to facilitate further discussion about how these two types of models could be combined to develop a more complete representation of prion disease dynamics within an actively growing and dividing yeast colony.

While current models of protein aggregation in yeast have successfully provided further insight into important mechanisms driving prion disease dynamics (i.e., conversion and fragmentation), they have been unable to recreate a number of important physiological characteristics including variable phenotype induction rates that result in sectored phenotypes among a single yeast colony (see Section 1 for details). One reason for this may be that a large number of models representing intracellular dynamics of protein misfolding diseases were developed for studying protein aggregation dynamics in isolation and these models disregard the contribution of individual cell behaviors within the growing yeast colony as a possible mechanism governing prion disease dynamics. A major open question in prion biology is to understand how prion aggregates spread between cells within a whole colony or tissue. Experimental observations such as sectoring provide compelling data that transmission mechanisms other than what is addressed by current aggregation-only-models must play a role in the presence and persistence of prion disease phenotypes in yeast colonies, i.e., processes such as conversion, fragmentation, nucleation, and even enzyme-mediated fragmentation alone cannot entirely explain the spread of prion disease throughout a yeast colony. Thus, in order to test hypotheses about the impact of individual cell behaviors on the spread of misfolded proteins, it is necessary to develop a novel modeling framework.

Developing a modeling framework for investigating prion disease dynamics within an entire yeast colony is challenging because it requires capturing the physical processes on an individual cell level that determine colony growth (i.e., budding and variable cell cycle length) as well as capturing the interplay of individual cell processes with protein aggregation dynamics (i.e., asymmetric protein distribution at the time of division, persistence of diseased phenotype/aggregate that was given to daughter while it grows to begin a new cell cycle, lineage-dependent protein propagation). Several models have already been developed to investigate physical mechanisms controlling patterns of yeast colony growth such as cell division polarity, mother-daughter size asymmetry, and cell-cell adhesion via budding of the new daughter cell [92, 93, 105]. Jönsson and Levchenko developed an off-lattice, center-based model in which cells are modeled as elastic spheres of variable size [92]. Their work showed that cell growth inhibition by neighboring cells and polar division growth patterns were the most significant factors driving appreciable differences in the shape and size of yeast colony development. In addition, Wang et al. built an off-lattice, center-based model that incorporates key biological processes in yeast cell colonies including budding, mating, mating type switch, changes in cell cycle length and cell size due to aging, and cell death [93]. The main results of their work include proposed mechanisms for how budding patterns in yeast cells affect colony growth. Both of these studies serve as excellent starting points for developing a modeling framework that includes the physical processes and individual cell behaviors of yeast cells that govern colony growth.

One way forward would be to extend current cell-based models for yeast colony growth by combining them with a detailed model of protein aggregation dynamics as well as implementing specific model components that capture the interplay of individual cell behaviors with protein aggregation dynamics. Specifically, the new modeling framework would need to include asymmetric protein distribution at the time of division and varied cell cycle lengths to represent how growth of a new daughter cell impacts persistence of aggregates that were inherited at time of division and track protein concentration through different cell lineages to investigate the impact of lineage on colony phenotypes in yeast. In the model, prion dynamics would need to be simulated within each individual cell using models of intracellular dynamics described in Section 2. In addition, using a center-based model would allow for the spatial arrangement of cells to account for the effect of biophysical properties such as increased adhesion between a mother and daughter cell during budding.

Preliminary simulations (**Figure** 7) show that individual cell behaviors indeed impact yeast colony structure as well as protein aggregation dynamics in a growing yeast colony. For our preliminary framework, we extended current cell-based models for yeast colony growth by combining them with a simplified model of protein aggregation dynamics as well as implementing asymmetric protein distribution at the time of division and strong adhesion between mother and daughter cells during budding. These are two cell behaviors we hypothesize to play an important role in protein aggregation dynamics in yeast colonies. Our simulations begin with a single founding cell that divides to give birth to the first daughter. All cells continue to grow and divide throughout the simulation and stop once the colony reaches about 1500 cells total. Successive daughter cells of the founder cell are enumerated in the order of their division. For example, the first daughter is the cell born after the first division of the founder cell, the second daughter is the cell born after the second division of the founder cell, and so on.

Figure 7A shows sub-colony structure for each of nine daughters cells born from the original founder cell of the colony represented in black. The first daughter sub-colony (grayish black) occupies a larger area than any of the other sub-colonies. In addition, the average location of each successive generation is located closer to the boundary (**Figure 7B**). Protein aggregation dynamics were modeled using a logistic growth function with a strong Allee effect. **Figure 7C** shows the number of aggregates per cell in a simulation that resulted in a sectored colony. The interplay between protein aggregation within each cell and individual cell behaviors results in complete loss of aggregates in an early daughter cell, and subsequently, the sub-colony generated by that daughter remains free of aggregates throughout the simulation. These results suggest that the unified model may have the potential to predict mechanisms underlying experimentally observed phenomena such as

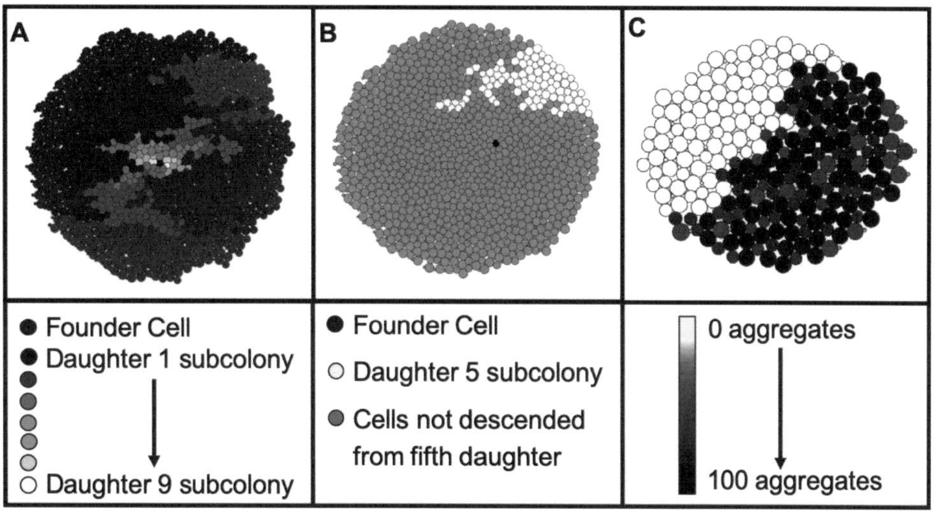

Figure 7.
Impact of individual cell behaviors on yeast colony structure and protein aggregation. In our preliminary simulations, we extended current cell-based modeling approaches for yeast colony growth by combining them with a simplified model of protein aggregation dynamics along with asymmetric protein distribution at cell division and strong adhesion between mother and daughter cells during budding. Simulations begin with a single founder cell and stop after 15 hours when the colony contains about 1500 cells total. (A) Example of sub-colony structure for each of nine daughter cells born from the original founder (black cell). Each sub-colony is depicted with a different color indicated below. (B) We show in white the fifth-generation sub-colony corresponding to the fifth daughter of the founder cell and all of its offspring. (C) Example of the interplay between protein aggregation dynamics and individual cell behaviors in a simulation leading to a sectored colony. Protein aggregation dynamics for individual cells were modeled using a logistic growth function with a strong Allee effect which results in stable extinction of protein aggregates in the sub-colony of an early daughter cell. The number of aggregates per cell is represented by different colors given in the key below.

sectored prion phenotypes in yeast colonies in addition to serving as a tool for future hypothesis generation and testing.

Although our preliminary studies appear promising, there is still a large gap to fill in terms of developing models that consider underlying microscopic processes of protein aggregation together with macroscopic properties of the environment in which they are taking place. Indeed, as mentioned in Section 2 there have been recent efforts to model the impact of tissue structures on the spread of protein misfolding diseases in mammalian systems [63, 64]. However, most of the large-scale models for protein aggregation lack details of components at the microscopic scale that would allow for the study of the interplay of different temporal and spatial scales. We hope that future modeling studies will begin to incorporate both scales in order to test hypotheses about mechanisms that can explain unresolved experimental data and yield new strategies for treating protein misfolding diseases.

Author details

Mikahl Banwarth-Kuhn and Suzanne Sindi*
Department of Applied Mathematics, University of California-Merced, Merced, USA

*Address all correspondence to: ssindi@ucmerced.edu

References

[1] Brundin P, Melki R, Kopito R. Prion-like transmission of protein aggregates in neurodegenerative diseases. Nature Reviews Molecular Cell Biology. 2010; **11**(4):301-307

[2] Knowles TPJ, Vendruscolo M, Dobson CM. The amyloid state and its association with protein misfolding diseases. Nature Reviews Molecular Cell Biology. 2014;**15**(6):384-396

[3] Wang H, Rhoads DD, Appleby BS. Human prion diseases. Current Opinion in Infectious Diseases. 2019;**32**(3): 272-276

[4] Zabel MD, Reid C. A brief history of prions. Pathogens and Disease. December 2015;**73**(9):ftv087. Available from: https://doi.org/10.1093/femspd/ftv087

[5] Ironside JW, Ritchie DL, Head MW. Prion diseases. Handbook of Clinical Neurology. 2018;**145**:393-403

[6] Sindi SS, Serio TR. Prion dynamics and the quest for the genetic determinant in protein-only inheritance. Current Opinion in Microbiology. 2009;**12**(6):623-630

[7] Inga Zerr KK, Summers DM, Romero C, Taratuto A, Heinemann U, Breithaupt M, et al. Updated clinical diagnostic criteria for sporadic creutzfeldt-jakob disease. Brain. 2009; **132**(10):2659-2668

[8] Medori R, Tritschler H-J, LeBlanc A, Villare F, Manetto V, Chen HY, et al. Fatal familial insomnia, a prion disease with a mutation at codon 178 of the prion protein gene. New England Journal of Medicine. 1992;**326**(7): 444-449

[9] Plummer PJG. Scrapie—A disease of sheep: A review of the literature. Canadian Journal of Comparative Medicine and Veterinary Science. 1946; **10**(2):49

[10] Sigurdson CJ, Aguzzi A. Chronic wasting disease. Biochimica et Biophysica Acta (BBA)-Molecular Basis of Disease. 2007;**1772**(6):610-618

[11] Collinge J, Whitfield J, McKintosh E, Beck J, Mead S, Thomas DJ, et al. Kuru in the 21st century—An acquired human prion disease with very long incubation periods. The Lancet. 2006;**367**(9528): 2068-2074

[12] Brown P, Will RG, Bradley R, Asher DM, Detwiler L. Bovine spongiform encephalopathy and variant creutzfeldt-jakob disease: Background, evolution, and current concerns. Emerging Infectious Diseases. 2001; **7**(1):6

[13] Brotherston JG, Renwick CC, Stamp JT, Zlotnik I, Pattison IH. Spread of scrapie by contact to goats and sheep. Journal of Comparative Pathology. 1968; **78**(1):9-17

[14] Manix M, Kalakoti P, Henry M, Thakur J, Menger R, Guthikonda B, et al. Creutzfeldt-jakob disease: Updated diagnostic criteria, treatment algorithm, and the utility of brain biopsy. Neurosurgical Focus. 2015;**39**(5):E2

[15] Goedert M, Masuda-Suzukake M, Falcon B. Like prions: The propagation of aggregated tau and α-synuclein in neurodegeneration. Brain. 2017;**140**(2): 266-278

[16] Tuite MF, Serio TR. The prion hypothesis: From biological anomaly to basic regulatory mechanism. Nature Reviews Molecular Cell Biology. 2010; **11**(12):823-833

[17] Jaunmuktane Z, Mead S, Ellis M, Wadsworth JDF, Nicoll AJ, Kenny J, et al. Evidence for human transmission

of amyloid-β pathology and cerebral amyloid angiopathy. Nature. 2015; **525**(7568):247

[18] Gaugler J, James B, Johnson T, Marin A, Weuve J. 2019 Alzheimer's disease facts and figures. Alzheimers & Dementia. 2019;**15**(3):321-387

[19] Shorter J, Lindquist S. Prions as adaptive conduits of memory and inheritance. Nature Reviews Genetics. 2005;**6**(6):435

[20] Chakravarty AK, Jarosz DF. More than just a phase: Prions at the crossroads of epigenetic inheritance and evolutionary change. Journal of Molecular Biology. 2 Nov 2018;**430**(23): 4607-4618. DOI: 10.1016/j.jmb.2018.07.017. Epub 2018 Jul 19

[21] Cox BS, Tuite MF, McLaughlin CS. The *psi* factor of yeast: A problem in inheritance. Yeast. 1988;**4**(3):159-178

[22] Goldberg AD, Allis CD, Bernstein E. Epigenetics: A landscape takes shape. Cell. 2007;**128**(4):635-638

[23] Liebman SW, Chernoff YO. Prions in yeast. Genetics. 2012;**191**(4): 1041-1072

[24] DiSalvo S, Derdowski A, Pezza JA, Serio TR. Dominant prion mutants induce curing through pathways that promote chaperone-mediated disaggregation. Nature Structural & Molecular Biology. 2011;**18**(4):486

[25] Pei F, DiSalvo S, Sindi SS, Serio TR. A dominant-negative mutant inhibits multiple prion variants through a common mechanism. PLoS Genetics. 2017;**13**(10):e1007085

[26] Byrne LJ, Cole DJ, Cox BS, Ridout MS, Morgan BJ, Tuite MF. The number and transmission of $[PSI^{+}]$ prion seeds (propagons) in the yeast saccharomyces cerevisiae. PLoS One. 2009;**4**(3):e4670

[27] Derdowski A, Sindi SS, Klaips CL, DiSalvo S, Serio TR. A size threshold limits prion transmission and establishes phenotypic diversity. Science. 2010; **330**(6004):680-683

[28] Gousset K, Schiff E, Langevin C, Marijanovic Z, Caputo A, Browman DT, et al. Prions hijack tunnelling nanotubes for intercellular spread. Nature Cell Biology. 2009;**11**(3):328

[29] Vilette D, Courte J, Peyrin JM, Coudert L, Schaeffer L, Andréoletti O, et al. Cellular mechanisms responsible for cell-to-cell spreading of prions. Cellular and Molecular Life Sciences. 2018;**75**(14):2557-2574

[30] Cipollina C, Vai M, Porro D, Hatzis C. Towards understanding of the complex structure of growing yeast populations. Journal of Biotechnology. 2007;**128**(2):393-402

[31] Pezza JA, Villali J, Sindi SS, Serio TR. Amyloid-associated activity contributes to the severity and toxicity of a prion phenotype. Nature Communications. 2014;**5**:4384

[32] Ginovart M, Carbó R, Blanco M, Portell X. Digital image analysis of yeast single cells growing in two different oxygen concentrations to analyze the population growth and to assist individual-based modeling. Frontiers in Microbiology. 2018;**8**:2628

[33] Suzanne S. Sindi. Mathematical modeling of prion disease. In: Prion—An Overview. Rijeka: IntechOpen; 2017

[34] Carbonell F, Iturria-Medina Y, Evans AC. Mathematical modeling of protein misfolding mechanisms in neurological diseases: A historical overview. Frontiers in Neurology. 2018;**9**:37

[35] Alper T, Cramp WA, Haig DA, Clarke MC. Does the agent of scrapie replicate without nucleic acid? Nature. 1967;**214**(5090):764

[36] Alper T, Haig DA, Clarke MC, et al. The exceptionally small size of the scrapie agent. Biochemical and Biophysical Research Communications. 1966;**22**:278-284

[37] Pattison IH, Jones KM. The possible nature of the transmissible agent of scrapie. Veterinary Record. 1967;**80**(1): 2-9

[38] Griffith JS. Nature of the scrapie agent: Self-replication and scrapie. Nature. 1967;**215**(5105):1043

[39] Masel J, Jansen VAA, Nowak MA. Quantifying the kinetic parameters of prion replication. Biophysical Chemistry. 1999;**77**(2):139-152

[40] Nowak MA, Krakauer DC, Klug A, May RM. Prion infection dynamics. Integrative Biology Issues News and Reviews. 1998;**1**(1):3-15

[41] Eigen M. Prionics or the kinetic basis of prion diseases. Biophysical Chemistry. 1996;**63**(1):A1-A18

[42] Davis JK. Suzanne S Sindi. Initial condition of stochastic self-assembly. Physical Review E. 2016;**93**(2):022109

[43] Yvinec R, D'Orsogna MR, Chou T. First passage times in homogeneous nucleation and self-assembly. The Journal of Chemical Physics. 2012; **137**(24):244107

[44] Greer ML, Pujo-Menjouet L, Webb GF. A mathematical analysis of the dynamics of prion proliferation. Journal of Theoretical Biology. 2006; **242**(3):598-606

[45] Prüss J, Pujo-Menjouet L, Webb G, Zacher R. Analysis of a model for the dynamics of prions. Discrete and Continuous Dynamical Systems—Series B. 2006;**6**(1):225-235

[46] Engler H, Prüss J, Webb GF. Analysis of a model for the dynamics of prions ii. Journal of Mathematical Analysis and Applications. 2006;**324**(1): 98-117

[47] Doumic M, Goudon T, Lepoutre T, et al. Scaling limit of a discrete prion dynamics model. Communications in Mathematical Sciences. 2009;**7**(4): 839-865

[48] Pöschel T, Brilliantov NV, Frömmel C. Kinetics of prion growth. Biophysical Journal. 2003;**85**(6): 3460-3474

[49] Davis JK, Sindi SS. A study in nucleated polymerization models of protein aggregation. Applied Mathematics Letters. 2015;**40**:97-101

[50] Calvez V, Lenuzza N, Doumic M, Deslys J-P, Mouthon F, Perthame B. Prion dynamics with size dependency–strain phenomena. Journal of Biological Dynamics. 2010;**4**(1):28-42

[51] Doumic M, Gabriel P. Eigenelements of a general aggregation-fragmentation model. Mathematical Models and Methods in Applied Sciences. 2010;**20**(05):757-783

[52] Gillam JE, MacPhee CE. Modelling amyloid fibril formation kinetics: Mechanisms of nucleation and growth. Journal of Physics: Condensed Matter. 2013;**25**(37):373101

[53] Leis E, Walker C. Existence of global classical and weak solutions to a prion equation with polymer joining. Journal of Evolution Equations. 2017;**17**(4): 1227-1258

[54] Davis JK, Sindi SS. A mathematical model of the dynamics of prion aggregates with chaperone-mediated fragmentation. Journal of Mathematical Biology. 2016;**72**(6):1555-1578

[55] Lemarre P, Pujo-Menjouet L, Sindi SS. Generalizing a mathematical model of prion aggregation allows strain coexistence and co-stability by

including a novel misfolded species. Journal of Mathematical Biology. 2019; **78**(1–2):465-495

[56] Helal M, Igel-Egalon A, Lakmeche A, Mazzocco P, Perrillat-Mercerot A, Pujo-Menjouet L, et al. Stability analysis of a steady state of a model describing alzheimer's disease and interactions with prion proteins. Journal of Mathematical Biology. 2019;**78**(1–2):57-81

[57] Cole DJ, Morgan BJT, Ridout MS, Byrne LJ, Tuite MF. Estimating the number of prions in yeast cells. Mathematical Medicine and Biology. 2004;**21**(4):369-395

[58] Olofsson P, Sindi SS, et al. A crump-mode-jagers branching process model of prion loss in yeast. Journal of Applied Probability. 2014;**51**(2):453-465

[59] Palmer KJ, Ridout MS, Morgan BJT. Kinetic models of guanidine hydrochloride-induced curing of the yeast [psi$^+$] prion. Journal of Theoretical Biology. 2011;**274**(1):1-11

[60] Ridout M, Giagos V, Morgan B, Naeimi W, von der Haar T, Tuite M. Modelling prion dynamics in yeast. In: Int. Statistical Inst.: Proc. 58th World Statistical Congress, Dublin (Session IPS020). 2011

[61] Banks HT, Flores KB, Langlois CR, Serio TR, Sindi SS. Estimating the rate of prion aggregate amplification in yeast with a generation and structured population model. Inverse Problems in Science and Engineering. 2018;**26**(2): 257-279

[62] Tanaka M, Collins SR, Toyama BH, Weissman JS. The physical basis of how prion conformations determine strain phenotypes. Nature. 2006;**442**(7102): 585-589

[63] Abdelnour F, Voss HU, Raj A. Network diffusion accurately models the relationship between structural and functional brain connectivity networks. NeuroImage. 2014;**90**:335-347

[64] Raj A, Kuceyeski A, Weiner M. A network diffusion model of disease progression in dementia. Neuron. 2012; **73**(6):1204-1215

[65] Gravenor MB, Cox DR, Hoinville LJ, Hoek A, McLean R. The flock-to-flock force of infection for scrapie in Britain. Proceedings of the Royal Society of London. Series B: Biological Sciences. 2001;**268**(1467):587-592

[66] Gubbins S. A modelling framework to describe the spread of scrapie between sheep flocks in Great Britain. Preventive Veterinary Medicine. 2005; **67**(2–3):143-155

[67] Touzeau S, Chase-Topping ME, Matthews L, Lajous D, Eychenne F, Hunter N, et al. Modelling the spread of scrapie in a sheep flock: Evidence for increased transmission during lambing seasons. Archives of Virology. 2006; **151**(4):735-751

[68] Miller MW, Hobbs NT, Tavener SJ. Dynamics of prion disease transmission in mule deer. Ecological Applications. 2006;**16**(6):2208-2214

[69] Sharp A, Pastor J. Stable limit cycles and the paradox of enrichment in a model of chronic wasting disease. Ecological Applications. 2011;**21**(4): 1024-1030

[70] Alber MS, Kiskowski MA, Glazier JA, Yi J. On cellular automaton approaches to modeling biological cells. In: Mathematical Systems Theory in Biology, Communications, Computation, and Finance. New York, NY: Springer New York; 2003. pp. 1-39

[71] Boon JP, Dab D, Kapral R, Lawniczak A. Lattice gas automata for reactive systems. Physics Reports. 1996; **273**(2):55-147

[72] Chopard B. Cellular Automata Modeling of Physical Systems. New York, NY: Springer New York; 2012

[73] Andreas Deutsch and Sabine Dormann. Cellular Automaton Modeling of Biological Pattern Formation: Characterization, Examples, and Analysis. Boston, MA: Birkhäuser; 2018

[74] Deutsch A, Dormann S, et al. Cellular Automaton Modeling of Biological Pattern Formation. Boston, MA: Springer; 2005

[75] Hwang M, Garbey M, Berceli SA, Tran-Son-Tay R. Rule-based simulation of multi-cellular biological systems—A review of modeling techniques. Cellular and Molecular Bioengineering. 2009; **2**(3):285-294

[76] Van Liedekerke P, Palm MM, Jagiella N, Drasdo D. Simulating tissue mechanics with agent-based models: Concepts, perspectives and some novel results. Computational Particle Mechanics. 2015;**2**(4):401-444

[77] Wolf-Gladrow DA. Lattice-Gas Cellular Automata and Lattice Boltzmann Models: An Introduction. Berlin Heidelberg: Springer-Verlag; 2004

[78] Alexander Anderson and Katarzyna Rejniak. Single-Cell-Based Models in Biology and Medicine. Boston, MA: Birkhäuser; 2007

[79] Hirashima T, Rens EG, Merks RMH. Cellular potts modeling of complex multicellular behaviors in tissue morphogenesis. Development, Growth & Differentiation. 2017;**59**(5):329-339

[80] Alber M, Chen N, Lushnikov PM, Newman SA. Continuous macroscopic limit of a discrete stochastic model for interaction of living cells. Physical Review Letters. 2007;**99**(16):168102

[81] Stevens A. The derivation of chemotaxis equations as limit dynamics of moderately interacting stochastic many-particle systems. SIAM Journal on Applied Mathematics. 2000;**61**(1): 183-212

[82] Chaturvedi R, Huang C, Kazmierczak B, Schneider T, Izaguirre JA, Glimm T, et al. On multiscale approaches to three-dimensional modelling of morphogenesis. Journal of the Royal Society Interface. 2005;**2**(3):237-253

[83] Drasdo D, Hoehme S, Block M. On the role of physics in the growth and pattern formation of multi-cellular systems: What can we learn from individual-cell based models? Journal of Statistical Physics. 2007;**128**(1–2):287

[84] Fletcher AG, Cooper F, Baker RE. Mechanocellular models of epithelial morphogenesis. Philosophical Transactions of the Royal Society B: Biological Sciences. 2017;**372**(1720): 20150519

[85] Galle J, Aust G, Schaller G, Beyer T, Drasdo D. Individual cell-based models of the spatial-temporal organization of multicellular systems-achievements and limitations. Cytometry Part A: The Journal of the International Society for Analytical Cytology. 2006;**69**(7): 704-710

[86] Sandersius SA, Weijer CJ, Newman TJ. Emergent cell and tissue dynamics from subcellular modeling of active biomechanical processes. Physical Biology. 2011;**8**(4):045007

[87] Farhadifar R, Röper J-C, Aigouy B, Eaton S, Jülicher F. The influence of cell mechanics, cell-cell interactions, and proliferation on epithelial packing. Current Biology. 2007;**17**(24):2095-2104

[88] Kursawe J, Brodskiy PA, Zartman JJ, Baker RE, Fletcher AG. Capabilities and

limitations of tissue size control through passive mechanical forces. PLoS Computational Biology. 2015;**11**(12): e1004679

[89] Newman TJ. Modeling multicellular systems using subcellular elements. Mathematical Biosciences and Engineering. 2005;**2**(3):613-624

[90] Newman TJ. Modeling multicellular structures using the subcellular element model. In: Anderson ARA, Chaplain MAJ, Rejniak KA, editors. Single-Cell-Based Models in Biology and Medicine. Basel: Birkhäuser Basel; 2007. pp. 221-239

[91] Drasdo D, Höhme S. A single-cell-based model of tumor growth in vitro: Monolayers and spheroids. Physical Biology. 2005;**2**(3):133

[92] Jönsson H, Levchenko A. An explicit spatial model of yeast microcolony growth. Multiscale Modeling & Simulation. 2005;**3**(2):346-361

[93] Wang Y, Lo W-C, Chou C-S. A modeling study of budding yeast colony formation and its relationship to budding pattern and aging. PLoS Computational Biology. 2017;**13**(11): e1005843

[94] Banwarth-Kuhn M, Nematbakhsh A, Rodriguez KW, Snipes S, Rasmussen CG, Venugopala Reddy G, et al. Cell-based model of the generation and maintenance of the shape and structure of the multilayered shoot apical meristem of arabidopsis thaliana. Bulletin of Mathematical Biology. Aug 2019;**81**(8):3245-3281

[95] Nematbakhsh A, Sun W, Brodskiy PA, Amiri A, Narciso C, Xu Z, et al. Multi-scale computational study of the mechanical regulation of cell mitotic rounding in epithelia. PLoS Computational Biology. 2017;**13**(5): e1005533

[96] Diaz de la Loza MC, Thompson BJ. Forces shaping the drosophila wing. Mechanisms of Development. 2017;**144** (Pt A):23-32

[97] Pathmanathan P, Cooper J, Fletcher A, Mirams G, Murray P, Osborne J, et al. A computational study of discrete mechanical tissue models. Physical Biology. 2009;**6**(3):036001

[98] Drasdo D. Coarse graining in simulated cell populations. Advances in Complex Systems. 2005;**8**(02):319-363

[99] Galle J, Loeffler M, Drasdo D. Modeling the effect of deregulated proliferation and apoptosis on the growth dynamics of epithelial cell populations in vitro. Biophysical Journal. 2005;**88**(1):62-75

[100] Drasdo D, Höhme S. Individual-based approaches to birth and death in avascu1ar tumors. Mathematical and Computer Modelling. 2003;**37**(11): 1163-1175

[101] Drasdo D, Forgacs G. Modeling the interplay of generic and genetic mechanisms in cleavage, blastulation, and gastrulation. Developmental Dynamics: An Official Publication of the American Association of Anatomists. 2000;**219**(2):182-191

[102] Dirk Drasdo RK, McCaskill JS. Monte carlo approach to tissue-cell populations. Physical Review E. 1995; **52**(6):6635

[103] Drasdo D, Loeffler M. Individual-based models to growth and folding in one-layered tissues: Intestinal crypts and early development. Nonlinear Analysis-Theory Methods and Applications. 2001;**47**(1):245-256

[104] Hoehme S, Brulport M, Bauer A, Bedawy E, Schormann W, Hermes M, et al. Prediction and validation of cell alignment along microvessels as order principle to restore tissue architecture in

liver regeneration. Proceedings of the National Academy of Sciences. 2010; **107**(23):10371-10376

[105] Aprianti D, Khotimah SN, Viridi S. Budding yeast colony growth study based on circular granular cell. Journal of Physics: Conference Series. 2016;**739**: 012026

[106] Byrne H, Drasdo D. Individual-based and continuum models of growing cell populations: A comparison. Journal of Mathematical Biology. 2009; **58**(4–5):657

[107] Fletcher AG, Osterfield M, Baker RE, Shvartsman SY. Vertex models of epithelial morphogenesis. Biophysical Journal. 2014;**106**(11): 2291-2304

[108] Fozard JA, Lucas M, King JR, Jensen OE. Vertex-element models for anisotropic growth of elongated plant organs. Frontiers in Plant Science. 2013; **4**:233

[109] Weaire D, Fortes MA. Stress and strain in liquid and solid foams. Advances in Physics. 1994;**43**(6): 685-738

[110] Odell GM, Oster G, Alberch P, Burnside B. The mechanical basis of morphogenesis: I. Epithelial folding and invagination. Developmental Biology. 1981;**85**(2):446-462

[111] Tanaka S. Simulation frameworks for morphogenetic problems. Computation. 2015;**3**(2):197-221

[112] Vogel V, Sheetz M. Local force and geometry sensing regulate cell functions. Nature Reviews Molecular Cell Biology. 2006;**7**(4):265

[113] Rauzi M, Lenne P-F, Lecuit T. Planar polarized actomyosin contractile flows control epithelial junction remodelling. Nature. 2010;**468**(7327): 1110

[114] Brodland GW, Chen HH. The mechanics of heterotypic cell aggregates: Insights from computer simulations. Journal of Biomechanical Engineering. 2000;**122**(4):402-407

[115] Odell G, Oster G, Burnside B, Alberch P. A mechanical model for epithelial morphogenesis. Journal of Mathematical Biology. 1980;**9**(3): 291-295

[116] Morse PM. Diatomic molecules according to the wave mechanics. II. Vibrational levels. Physical Review. 1929;**34**(1):57

[117] Sandersius SA, Newman TJ. Modeling cell rheology with the subcellular element model. Physical Biology. 2008;**5**(1):015002

[118] Wottawah F, Schinkinger S, Lincoln B, Ananthakrishnan R, Romeyke M, Guck J, et al. Optical rheology of biological cells. Physical Review Letters. 2005;**94**(9):098103

[119] Schiff. Quantum Mechanics. 3rd ed. American Journal of Physics. 1973;**41** (4):599-600,

[120] Amiri A, Harvey C, Buchmann A, Christley S, Shrout JD, Aranson IS, et al. Reversals and collisions optimize protein exchange in bacterial swarms. Physical Review E. 2017;**95**(3–1):032408

[121] Christley S, Alber MS, Newman SA. Patterns of mesenchymal condensation in a multiscale, discrete stochastic model. PLoS Computational Biology. 2007;**3**(4):e76

[122] Christley S, Lee B, Dai X, Nie Q. Integrative multicellular biological modeling: A case study of 3D epidermal development using GPU algorithms. BMC Systems Biology. 2010;**4**:107

[123] Gord A, Holmes WR, Dai X, Nie Q. Computational modelling of epidermal stratification highlights the importance

of asymmetric cell division for predictable and robust layer formation. Journal of the Royal Society Interface. 2014;**11**(99):20140631

[124] Sweet CR, Chatterjee S, Xu Z, Bisordi K, Rosen ED, Alber M. Modelling platelet–blood flow interaction using the subcellular element langevin method. Journal of the Royal Society Interface. 2011;**8**(65): 1760-1771

[125] Wu Z, Xu Z, Kim O, Alber M. Three-dimensional multi-scale model of deformable platelets adhesion to vessel wall in blood flow. Philosophical Transactions of the Royal Society A: Mathematical, Physical and Engineering Sciences. 2014;**372** (2021):20130380

[126] Wayne Brodland G. How computational models can help unlock biological systems. Seminars in Cell and Developmental Biology. 2015;**47-48**: 62-73

[127] Milde F, Tauriello G, Haberkern H, Koumoutsakos P. SEM++: A particle model of cellular growth, signaling and migration. Computational Particle Mechanics. 2014;**1**(2):211-227

Chapter 5

Zeta Potential as a Diagnostic Tool to Determine the Angina Risk

Swati S. Gaikwad, Jasmine G. Avari and Mansi Liladhar Patil

Abstract

Red blood cells (RBCs) moving in circulation are continuously exposed to the reactive oxygen species (ROS) that are circulating within the vascular system of body. The objective of the study is to determine the erythrocyte zeta potential and its morphological changes caused by oxidative stress in hypertensive patients in relation to the risk of development of cardiovascular disease (myocardial infarction). In this retrospective study, we investigated 186 samples, which include hypertension patients (n = 64), myocardial infarction patients (n = 52), treated myocardial infarction patients (n = 20), and normal healthy volunteers (n = 50). Morphological and electrochemical characteristics of RBCs in patients were analyzed using scanning electron microscopy and zeta potential study. These parameters were also statistically analyzed applying one way ANNOVA (Tukey's comparison) using Graph Pad. Statistical analyses of data showed that Hypertensive (ZP-16.06 mV) and cardiac patients (ZP-9.94 mV) RBCs possessed significantly reduced zeta potential relative to that of RBCs from healthy individuals (ZP-23.39 mV) (P-value <0.0001). Microscopic imaging of Hypertensive RBCs revealed increased anisocytes and poikilocytes. These parameters were found exacerbated in patients suffering from Myocardial infarction. Hence, it can be concluded that ZP can be used as an effective diagnostic tool for detection of hypertension associated cardiovascular disorder risk.

Keywords: red blood cells (RBCs), zeta potential, erythrocyte fragility, lipid peroxidation, hypertension, myocardial infarction

1. Introduction

Hypertension is an emerging public health problem of this millennium and it is a major challenge to disclose the mechanism involved in the coexistence of hypertension and cardiovascular disease to improve the health of the Hypertensive patients. Patients with symptoms of a cardiovascular disease frequently present without striking evidence of cardiac specific enzymes in blood laboratory assessments or specific electrocardiogram findings. Recently, researchers have reported higher mortality risk associated with higher RDW in patient populations with cardiovascular disease (CVD) [1]. Nowadays, also there is a growing interest in characterizing RBC membrane defects in several diseases, as changes in membrane structure also contribute to the pathophysiology of the disease process.

Human erythrocyte contains about 95% of the glutathione which is responsible for scavenging reactive oxygen species [2]. Sialic acid is an important factor for maintenance of the surface electrical charge and stability of biological cellular system [3]. Erythrocytes and Erythrocyte membrane are more vulnerable to peroxidation due to constant exposure to high oxygen tension and richness in polyunsaturated fatty acid, respectively [4]. The interaction of Reactive oxygen species (ROS) with particularly with fatty acids available in membrane can result in undesirable irreversible changes in cellular membrane. The membranes are therefore naturally protected by available anti-oxidative enzymes like superoxide dismutase, catalase, and glutathione peroxidase and vitamins E and A from oxidative damages [5]. By-products of peroxidation have been shown to cause profound alterations in the structural organization and functions of the cell membrane including decreased membrane fluidity, increased membrane permeability, inactivation of membrane-bound enzymes and loss of essential fatty acids [6]. Reactive oxygen species are the responsible moiety that are involved in the generation and progression of atherosclerosis and that contribute to the development of plaque instability in acute MI [7].

Blood viscosity is strongly affected by the surface charge of RBCs and is responsible for the spacing between them. A higher surface charge causes repulsive forces to increases distant RBCs, preventing their close aggregation, lowers the viscosity, and results into very low peripheral resistance to flow. Therefore, it can be hypothesized that stress like condition develops in hypertension generate ROS which can induce membrane deformity and as a result affect the membrane surface charge. The occlusive arterial disease resulting from RBC aggregation may develop due to comparable membrane deformity [8]. Besides vascular and cardiac tissue integrity, blood and especially the RBC, is critical to performing this critical assignment of CV risk assessment considering it makes up over 90% of formed elements within blood. In cardiovascular diseases, impaired blood rheology has been observed. A direct relationship has been found between increased RBC deformability and increased risk for arterial hypertension [9]. Accurate risk stratification of patients with chronic heart failure is critically important to efficiently target the use of evidence-based therapies and identify high-risk patients who may benefit from advanced treatments [10].

We tested the hypothesis that variations in zeta potential and deformation of erythrocytes were associated with risk of adverse cardiovascular outcomes in a population with hypertension that were free of symptomatic heart failure [11]. In the present work we envisaged to study and evaluate morphological changes taking place in RBCs, erythrocyte fragility, lipid per oxidation and zeta potential which can act as invaluable aid in the diagnosis of a hypertension and risk of Cardiovascular Disorder in hypertension patients. Hence, the aim of this study was to test the hypothesis of association of variation in RBC morphology, erythrocyte fragility and zeta potential in hypertension and its relation with the risk of cardiovascular disorder in hypertension patients.

2. Morphological characterization

2.1 Preparation of blood Smear

Blood smear was prepared with the aid of wedge method [12]. In this method a drop of blood was placed on base closed to one cease of the slide at least 1 cm away from the edged of the slide. Another slide with the smooth end was used as spreader and smear was prepared by moving spreader inclined at 30–45° angle to the blood.

This Smear was air dried and fixed with Leishman stain and located and observed beneath a trinocular research microscope (RXT4, Radical).

2.2 Scanning electron microscopy

To observe the morphological variations in the erythrocyte membrane structure in Erythrocytes of patients suffering with hypertension and MI, erythrocytes were analyzed by scanning electron microscopy. With this motive blood sample was taken in Eppendorf tube containing 10 μl of Heparin (5000 UI/ml) in 900 μl of pH 7.4 phosphate buffer saline. The blood suspension was then centrifuged (1000 rpm for 10 min) and washed with buffer three times. The supernatant was removed and replaced by same volume of buffer. One drop of these separated erythrocytes were then exposed to 500 μl of 2.5% Glutaraldehyde in distilled water overnight at 4°C to fix. Again samples were washed thrice with distilled water and centrifuged. About 40 μl of each sample was placed on glass covered studs and air dried at room temperature. The Scanning electron microscopy (SEM) analysis of prepared samples was accomplished using Jeol, Japan (Model—JSM 5610LV).

2.3 Preparation of iso-osmotic dextrose solution

About 5 g of anhydrous Dextrose (Merck) was solubilized in 100 ml of distilled water to prepare a 5% w/v iso-osmotic Dextrose solution.

2.4 Preparation of blood sample for zeta potential measurement

About 0.1 ml of blood sample was transferred into 50 ml of freshly prepared 5% w/w dextrose solution which is isotonic with the red blood cells [12].

2.5 Zeta potential measurement by zeta meter system 4.0

Zeta meter System 4.0 instrument was used to measure the zeta potential of the Erythrocytes [12]. Zeta potential of the system is measured by applying the electric field to the samples using electrodes and determining the mobility/velocity of the particle under the applied field. Zeta potential was calculated according to the simplified Helmholtz-Smoluchowski equation as follows: Mean velocity of the 10 readings was used to calculate the zeta potential Eq. (1).

$$\text{Zeta Potential, } \zeta = \frac{113000 \times v \times EM}{D} \tag{1}$$

where ν = viscosity of sample in poise at temperature “t”; EM = electrophoretic mobility; D = dielectric constant.

The electrophoretic cell consists of capillary which is embedded inside a chamber having electrodes at both ends having cavity for sample connection with electrodes. From the cavity of any one end of the electrophoresis cell the Sample is introduced into the capillary to fill it completely. Electrodes are connected to the cell with the applied electric field at specific voltage of 200 V. Due to the applied electric field charged particles move towards oppositely charged electrode. Their velocity under the applied electric field is measured and expressed in terms of electro kinetic potential/zeta potential. Nowadays, this system is preferably used for determining the zeta potential of various types of biological membranes.

In this method, fresh blood samples were collected from volunteer and blood suspension was prepared as described in preparation of sample. Prior measurement of zeta potential temperature of and other parameters for ZP measurements were adjusted such as light intensity; focal plane and tracking duration were optimized for stable data collection with minimal error. The collected and prepared samples were then added to the previously cleaned zeta-meter cell, which were then connected to electrodes and placed under the lens over stage. The Erythrocytes were tracked by using remote by microscopically-acquired video images, values were recorded 10 times for each sample and average zeta-potential in mv was obtained standard deviation from software Zeta meter-ZM4DAQ software.

2.6 Statistical analysis

The experimental results for ZP are expressed as mean ± standard (SD). Investigated groups were compared by the statistical one way analysis of variance (ANOVA) and evaluated by Tukey's multiple comparison tests using Graph Pad Prism version 5.00 for Windows. Experimental Results were considered significant at $p < 0.05$ for the different groups which indicates that the control and other patient groups differ significantly from one another in all situations.

3. Results and discussion

Morphological study using the smear reveals that there is deformation in RBC shape (Anisocytosis) in majority of patients. In Hypertension and patients with MI showed RBC deformation from spherical biconcave shape to tear shape which further continues to Rouleaux formation (**Figure 1a–f**). Also, it was observed that there is increased aggregation of Erythrocytes in patients suffering from Hypertension and Myocardial Infarction compared to normal volunteers.

SEM studies revealed that there are number of deformed erythrocytes observed in myocardial infarction patients (**Figure 2c**) in aggregated form. It was observed in patients with hypertension, Erythrocytes appears with uneven surface of cell membrane which indicates the oxidative damage in the cell membrane (**Figure 2b**) as compared to normal human erythrocyte as shown in (**Figure 2a**).

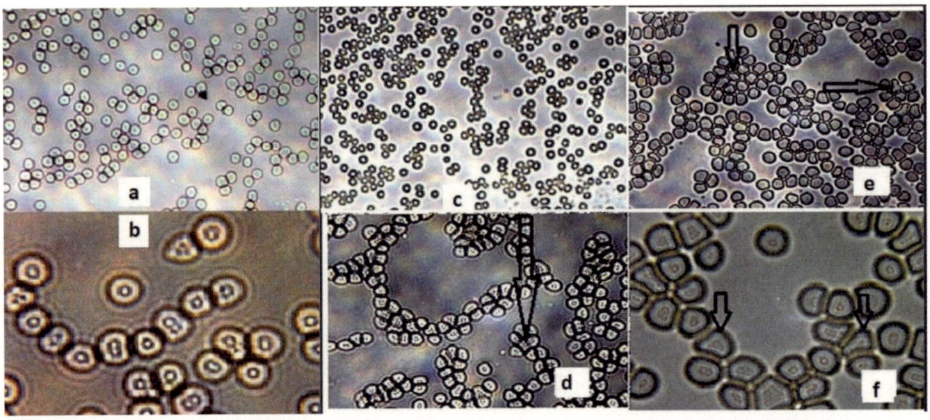

Figure 1.
(a, b) Erythrocytes of normal human erythrocytes. (c, d) Erythrocytes of patient suffering from Hypertension, blood smear showing Rouleaux formation. (e, f) Erythrocytes from patient suffering from myocardial infarction, blood smear showing anisocytes.

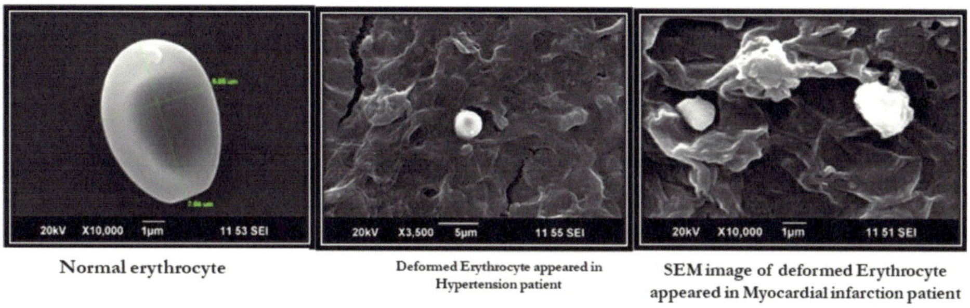

Figure 2.
Scanning electron microscopic images of normal human erythrocytes, patients suffering from hypertension and myocardial infarction showing deformed erythrocytes.

Similarly, the result for erythrocyte membrane osmotic fragility in test samples reveals that erythrocytes of patients with MI and hypertension become more fragile compared to erythrocytes of control group (**Figure 3**). The erythrocytes are very much susceptible to oxidative cellular damage on exposure to the excessive oxidative stress.

By using a Zeta meter system 4.0, the zeta potential of human Erythrocytes was obtained. The results obtained showed ZP values ranged from −20.13 mV to −26.46 (+1.87) mV for the healthy individuals with a mean value of −23.39 mV. But the ZP values obtained from the Patients with MI is much lower and ranges from −2.58 mV to −22.76 (+3.57) mV compared to patients suffering from only Hypertension values ranges from −12.13 mV to −19.61(+1.20) mV. Also it was observed that Patients with MI who underwent CABG and angioplasty and receiving medications the zeta potential of such patients was ranging from −21.53 mV to −32.59 mV(+2.95) indicating higher stability of erythrocytes in blood vessels (**Figure 4**).

It was specifically observed that the patients with Hypertension had a mean ZP value −16.06 mV and patients with MI mean ZP value was obtained as −9.938 mV ($p < 0.001$). Data obtained from the analysis of blood samples from patients with hypertension and healthy volunteers was subjected to one-way ANNOVA with the application of Tukey's multiple comparison tests, and results indicated that there is a significant lowering of zeta potential of Patients with Hypertension from that

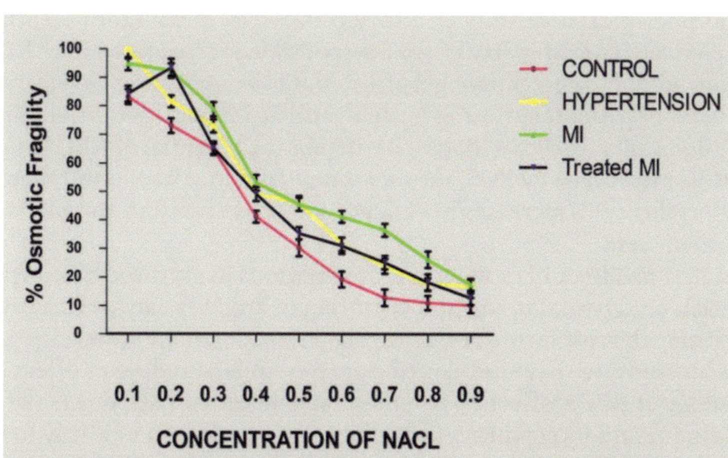

Figure 3.
Comparison between mean erythrocyte fragility of erythrocytes in normal volunteers (CONTROL), patients with hypertension and myocardial infarction and myocardial infarction patients on treatment.

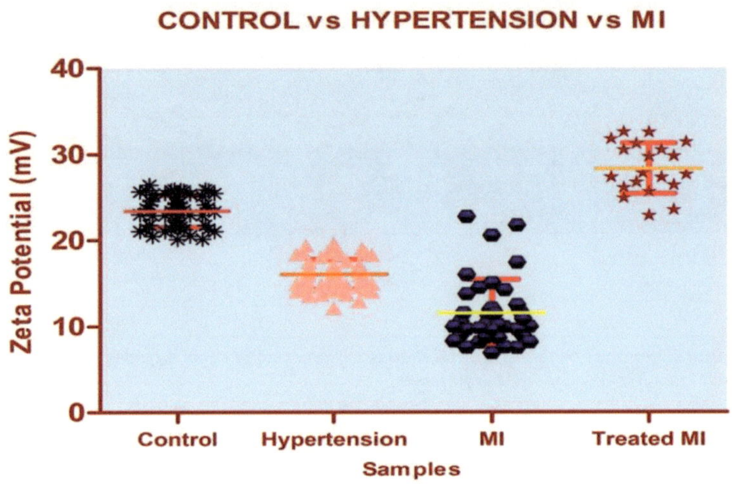

Figure 4.
Comparison between mean ZP values of normal volunteers, patients with hypertension and myocardial infarction patient and myocardial infarction patients on treatment.

of the healthy volunteers. Also, there is greater decrease in ZP values of patients with Hypertension suffering from MI indicate the exacerbation in RBC deformity in patients. This may be due to hypertension induced complications. Higher ZP in treated patients indicates the increased stability of erythrocytes due to reduced oxidative stress.

4. Conclusion

The exact mechanism of development of CVD is complex and is not yet fully understood. But from the literature survey it was clear that ROS plays an important role in the progression and development of CVD. Also, it was known that there is a strong relation between ROS and the pathophysiology of CVD. From the present research work it can be concluded that due ROS in Patients with Hypertension the erythrocytes are affected, their membrane gets oxidized resulting in various types of morphological deformity (Anisocytosis). Also, membrane potential (ZP) which is a characteristic property of RBC responsible for free flowing of RBC in the blood stream without aggregation, get affected. These conditions get exacerbated in erythrocytes of patients suffering from myocardial infarction. Development of membrane deformity directly reduces the membrane potential of RBC. Due to oxidation of RBC membrane by ROS, the membrane becomes fragile and therefore the fragility of erythrocytes increases in Patients with Hypertension and MI compared to healthy volunteers.

Results obtained in our lab suggest that variations in erythrocyte morphology, zeta potential, lipid peroxidation and erythrocyte Fragility can act as a key indicator to determine the risk factor of myocardial infarction in hypertensive patients. In the present study, we have developed a greater understanding of effect of ROS on morphology of RBCs, effect on its membrane potential (ZP) due to deformity in the membrane and its erythrocyte fragility. This could be a new way to realize a better treatment in hypertensive patients and a prevention of cardiovascular complications (i.e.: myocardial infarction, TIA, etc.). However, more works in the near future are necessary to improve the detection and treatment of the ROS mediated dysfunction.

Acknowledgements

We are grateful to Dr. Varma Pathology Laboratory, Platina heart hospital, Nagpur and healthy voluntary donors of the Department of Pharmaceutical Sciences, Nagpur University, Nagpur.

Author details

Swati S. Gaikwad[1*], Jasmine G. Avari[2] and Mansi Liladhar Patil[3]

1 Nagpur College of Pharmacy, Wanadongri, Nagpur, India

2 Department of Pharmaceutical Sciences, RTMNU, Nagpur, India

3 Dr. D. Y. Patil Institute of Pharmaceutical Sciences and Research, Pune, India

*Address all correspondence to: swati.gaikwad05@gmail.com

References

[1] Patel KV, Ferrucci L, Ershler WB, Longo DL, Guralnik JM. Red blood cell distribution width and the risk of death in middle-aged and older adults. Archives of Internal Medicine. 2009;**169**(5):515-523. DOI: 10.1001/archinternmed.2009.11

[2] Isaji Y. Use of erythrocyte adhesion assay to predict the risk of diabetic complications. Biochemical Engineering Journal. 2009;**439**:178-184

[3] Riddick TM. Chapter 22: The application of basic concepts of zeta potential to cardiovascular disease. Vol. 1. In: Control of Colloid Stability through Zeta Potential. Wynnewood, PA: Published for Zeta-Meter, Inc., by Livingston Pub. Co; 1968

[4] Gurel A, Armutcu F, Damatoglu S, Unalacak M, Demircan N. Evaluation of erythrocyte NA+, K+ -ATPase and superoxide dismutase activities and malondialdehyde level alteration in coal miners. European Journal of General Medicine. 2004;**1**(4):22-28

[5] Birben E, Sahiner UM, Sackesen C, Erzurum S, Kalayci O. Oxidative stress and antioxidant defense. The World Allergy Organization Journal. 2012;**5**(1): 9-19. DOI: 10.1097/WOX.0b013e3182439613

[6] Ayala A, Muñoz MF, AruUelles S. Lipid peroxidation: Production, metabolism, and signalling mechanisms of malondialdehyde and 4-hydroxy-2-nonenal. Oxidative Medicine and Cellular Longevity. 2014:360438. DOI: 10.1155/2014/360438. Epub 8 May 2014. PMID: 24999379; PMCID: PMC4066722

[7] Sugamura K, Keaney JF. Reactive oxygen species in cardiovascular disease. Free Radical Biology & Medicine. 2011;**51**(5):978-992. DOI: 10.1016/j.freeradbiomed.2011.05.004

[8] Chevalier G, Sinatra ST, Oschman JL, Delany RM. Earthing (grounding) the human body reduces blood viscosity—A major factor in cardiovascular disease. Journal of Alternative and Complementary Medicine. 2013;**19**(2):102-110. DOI: 10.1089/acm.2011.0820

[9] Kuhn V, Diederich L, Keller TCS, Kramer CM, Luckstadt W, Panknin C, et al. Red blood cell function and dysfunction: Redox regulation, nitric oxide metabolism. Anemia. Antioxidants & Redox Signalling. 2017;**26**(13):718-742. DOI: 10.1089/ars.2016.6954

[10] Felker GM, Allen LA, Pocock SJ, et al. Red cell distribution width as a novel prognostic marker in heart failure. Journal of the American College of Cardiology. 2007;**50**:40-47

[11] Tonelli M, Sacks F, Arnold M, et al. Relation between red blood cell distribution width and cardiovascular event rate in people with coronary disease. Circulation. 2008;**117**:163-168

[12] Buys AV, Van Rooy MJ, Soma P, Van Papendorp D, Lipinski B, Pretorius E. Changes in red blood cell membrane structure in type 2 diabetes: A scanning electron and atomic force microscopy study. Cardiovascular Diabetology. 2013;**12**:25. DOI: 10.1186/1475-2840-12-25

Section 2

Triglycerides

Chapter 6

Postprandial Triglycerides, Oxidative Stress, and Inflammation

Ngoc-Anh Le

Abstract

Among the most common non-communicable diseases are obesity, cardiovascular disease, and diabetes, which are responsible for the major cardiometabolic phenotypes. Together with mitochondrial alterations, oxidative stress and inflammation are key molecular mechanisms that contribute to the onset and development of these conditions. Meal consumption is a recurring daily activity that is directly linked to oxidative stress and inflammation. Acute increases in lipids, notably triglycerides, during the postabsorptive period have been suggested to induce a state of inflammation with stimulation of adhesion molecules, cytokines, oxidative stress, and leukocyte activation. Not only lipids but also meal-induced elevations in glucose have also been linked to postprandial oxidative stress and inflammation. The impact of postprandial hypertriglyceridemia and hyperglycemia on oxidative stress and inflammation is not only independent but may be cumulative. It is our hypothesis that, in a system that could not maintain homeostasis to continuous changes of the environment, repeated exposures to meals that provide modest doses of fat and glucose could potentially elicit abnormal responses that contribute to the onset and development of chronic cardiometabolic phenotypes.

Keywords: postprandial triglycerides, oxidative stress, inflammation, hypertriglyceridemia, hyperglycemia, cardiometabolic phenotypes

1. Introduction

Non-communicable diseases refer to chronic conditions that are non-infectious and non-transmissible and include obesity, cardiovascular disease, diabetes, cancer, as well as respiratory and neurological diseases. They represent the most common cause of death and disability in developed as well as developing countries [1]. In addition to age-related mitochondrial alterations, oxidative stress and inflammation are key mechanisms in the onset and development of these conditions [2].

Lipids, in particular esterified lipids such as triglycerides (TG), phospholipids, and cholesteryl esters are essential metabolites for energy, cell membrane integrity as well as regulatory hormones. Reactive oxygen species (ROS) are produced by normal physiological processes and play an important role in cell signaling and tissue homeostasis. In view of their high content of polyunsaturated fatty acids (PUFA), lipids transported in plasma lipoproteins as well as those in cellular membranes are especially susceptible to ROS damage, i.e. lipid peroxidation. An imbalance between the formation and detoxification of ROS can lead to the oxidative stress and the nonenzymatic modification of biomolecules, such as proteins, carbohydrates, nucleic acids and lipids. Lipid peroxidation damages surface phospholipids directly and delivery of oxidized fatty acids to the cells play an important role in many inflammatory diseases and can mediate proinflammatory changes [3]. With meal consumption, the intermittent influx of newly absorbed dietary fat in the form of TG-rich chylomicrons is associated with delayed clearance of plasma lipids. This delay is further exacerbated in individuals with metabolic syndrome, type 2 diabetes as well as in obese individuals resulting in greater ROS damage.

Chronic hyperglycemia in diabetes is associated with concomitant increase in the level of ROS and a reduction in enzymatic and nonenzymatic cell antioxidant defenses [4, 5]. In vitro study has demonstrated that intermittent exposure to high glucose environment could stimulate superoxide production and enhance endothelial cell apoptosis to a greater extent than exposure to constant high glucose [6]. In vivo study also noted that intermittent meal-induced elevations in plasma glucose could also play a role in inducing inflammation and oxidative stress [7]. Ingestion of glucose has been reported to result in increased production of tumor necrosis factor alpha (TNFα) and interleukin-6 (IL-6) by peripheral blood mononuclear cells, and increased formation of ROS [8, 9]. Furthermore, addition of glucose to a fatty meal delays the metabolism of intestinal chylomicrons in healthy subjects, thus enhancing lipid peroxidation. In fact, there is ample evidence that the impact of postprandial hyperlipidemia and hyperglycemia on inflammation and oxidative stress is independent and cumulative [10].

2. Lipoprotein-associated oxidative stress and inflammation

Oxidative stress refers to an imbalance between the generation of reactive oxygen/nitrogen species (ROS/RNS) and the capacity of antioxidative protection systems to counteract them. All substrates in the cells are potential targets of ROS with lipids being most susceptible to undergo oxidative modification, in particular polyunsaturated fatty acids (PUFA) with the multiple double bonds [11]. Peroxidation of PUFA leads to the formation of isoprostanes [12] and their levels are considered to accurately reflect the oxidative status of the environment [13]. Oxidative modification of lipids can also result in the formation of reactive aldehydes, such as malondialdehyde (MDA) and 4-hydroxynonenal (HNE) [14]. MDA and HNE modified metabolites can be detected in plasma [15].

Serving as both a source of energy and structural components of cells, lipids are transported in the circulation by a network of globular particles of varying in size (from 5 to 1200 nm in diameter) and composition. Independent of the size, these so-called lipoproteins consist of a hydrophilic outer layer and a hydrophobic inner core [16]. The surface of lipoproteins consists of an amphipathic phospholipid bilayer stabilized by non-esterified cholesterol and specific proteins known as apolipoproteins. The inner core includes nonpolar lipids such as triglycerides and cholesteryl esters in various amounts. The key lipoproteins are the triglyceride-rich lipoproteins (TRL), the cholesterol-rich atherogenic lipoproteins, low-density lipoproteins (LDL), and the protective, anti-oxidative, anti-atherogenic lipoproteins, high-density lipoproteins (HDL). TRL are responsible for the delivery of TG to peripheral tissues can be of intestinal origin, chylomicrons, or of hepatic origin, very-low density lipoproteins (VLDL). While chylomicrons which carry dietary fats are present only in the postprandial state, VLDL are constantly being produced by the liver to transport endogenous triglycerides and are metabolic precursors of

LDL which are delivered to peripheral tissues as cholesterol-rich particles [16]. In addition to their anti-inflammatory and anti-oxidative properties, HDL serves as the primary vehicle for reverse cholesterol transport [16].

In the circulation, very-low density lipoproteins (VLDL) are triglyceride-rich (TG-rich) lipoproteins synthesized in the liver for the export of new formed TG to peripheral tissues [17]. VLDL unloads their cargo of triglycerides via interactions with lipoprotein lipase that is anchored to the endothelium via heparan sulfate proteoglycans [18]. As part of this process, lipoprotein particles form tight junction with the endothelium allowing released free fatty acids and monoglycerides to be delivered to the underlying cells. The neutral and oxidized free fatty acids released by the hydrolysis of TG-rich lipoproteins can induce endothelial inflammation [19]. In the same process, excess ROS generated by activated macrophages in the arterial wall can attach themselves to lipoproteins in the circulation (**Figure 1**). Plasma lipoproteins are thus seeded with oxidants that can initiate the propagation of oxidative damage, unless the oxidative epitopes are quenched by antioxidant mechanisms [17, 20].

Following meal consumption, newly absorbed dietary fats are packaged in the intestine and secreted into the circulation as chylomicrons which contain over 90% TG by mass and range in diameter from 100 to 1200 nm [16]. These intestine-derived TG-rich lipoproteins compete with hepatic VLDL for interactions with lipoprotein lipase to deliver TG to peripheral tissues. Depending on the TG contents in the meal as well as the capacity of the lipolytic system, retention times for both VLDL and chylomicrons at the endothelium could be significantly extended. The net result is a greater propensity for plasma lipoproteins to be seeded with ROS generated in the arterial wall, especially when the endothelium is under chronic inflammation. Indeed using immunoassay for IgG specific for oxidative epitopes we have reported acute reduction in circulating IgG levels following meal consumption in patients with documented coronary artery disease [21]. Furthermore, this transient reduction was observed only when the meal was enriched in polyunsaturated fatty acids which are most susceptible to modification by ROS [22].

With the hydrolysis of VLDL and chylomicron TG to free fatty acids and glycerol, smaller lipoprotein particles are formed relatively enriched in cholesterol, the so-called VLDL and chylomicron remnants. While most chylomicron remnants are cleared by the liver via receptor-mediated uptake, only 50% of the VLDL remnants are directly removed from plasma. The remaining 50% are converted to cholesterol-rich low-density lipoproteins (LDL), the major cholesterol carrying lipoproteins in normal human plasma [23–25]. If the oxidative epitopes transferred from the arterial wall to the TG-rich VLDL are not efficiently removed or quenched, we have suggested that this could represent an alternate pathway for the formation of oxidatively modified LDL (**Figure 1**) [20]. This is different from the traditional concept which postulates that LDL oxidation must take place in the arterial wall [26]. The presumption is that plasma lipoprotein lipids are well protected by inherent antioxidant defenses in particular, LDL itself contains most of the antioxidant alpha-tocopherol in plasma [26]. Recent data would suggest that under certain conditions, oxidative modification of plasma lipoproteins could be initiated in the circulation. By using the *in vitro* assay described by Esterbauer et al. [27] to assess the oxidative susceptibility of plasma LDL, we can show that, while LDL isolated from fasting plasma would undergo oxidative modification only in the presence of Cu^{++} as a catalyst, LDL isolated from non-fasting plasma was susceptible to auto-oxidation in the absence of Cu^{++} (**Figure 2A**) [28]. Furthermore, in some patients with metabolic syndrome, auto-oxidation of fasting LDL could be prevented following management with ABT-335 (choline salt of fenofibric acid) (**Figure 2B**) [28].

While endothelial cells are continuously exposed to high concentrations of free fatty acids in the fasted state generated by the hydrolysis of hepatic VLDL,

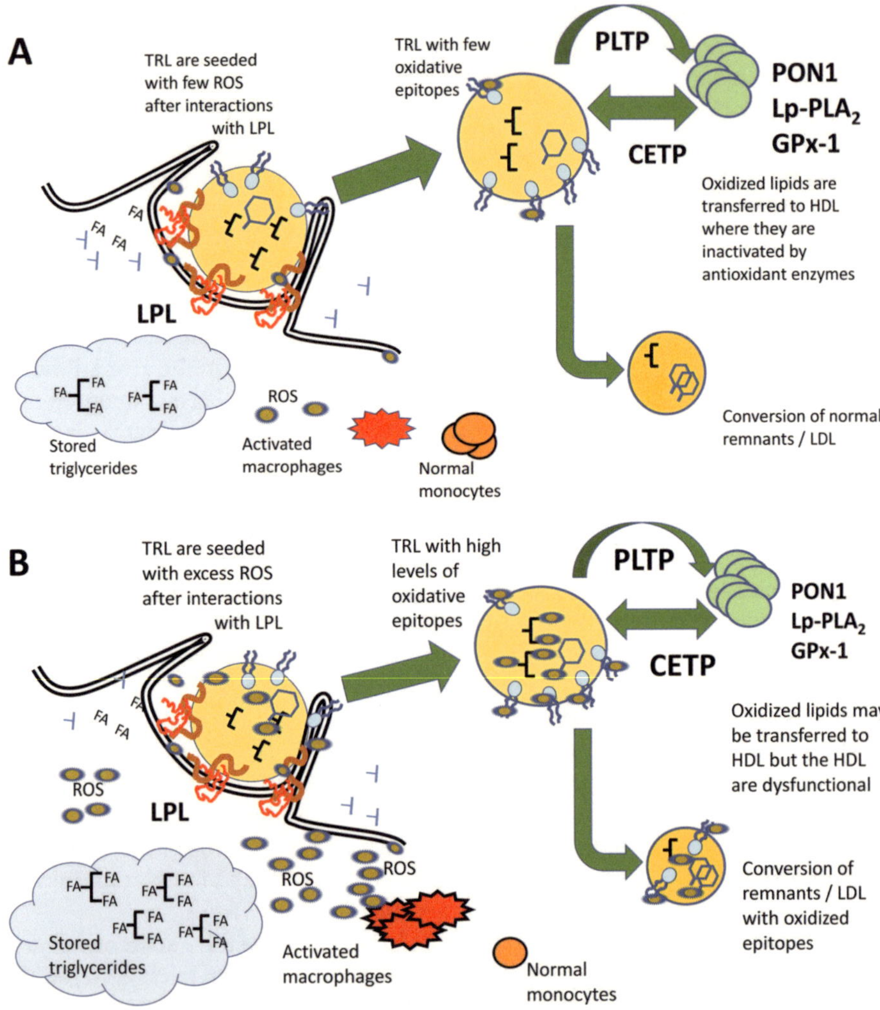

Figure 1.
Interactions of TG-rich lipoproteins with the endothelium. (A) In the case of the healthy endothelium, as TG-rich lipoproteins (TRL) are attached to the arterial wall via heparan sulfate proteoglycans, lipoprotein lipase (LPL) hydrolyzes TG into free fatty acids (FA) and monoglycerides that can penetrate the endothelium and be reconstitute as stored TG. With modest quantity of activated macrophages, only minimal amount of ROS is generated to seed plasma lipoproteins. In the circulation, lipoproteins undergo remodeling via the action of phospholipid lipid transfer protein (PLTP) and cholesteryl ester transfer protein (CETP) with the transfer of oxidized lipids to high-density lipoproteins (HDL) where they are inactivated by antioxidant enzymes, including paraoxonase (PON-1), lipoprotein-associated phospholipase A2 (Lp-PLA2) and glutathione peroxidase (GPx-1). With the depletion of TG, TG-rich lipoproteins are converted to normal chylomicron remnants and LDL. (B) In the case of the inflamed endothelium, there are excess generation of ROS by increasing number of activated macrophages. Together with impaired hydrolysis by LPL and greater flux of TG-rich lipoproteins, the retention time of TRL along the arterial wall is increased resulting to a greater opportunity for ROS to seed plasma lipoproteins. In the circulation, while oxidized lipids may still be transferred to HDL, the antioxidative properties of HDL have been reduced. The net result is the generation of chylomicron remnants and LDL with significant number of oxidative epitopes that could undergo spontaneous auto-oxidation.

intermittent entries of intestinal chylomicrons carrying dietary fats can further increase this concentration. Depending on the type of fat, interactions of these free fatty acids with endothelial cells could have deleterious effect on endothelial function [29, 30]. In vitro studies with human aortic endothelial cells have demonstrated that high levels of the polyunsaturated fatty acid linoleic acid increase endothelial cell permeability [31] whereas increased levels of the saturated fatty acid palmitate and the monounsaturated fatty acid oleate tend to decrease the activity of

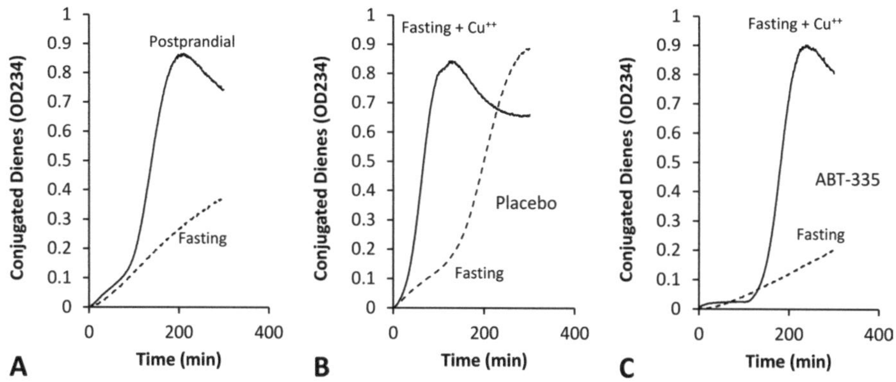

Figure 2.
Continuous monitoring of in vitro oxidation of LDL from two patients with metabolic syndrome. (A) In the absence of the catalyst Cu++, LDL isolated from Patient A at 4 h after meal undergoes spontaneous auto-oxidation while LDL isolated from fasting plasma does not. (B) During the placebo period, fasting LDL isolated from Patient B undergoes spontaneous auto-oxidation in the absence of Cu++ and the lag time is significantly reduced in the presence of Cu++. (C) After 8 weeks of therapy with ABT-335, fasting LDL from Patient B no longer undergoes auto-oxidation in the absence of Cu++ and demonstrates the typical oxidation kinetics in the presence of Cu++.

endothelial nitric oxide synthase [32]. However, if these newly released free fatty acids have been previously oxidized, they can elicit additional pro-inflammatory responses in endothelial cells. Oxidized free fatty acids can have direct effect on toll-like receptors, insulin resistance, upregulation of NF-κB, expression of adhesion molecules and macrophage cytotoxicity [33, 34].

In addition to the free fatty acids released as part of the hydrolysis of TG-rich lipoproteins, LDL is direct product of the metabolism of TG-rich VLDL [17]. The presence of oxidized phospholipids on the surface of plasma lipoproteins can also contribute to oxidative and inflammatory status [35–37]. As illustrated (**Figure 1**), a number of intrinsic enzymes on circulating HDL can blunt the deleterious effect of oxidized phospholipids, namely paraoxonase-1, lipoprotein-associated phospholipase A2 and glutathione peroxidase-1 [38]. However, in individuals with hypertriglyceridemia, excess generation of LDL with oxidized phospholipids [26, 39] and impaired anti-oxidant enzymes on HDL may exacerbate the oxidative process.

3. Glucose-associated oxidative stress and inflammation

Exposure to elevated plasma glucose in patients with type 2 diabetes mellitus is associated with both an increased level of ROS and a drop in cell antioxidant defense, enzymatic as well as non-enzymatic [4, 5]. In addition to serving as a barrier to control movements of fluid, solutes and cells between blood and tissue, the endothelial layer also play a key role in regulating vascular tone and inflammatory processes [40]. While the impact of hyperglycemia on endothelial cell metabolism is beyond the scope of this presentation it should be noted that, at normal fasting plasma glucose concentrations (4–6 mM, 72–108 mg/dL), the glucose transporter GLUT-1 is already functioning at saturation level [41]. GLUT-1 is the primary glucose carrier responsible for the uptake of glucose by endothelial cells [42]. In vitro studies with human umbilical vein endothelial cells have indicated that exposure to high concentration of glucose is associated with increased oxidative stress as assessed by increased levels of nitrotyrosine and 8-hydroxy-2′-deoxyguanosine [6]. In fact, the increase in oxidative stress was even more pronounced when endothelial cells were exposed intermittently to high and low glucose concentrations [6]. It has also been demonstrated that high glucose levels stimulate ROS production through protein kinase C (PKC)-dependent activation of NAD(P)H oxidase [43, 44]. The net result of constant and intermittent hyperglycemia is enhancement of endothelial cell apoptosis [6, 44].

In both normal and type 2 diabetic patients, intermittent hyperglycemia has been demonstrated to be more deleterious to endothelial cell metabolism and oxidative stress than constant elevations in plasma glucose [45]. In fact, reduction in the mean amplitude of glycemic excursion with dipeptidyl peptidase-IV inhibition in patients with type 2 diabetes is associated with reduction in oxidative stress and biomarkers of inflammation [46].

In addition to the stimulation of ROS production, exposure to high glucose concentrations can also lead to nonenzymatic modifications of proteins with the formation of glycated proteins or advanced glycation end-products (AGE) [47]. Glycation of specific proteins can severely alter their function as in the case of glycated hemoglobin C with increased affinity for oxygen and subsequent decreased oxygen delivery to tissues [48]. Glycation of plasma LDL by methylglyoxal, a side product of glycolysis, results in enhanced delivery of these pro-atherogenic particles to the arterial wall leading to increased risk for atherosclerosis in patients with type 2 diabetes [49]. Glycation of insulin has also been suggested to contribute to insulin resistance leading to more severe hyperglycemia [50].

4. Meal-induced oxidative stress and inflammation

The fat content of a typical Western meal ranges from 20 to 40 g of fat which corresponds to between 5 and 8 times the total pool of TG in plasma. In a healthy individual with normal metabolism, plasma TG level reaches peak level by 2 h after meal and returns to pre-meal level by 6 h. As many individuals customarily consume 3 meals a day, they spend the majority of their day in the postprandial state. Several processes contribute to the increased oxidative stress and inflammatory state associated with meal consumption [51]. The magnitude and timing of postprandial inflammatory response to a high-fat meal may depend on fat content, caloric intake, body-mass index [52] as well as age of the individual [53, 54].

Firstly, as noted in earlier section, the intermittent secretion of intestinal chylomicrons after each meal has a direct effect on the metabolism of circulating hepatic VLDL. While this may not be the case for individuals with normal TG levels, the lipolytic system may be saturated in individuals with elevated TG as in the case of obesity and diabetes. The net result is a delayed clearance of TG-rich lipoproteins, prolonged interactions of TG-rich particles with the endothelium, and greater potential for the transfer of ROS from the arterial wall to circulating plasma lipoproteins (**Figure 1B**). Treatment of peripheral blood mononuclear cells with lipolysis products of postprandial TG-rich lipoproteins resulted in increased expression of TNFa, IL-1b, and IL-8 [55].

Secondly, the meal-induced change in free fatty acids can contribute to the deleterious impact on the endothelium not only with respect to the increased in concentrations but also to the type of free fatty acids. In patients with insulin resistance, the impact of postprandial free fatty acids is further exacerbated by the failure of postprandial insulin to inhibit the activity of hormone-sensitive lipase (HSL) in tissues. In the normal individuals, the increase in insulin concentration during postprandial lipemia would inhibit the activity of HSL and shut off the mobilization of free fatty acids from peripheral tissues [56–58]. With insulin resistance, HSL remains active during postprandial lipemia and continues to mobilize free fatty acids from intracellular stores contributing to even higher plasma free fatty acid concentrations.

Thirdly and most importantly is the independent and cumulative effect of postprandial hypertriglyceridemia and hyperglycemia on oxidative stress and inflammation [10]. In both normal and diabetic individuals, compared to high fat meal, the inclusion of a glucose dose equivalent to an oral glucose tolerance test resulted in greater increase in nitrotyrosine and circulating adhesion molecules, including E-selectin, ICAM-1, and VCAM-1 [59]. Management with simvastatin in the short term (3 days) did not affect lipids but reduced the effect on nitrotyrosine and adhesion molecules. Extended therapy with simvastatin (3 months) blunted the meal-induced hypertriglyceridemia as well as the post-prandial responses in nitrotyrosine and adhesion molecules [10, 59]. In patients with type 2 diabetes, meal-induced increases in TG and glucose were attenuated by prandial + basal insulin [60]. Post-meal increased in C-reactive protein (hs-CRP), IL-6, and TNFa were also reduced [60] with the addition of prandial insulin.

5. Future directions

In view of the fact that individuals are in a postprandial state throughout the day, assessment of oxidative and inflammatory status in the fasted might not provide an accurate snapshot. It is important to understand the processes that affect oxidative stress and inflammatory status in a non-fasted state [61]. Additional research is needed to understand how nutrients, with respect to quality, quantity, and frequency, could be managed to attenuate the deleterious effect of postprandial hypertriglyceridemia and hyperglycemia on oxidative stress and inflammation [62]. Blunting the acute daily fluctuations in plasma glucose and triglycerides may be a novel mode of management to reduce oxidative stress and inflammation in high-risk individuals. Furthermore, special attention should be placed on the timing of antioxidant ingestion in relation to meal consumption as that might have a direct impact on the time course of postprandial response [63].

6. Conclusions

Assessment of oxidative and inflammatory status in the fasted state may not provide an accurate picture of the metabolic status of an individual. Intermittent elevations in plasma triglycerides and glucose associated with meal consumption throughout the day may be associated with considerable increase in oxidative stress and inflammation depending on the quantity and quality of the meals.

Abbreviations

AGE	advanced glycated end-products
CRP	(hs-CRP) high-sensitive C-reactive protein
FFA	free fatty acid
HDL	high-density lipoproteins
IL	interleukin
LDL	low-density lipoproteins
LPL	lipoprotein lipase
HSL	hormone sensitive lipase
ROS	reactive oxygen species
TG	triglycerides
TRL	TG-rich lipoproteins
VLDL	very-low density lipoproteins

Author details

Ngoc-Anh Le
Biomarker Core Laboratory, Foundation for Atlanta Veterans Education and Research, Atlanta Veterans Affairs Health Care System, Atlanta, Georgia, USA

*Address all correspondence to: anh.le@va.gov

References

[1] Ebrahim S, Pearce N, Smeeth L, Casas J, Jaffar S, Piot P. Tackling non-communicable diseases in low- and middle-income countries: Is the evidence from high income countries all we need? PLoS Medicine. 2013;**10**:e1001377

[2] Camps J, editor. Oxidative Stress and Inflammation in Non-Communicable Diseases—Molecular Mechanisms and Perspectives in Therapeutics. Switzerland: Springer; 2014

[3] Que X, Hung M, Yeang C. Oxidized phospholipids are proinflammatory and proatherogenic in hypercholesterolaemic mice. Nature. 2018;**558**(7709):301-306

[4] Catherwood M, Powell L, Anderson P, McMaster C, Sharpe P, Trimble E. Glucose-induced oxidative stress in mesangial cells. Kidney International. 2002;**61**:599-608

[5] Bonnefont-Rousselot D, Bastard JP, Jaudon MC, Delattre J. Consequences of the diabetic status on the oxidant/antioxidant balance. Diabetes & Metabolism. 2000;**26**:163-176

[6] Piconi L, Quagliaro R, Assaloni R, et al. Constant and intermitent high glucose enhances epithelial apoptosis through mitochondrial superoxide overproduction. Diabetes/Metabolism Research and Reviews. 2006;**22**:198-203

[7] Vries Md KB, Janssen H, Njo T, Westerman E, Cabezas MC. Postprandial inflammation: Targeting glucose and lipids. Advances in Experimental Medicine and Biology. 2014;**824**:161-170

[8] Deopurkar R, Ghanim H, Friedman J, Abuaysheh S, Sia C, Mahanty P. Differential effects of cream, glucose, and orange juice on inflammation, endotoxin, and the expression of toll-like receptor-4 and suppressor of cytokine signaling-3. Diabetes Care. 2010;**33**:991-997

[9] Wang L, Guo L, Zhang L, Zhou Y, He Q, Zhang Z. Effects of glucose load and nateglinide intervention on endothelial function and oxidative stress. Journal Diabetes Research. 2013;**2013**:849295

[10] Ceriello A, Taboga C, Tonutti L, et al. Evidence for an independent and cumulative effect of postprandial hypertriglyceridemia and hyperglycemia on endothelial dysfunction and oxidative stress generation. Effects of short- and long-term simvastatin treatment. Circulation. 2002;**106**:1211-1218

[11] Niki E, Yoshida Y, Saito Y, Noguchi N. Lipid peroxidation: Mechanisms, inhibition, and biological effects. Biochemical and Biophysical Research Communications. 2005;**338**:668-676

[12] Liu T, Stern A, Roberts L, Morrow J. The isoprostanes: Novel prostaglandin-like products of the free radical catalyzed peroxidation of arachidonic acid. Journal of Biomedical Science. 1999;**6**:226-235

[13] Milne G, Musiek E, Morrow J. F2 isoprostanes as markers of oxidative stress in vivo: An overview. Biomarkers. 2005;**10**(Suppl 1):S10-S23

[14] Esterbauer H, Schaur R, Zollner H. Chemistry and biochemistry of 4-hydroxynonenal, malondialdehyde and related aldehydes. Free Radical Biology and Medicine. 1991;**11**:81-128

[15] Holvoet P, Perez G, Zhao Z. MDA-modified LDL in patients with atherosclerotic disease. The Journal of Clinical Investigation. 1995;**95**:2611-2619

[16] Dominiczak M. Apolipoproteins and lipoproteins in human plasma. In: Rifai N, Warnick G, Domniczak M, editors. Handbook of Lipoprotein Testing. Washington, DC: American Association for Clinical Chemistry; 2000. pp. 1-29

[17] Le N-A, Walter M. The role of hypertriglyceridemia in atherosclerosis. Current Atherosclerosis Reports. 2007;**9**(2):110-115

[18] Weisgraber K, Rall SC Jr. Human apolipoproteinB-100 heparin-binding sites. The Journal of Biological Chemistry. 1987;**262**(23):11097-11103

[19] Wang L, Gill R, Pedersen T, Higgins L, Newman J, Rutledge J. TG-rich lipoprotein lipolysis releases neutral and oxidized FFAs that induce endothelial cell inflammation. Journal of Lipid Research. 2009;**50**:204-213

[20] Le N-A. Lipoprotein-associated oxidative stress: A new twist to the postprandial hypothesis. International Journal of Molecular Sciences. 2015;**16**:401-419

[21] Le NA, Li X, Kyung S, Brown WV. Evidence for the in vivo generation of oxidatively modified epitopes in patients with documented CAD. Metabolism. 2000;**49**:1271-1277

[22] Gradek Q, Harris M, Yahia N, Davis W, Le N-A, Brown W. Polyunsaturated fatty acids acutely suppress antibodies to malondialdehyde-modified LDL in patients with vascular disease. The American Journal of Cardiology. 2004;**93**:881-885

[23] Melish JS, Le NA, Ginsberg HN, Steinberg D, Brown WV. Dissociation of apoprotein B and triglyceride production in very low density lipoproteins. American Journal of Physiology. 1980;**239**:E354-EE62

[24] Ginsberg HN, Le NA, Melish JS, Steinberg D, Brown WV. Effect of a high carbohydrate diet on apoprotein B catabolism in man. Metabolism. 1981;**30**:347-353

[25] Welty F, Lichtenstein A, Barrett P. Hupamn apolipoprotein B-48 and apoB-100 kinetics with stable isotopes. Arteriosclerosis, Thrombosis, and Vascular Biology. 1999;**19**(12):2966-2974

[26] Yoshida H, Kisugi R. Mechanisms of LDL oxidation. Clinica Chimica Acta. 2010;**411**:1875-1882

[27] Esterbauer H, Striegl G, Puhl H, Rotheneder M. Continuous monitoring of in vitro oxidation of human LDL. Free Radical Biology & Medicine. 1989;**6**:67-75

[28] Le N-A, Farkas-Epperson M, Sweeney M, Wilson P, Brown W. Effect of ABT-335 (fenofibric acid) on meal-induced oxidative stress in patients with metabolic syndrome. Atherosclerosis. 2013;**231**:268-273

[29] Steinberg H, Tarshoby M, Modestel R, et al. Elevated circulatingfree fatty levels impair endothelium-dependent vasodilation. The Journal of Clinical Investigation. 1997;**100**:1230-1239

[30] Chung B, Hennig B, Cho B, Darnell B. Effect of the fat composition of a single meal on the composition andcytotoxic properties of lipolytically-releasable freefatty acids in postprandial lipemia. Atherosclerosis. 1998;**141**:321-332

[31] Hennig B, Watkins B. Linoleic acid and linolenic acid effect on permaebility properties of cultured endothelial cell monolayers. The American Journal of Clinical Nutrition. 1989;**49**:301-309

[32] Halle M, Eriksson P, Tornvall P. Effects of free fatty acids and

triglyceride-rich fat emulsion o endothelial nitric oxide synthase. European Journal of Clinical Investigation. 2005;**35**:154-155

[33] Goldberg I, Eckel R, McPherson R. Triglycerides and heart disease: Still a hypothesis? Arteriosclerosis, Thrombosis, and Vascular Biology. 2011;**31**:1716-1725

[34] Rask-Madsen C, Kahn C. Tissue-specific insulin signaling, metabolic syndrome and cardiovascular disease. Arteriosclerosis, Thrombosis, and Vascular Biology. 2012;**32**:2052-2059

[35] Stamenkovic A, Pierce G, Ravandi A. Oxidized lipids: Not just another brick in the wall. Canadian Journal of Physiology and Pharmacology. 2019;**97**:473-485

[36] Freigang S. The regulation of inflammation by oxidized phospholipids. European Journal of Immunology. 2016;**46**:1818-1825

[37] Leitinger N. Oxidized phspholipids as modulators of inflammation in atherpsclerosis. Current Opinion in Lipidology. 2003;**14**:421-430

[38] Le N-A, Walter M. Intrinsic enzymes of HDL. Journal of Clinical Lipidology. 2007;**1**:20-30

[39] Matsuura E, Hughes G, Khamashta M. Oxidation of LDL and its clinical implication. Autoimmunity Reviews. 2008;**7**:558-566

[40] Hansen N, Hansen A, Sams A. The endothelial border to health: Mechanistic evidence of the hyperglycemic culprit of inflammatory disease acceleration. International Union of Biochemistry and Molecular Biology Life. 2017;**69**(3):148-161

[41] Mann G, Yudilevich D, Sobrevia L. Regulation of amino acid and glucose transportersin endothelial and smooth muscle cells. Physiological Reviews. 2003;**83**:183-252

[42] Olson A, Pessin J. Structure, fuction, and regulation of the mammalian facilitative glucose transporter gene family. Annual Review of Nutrition. 1996;**16**:235-256

[43] Inoguchi T, Li P, Umeda F, et al. High glucose level and free fatty acid stimulate reactive oxygen species production through protein kinase C-dependent activation of NAD(P) H oxidase in cultured vascular cells. Diabetes. 2000;**49**:1939-1945

[44] Quagliaro L, Piconi L, Assaloni R, Martinelli L, Motz E, Ceriello A. Intermittent high glucose enhances apoptosis related to oxidative stress in human umbilical vein endothelial cells in culture: The role of protein kinase C and NAD(P)H-oxidase activation. Diabetes. 2003;**52**:1795-2804

[45] Ceriello A, Esposito K, Piconi L, et al. Oscillating glucose is more deleterious to endothelial function and oxidative stress than mean glucose in normal and type 2 diabetic patients. Diabetes. 2008;**57**:1349-1354

[46] Rizzo M, Barbieri M, Marfella R, Paolisso G. Reduction of oxidative stress by blunting daily acute glucose fluctuations in patients with type 2 diabetes. Diabetes Care. 2012;**35**:2076-2082

[47] Eble A, Thorpe S, Baynes J. Nonenzymatic glucosylation and glucose-dependent cross-linking of protein. The Journal of Biological Chemistry. 1983;**258**:9406-9412

[48] Coletta M, Amiconi G, Bellelli A, Betollini A, Carsky J. Alteration of t-state binding properties of naturally glycated hemoglobin HbA1c. Journal of Molecular Biology. 1988;**5**:233-239

[49] Rabbini N, Godfrey L, Xue M, et al. Glycation of LDL by methylglyoxal increases arterial atherogenicity: A possible contributor to increased ris of cardiovascular disease in diabetes. Diabetes. 2011;**60**(7):1973-1980

[50] Song F, Schmidt A. Glycation and insulin resistance: Novel mechanisms and unique targets. Arteriosclerosis, Thrombosis, and Vascular Biology. 2012;**32**:1760-1765

[51] Herieka M, Erridge C. High-fat meal induced postprandial inflammation. Molecular Nutrition & Food Research. 2014;**58**:136-146

[52] Wang Y, Schulze J, Raymond N, et al. Endothelial inflammation correlates with subject triglycerides and waist size after high-fat meal. American Journal of Physiology. Heart and Circulatory Physiology. 2011;**300**(3):H784-HH91

[53] Emerson S, Kurti S, Harms C, et al. Magnitude and timing of the postprandial inflammatory response to a high-fat meal in healthy adults: A systematic review. Advances in Nutrition. 2017;**8**:213-225

[54] Milan A, Pundir S, Pileggi C, Markworth J, Lewandowski P, Cameron-Smith D. Comparisons of the postprandial inflammatory and endotoxaemic responses to mixed meals in yound and loder individuals: A randomized trial. Nutrients. 2017;**9**:354-370

[55] den Hartigh L, Altman R, Norman JE, Rutledge J. Postprandial VLDL lipolysis products increase monocyte adhesion and lipid droplet formation via activation of ERK2 and NFkB. American Journal of Physiology. Heart and Circulatory Physiology. 2014;**306**(1):H109-H120

[56] Lampidonis A, Rogdakis E, Voutsinas G, Stravopodis D. The resurgence of hormone-sensitive lipase in mammalian lypolysis. Gene. 2011;**477**:1-11

[57] Turner N, Cooney G, Kraegen E, Bruce C. Fatty acid metabolism, energy expenditure and insulin resistance in muscle. The Journal of Endocrinology. 2014;**220**:T61-T79

[58] Papackova Z, Cahova M. Fatty acid signaling: The new function of intracellular lipases. International Journal of Molecular Sciences. 2015;**16**:3831-3855

[59] Ceriello A, Quagliaro L, Piconi L, et al. Effect of postprandial hypertriglyceridemia and hyperglycemia on circulating adhesion molecules and oxidative stress generation and the possible role of simvastatin treatment. Diabetes. 2004;**53**:701-710

[60] Beisswenger P, Brown W, Ceriello A, et al. Meal-induced increases in C-reactive protein, IL6, and TNFa are attenuated by prandial + basal insulin in patients with type 2 diabetes. Diabetic Medicine. 2011;**28**:1088-1095

[61] Meessen E, Warmbrunn M, Nieuwdorp M, Soeters M. Human postprandial nutrient metabolism and low-grade inflammation: A narrative review. Nutrients. 2019;**11**:3000-3021

[62] Reverri E, Randolph J, Steinberg F, Kappagoda C, Edirisinghe I, Burton-Freeman B. Black beans, fiber, and antioxidant capacity pilot study: Examination of whole foods vs. functional components on postprandial metabolic, oxidative stress, and inflammation in adults with metabolic syndrome. Nutrients. 2015;**7**:6139-6154

[63] Carroll M, Schade D. Timing of antioxidant vitamin ingestion alters postprandial proatherogenic serum markers. Circulation. 2003;**108**:24-31

Chapter 7

Glycerol Conversion to Diglycerol via Etherification under Microwave Irradiation

Muhammad Ayoub, Wan Jie Wei, Manzoor Ahmad, Ranitha Mathialagan, Sarah Farrukh, Mohammed Danish, Sami Ullah and Salman Raza Naqvi

Abstract

According to Grand View Research in polyglycerol market size, demand for diglycerol is expected to grow by 50% from 2012 to 2022 due to its extensive use in various industries, thus validating the importance and value addition of producing diglycerol. Due to the volatility of refined glycerine market price and increasing demand of diglycerol, research has been conducted to upgrade glycerol via various processes. Etherification is a single-step process of catalytic conversion of glycerol into polyglycerols, involving the condensation of two glycerol molecules to form the simplest oligomer which is diglycerol with linear, branched, or cyclic isomers. Thus, this chapter will discuss on the methods of synthesizing diglycerol followed by the type of catalyst to be used. These include homogenous and heterogenous catalyst with their subdivision of acid and base type, respectively. Besides, this chapter does include on the method for the etherification process where it highlighted the advantage of advance technology microwave irradiation over conventional heating.

Keywords: diglycerol, etherification, microwave, yellow glycerine, industrial

1. Introduction

Glycerine, also called glycerol (propane-1,2,3-triol, $C_3H_8O_3$), occurs as the backbone in triglycerides which are the main constituents of all vegetable and animal fats and oils. It is colorless and odorless, has sweet taste, and is very viscous and hygroscopic [1, 2]. Glycerol is a key ingredient in personal care product due to its property of acting as humectants, solvents, moisturizer, as well as an additive to make coatings. Glycerol is currently produced from biodiesel transesterification, saponification, and hydrolysis processes in fatty acid and soap production [3]. Among these, biodiesel transesterification is the biggest crude glycerol generation sector, contributing up to 67% of the total crude glycerol production, followed by high pressure splitting (17%) and soap production (<10%) [4].

Due to the growing biodiesel demand as substitution to petroleum-based diesel, the production of crude glycerol as by-product from this process is also increased. For every 1 ton of biodiesel produced, approximately 100 kg of glycerol will be generated [5]. The crude glycerol generated contains many impurities such as methanol, organic and inorganic salts, water, unreacted triglycerides, soap, fatty acid, etc. Hence, for large-scale biodiesel manufacturers, the plants are normally equipped with crude glycerol refining facilities to produce glycerol with purities up to 95.5 and 99% [3].

Generally, there are three types of glycerol, namely, crude glycerine, technical grade glycerine, and refined glycerine. The process of refining glycerine is a complex process which involves distillation, bleaching, acidification, and several intermediate steps. In the distillation step, refined glycerine is obtained from the side draw of the distillation column and then subsequently bleached and deodorized to achieve the refined grade. Technical grade glycerine, also called yellow glycerine, is obtained from the top of the column with some impurities such as water, fatty acids, soap, etc. [2]. **Figure** 1 shows the general process flow of glycerine refining.

According to MarketWatch 2018, the global glycerine market is expected to reach approximately 6200 metric tons by 2024 from 3550 metric tons in 2016, resulted from the growth in bio-renewable chemicals, biodiesel production, and wide range of applications. However, with regard to the market price, technical grade glycerine is cheaper than the refined glycerine which are approximately RM4400 and RM5400 per metric ton, respectively, as reported by Oleoline (2017).

Furthermore, glycerol can be used as a starting material for the synthesis of value-added chemicals via catalytic conversion such as hydrogenolysis to propanediol and ethylene glycol, dehydration to acrolein, fermentation to propane-1,3-diol, thermal reduction into syngas, etherification to fuel oxygenates, conversion into glycerol carbonate, and synthesis of epichlorohydrin [6]. Among these, the etherification of glycerol into polyglycerols particularly diglycerol (DG) and triglycerol (TG) is gaining more interest in the recent research due to their possibility in controlling the hydrophilic–lipophilic balance (HLB) which is highly important as additive to food and pharmaceutical industries, lubricants, stabilizers, dispersants, plasticizers, etc. [6]. According to Grand View Research in polyglycerol market size, the demand for diglycerol and triglycerol is expected to grow by 50% from 2012 to 2022. The global diglycerol demand is expected to grow at a CAGR of 5.3% due its extensive use in various industries, thus validating the importance and value addition of producing diglycerol.

Due to the volatility of refined glycerine market price and increasing demand of polyglycerol especially the shortest chain oligomer which is diglycerol, research has been conducted to upgrade glycerol via various processes such as pyrolysis, epichlorohydrin, and catalytic etherification to produce polyglycerols. Etherification is a single-step process of catalytic conversion of glycerol into polyglycerols, involving the condensation of two glycerol molecules to form the simplest oligomer which is diglycerol with linear, branched, or cyclic isomers. Further reaction yields tri-, tetra-, and higher oligomers. Short-chain oligomers are preferred. Therefore, the

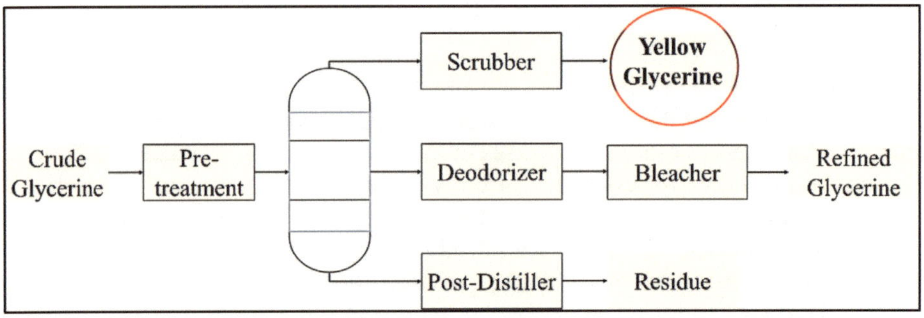

Figure 1.
General process flow of glycerine refining.

etherification of glycerol under controlled conditions is required [7]. The etherification method has potential to replace the conventional epichlorohydrin route of producing polyglycerols which is relatively complex and involves in the production of toxic intermediate. Polyglycerols can be produced from different raw materials such as crude glycerine, yellow glycerine, as well as pure glycerine [7–9]. Hence, the economical production of glycerol derivatives is directly related with the quality of glycerine feedstocks and prices.

With regard to the heating method, microwave irradiation has been increasingly popular as a heating method for organic reaction. It had been proven to be more effective than the conventional water bath heating [10]. Various benefits of microwave irradiation included internal rapid heating that reduced the reaction time which compensates its higher power consumption. Besides, efficient and uniform heating that enables good temperature control can be achieved via microwave heating.

2. Glycerine conversion

2.1 Existing diglycerol synthesis method

2.1.1 Laboratory-scale routes

On small scale, pure diglycerol is produced via direct synthesis methods in which diallyl ether is used as a primary reactant [1]. Diallyl ether is accessible by the reaction between allyl chloride and allyl alcohol in inert solvents under hydrogen chloride release. Direct hydroxylation of this product can be performed with peroxyformic acid, CH_2O_3, or permanganate at 40°C under safety precautions for 4.5 hours. However, several additional steps are needed for neutralization, filtration, derivatization, and fractional distillation which are required for the isolation of diglycerol and triglycerol. The isolation of diglycerol can be done using neutralization with barium hydroxide solution, centrifugation to separate the solid, digestion of the product in absolute ethanol, and fractional distillation under reduced pressure.

2.1.2 Thermal conversion of glycerol

The thermal reaction for glycerol oligomerization is conducted at a certain temperature under inert protecting reactor condition [1]. For selective reaction, pure glycerol up to 99% purity should be used. Commonly, before the use of the oligomeric products for further reactions, a distillation is needed to remove unconverted glycerol. Reaction temperature, basicity, and organic impurities would affect the glycerol oligomerization. For pure thermal conversion of glycerol without the addition of catalyst, temperature is normally set above 200°C. However, at reaction temperature higher than 290°C, side products with strong smelling are formed. At 180°C (low temperature) with the addition of alkali catalyst, the formation of diglycerol from glycerol is observed but at low conversion of glycerol.

2.1.3 Epichlorohydrin method

The epichlorohydrin method of producing polyglycerol is commonly applied. It involves basic hydrolysis by sodium hydroxide, led to the formation of intermediate product, namely, glycidol and glycerol, and it will react with unconverted epichlorohydrin or glycerol to diglycerol [6]. Further separation and purification of products are required. Residue glycerol will be separated, and then water has to be removed from the raw diglycerol, and, finally, the product has to be subjected to a fine distillation. The reaction of glycidol or epichlorohydrin with glycerol has in common that the coupling OH groups are not confined to the terminal positions but also the middle OH groups of glycerol can be involved as well. This leads to the formation of primary or secondary dimer of glycerol.

2.1.4 Catalytic reaction method

The etherification reaction of glycerol to produce value-added diglycerol has been extensively studied either using homogeneous or heterogenous alkali or acid catalysts. The etherification reaction is normally conducted under solvent-free condition for economical and environmental reasons [7]. Etherification is a condensation process whereby two glycerol molecules will react to form ether bond in between the molecules by removing water molecule. This reaction is also called etherification since the final product is in ether form as shown in **Figure 2**. Glycerol oligomers can be in linear, branched, or cyclic form [11]. The formation of different isomers is affected by the reaction conditions such as temperature, time, catalyst type, catalyst loading as well as the starting raw material for diglycerol synthesis. Diglycerols are mostly obtained from the oligomerization of glycerol catalyzed by acid or based catalysts.

2.2 Homogeneous catalyzed reaction for selective diglycerol

2.2.1 Homogeneous acid catalyst

The reaction by homogeneous acid catalyst is promoted by the protonation of one of the glycerol hydroxyl groups: a mechanism of SN_1 as shown in **Figure 3**. A carbocation is formed by splitting off a water molecule, followed by nucleophilic attack of a hydroxyl group of another glycerol molecule. Then, the formed ether is

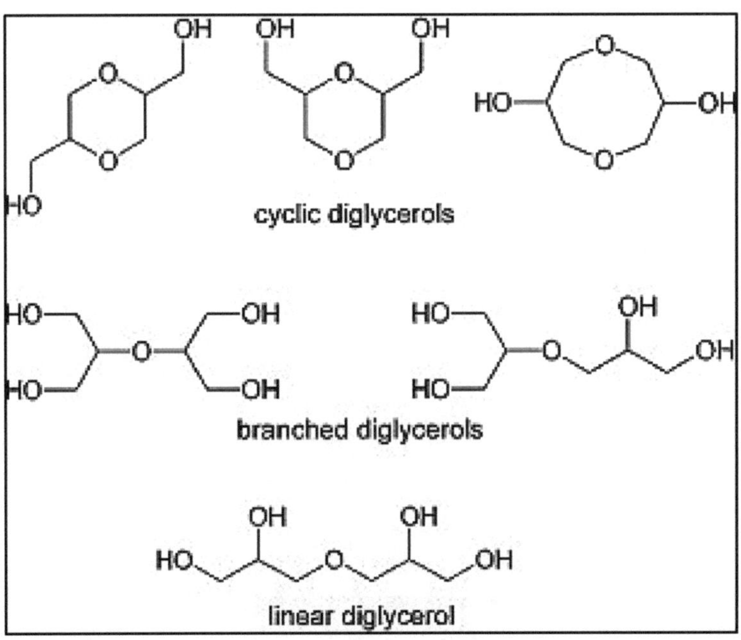

Figure 2.
Etherification of glycerol to polyglycerols.

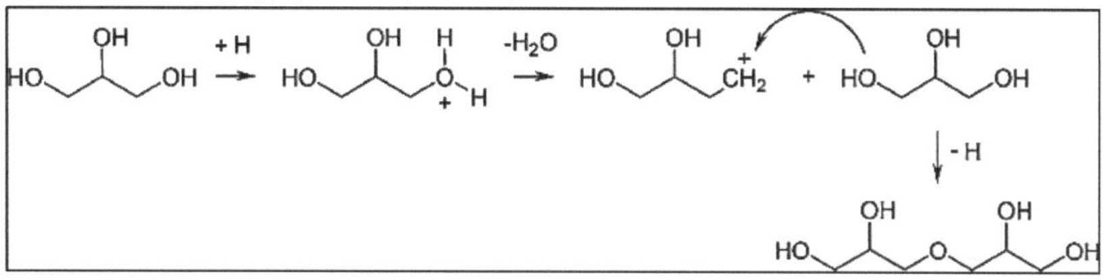

Figure 3.
SN_1-type mechanism for glycerol oligomerization in acid-catalyzed homogeneous reaction.

deprotonated, yielding the respective diglycerol [1]. Acid-catalyst oligomerization used sulfuric acid at 280°C in 2 hours giving more than 90% conversion of glycerol, and the main oligomers were triglycerol and tetraglycerol which only make the 20% of overall component, showing other side products were dominant in the reaction which were hardly identified [6].

Reported studies suggest that homogeneous acid-catalyzed reaction is generally fast but not selective for diglycerol. This could be due to the dehydration or oxidation of glycerol as secondary reactions to other undesired products. These secondary reactions may also result in the deterioration of the main product quality by changing its color, making the final product darker in appearance.

2.2.2 Homogeneous base catalyst

The reaction with glycerol conversion under basic homogeneous catalyst is proposed to follow an SN_2 mechanism as shown in **Figure 4**. In SN_2, the interaction of the base OH with glycerol weakens one of the glycerol OH bonds and enhances nucleophilic character of the hydroxyl oxygen. The attack of this polarized glycerol to a carbon of a second glycerol with simultaneous split off water results in diglycerol.

Several homogeneous bases have also been studied as catalysts in glycerol etherification. Depending on their basicity and solubility in glycerol, the following order of activity was reported under reaction temperature of 260°C at 4 hours with 2.5 mol% of catalyst:

K2CO3 > Li2CO3 > Na2CO3 > KOH > NaOH > CH3ONa > LiOH > MgCO3 > CaCO3.

Based on solubility measurements, the higher activity of carbonates than that of hydroxides was indeed ascribed to a better solubility of the former in glycerol and in the polyglycerol mixture at elevated temperature [6]. However, there are several studies presented that contradict the results from the proposed theory on solubility. A study conducted by [9] of glycerol conversion using Cs_2CO_3, CsOH, and $CsHCO_3$ showed 20% glycerol conversion and 100% diglycerol selectivity; hence, different anions did not alter the reaction characteristics. From the research carried out by

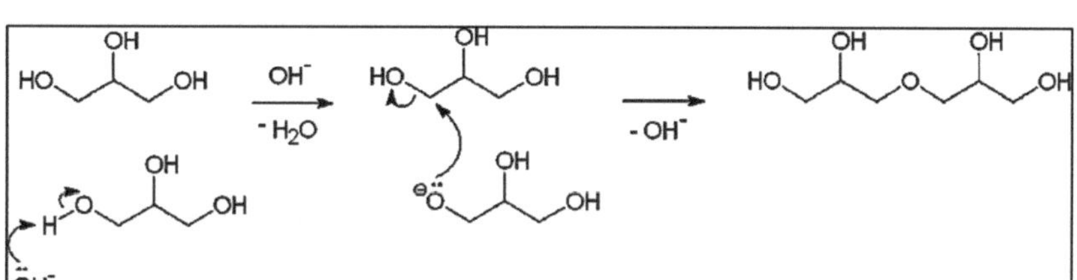

Figure 4.
SN_2-type mechanism for glycerol oligomerization in base-catalyzed homogeneous reaction.

[7], their studies reflected that the pH of the mixture of catalyst with glycerol which increased in the order LiOH, NaOH, KOH, and Na_2CO_3 was the main factor of high glycerol conversion and diglycerol selectivity. One hundred percent glycerol conversion was achieved for LiOH and NaOH catalysis system with 21 and 18% diglycerol selectivity, respectively. This could be resulted from the nature of lithium metal as the most active metal due to highest alkalinity, smallest ionic size, and highest atomic electronegativity [7].

Raw material choice on polyglycerol production is also very crucial in polyglycerol generation. A study done by [8] argued that polyglycerol production can be carried by the self-polymerization of crude glycerol which contained different amounts of soap (as homogeneous-based catalyst) which was originally presented in the crude glycerol samples. The experiment was carried out at 260°C under microwave heating; within 1 hour, the sample of soap with a highest soap content of 12.5% gave the highest conversion (94.94%) with 47.99% diglycerol. This study supported the claim by [7], whereby the higher the pH, the higher is the conversion and diglycerol selectivity. Besides, they also found that the processes with different feedstocks affected the amount of soap contained in the crude glycerol, preferably the refined, bleached, deodorized palm oil (RBDPO) which generated the highest amount of soap, creating a favorable condition for self-polymerization.

The shortcomings from homogeneous catalysis system include low selectivity to small-size polyglycerols which is the desired product of polyglycerols; uncontrollable oligomerization of glycerol which produces larger polyglycerols together with smaller polyglycerols would result in separation issue.

2.3 Heterogeneous catalyzed reaction for selective diglycerol

Recently, the main focus on catalytic etherification is on the implementation of heterogeneous catalyst in the etherification reaction to overcome the shortcomings of using homogeneous catalyst whereby the reactions are performed on solid materials of limited solubility [1]. Mesoporous solids modified by cesium impregnation or exchange lead to the best selectivity and yield to diglycerol in which 100% of diglycerol was produced after 10 hours reaction time [12]. The etherification activity was improved by the impregnation of LiOH with acid-treated montmorillonite K-10 (clay MK-10) [13]. Researchers reported that the clay Li/MK-10 catalyst gave a favorable result in selectively producing diglycerol which yielded 53% with high glycerol conversion of about 98%. However, the drawbacks of using heterogeneous catalyst are the leaching of metal from solid support into the liquid mixtures and instability of the solid support, resulting in dropping in efficiency to cater the reaction.

2.4 Microwave and conventional heating

Microwaves are electromagnetic waves which consist of both electric and magnetic fields that fall within 0.3–300 GHz on electromagnetic spectrum with wavelengths ranging from 0.01 to 1 m. Within the specific frequency range, the microwave transmits energy between 1.24×10^{-6} and 1.24×10^{-3} eV. **Table 1** shows the energy of different chemical bonds and microwaves [14]. With the energy lower than the Brownian motion, microwave will not be able to induce chemical reactions through bond breaking.

Microwave heating is a direct heating method which is highly dependent on the properties of materials to convert electromagnetic to heat energy. The two ways of inducing heat into a systems via microwave are dielectric heating and magnetic loss heating; however, based on the publications, researchers are mainly focus on studying the dielectric heating effects and mechanisms because this heating effect is more significant. Dielectric heating occurs due to the dielectric properties on the reacting medium. For example, materials with high dielectric constant tend to absorb microwave; however, poor dielectric property compounds are poor electromagnetic wave absorber. Hence, microwave is highly applied in organic reactions.

In conventional heating, heat is transferred through conduction and convection in which the heat is transferred from the sides/surface of a substance, then only travel to the center of targeted heating medium. This heating method is dependent on the heating flask properties such as thermal conductivity, specific heat capacity, density, etc. Hence, the portion of heat will be loss to the environment during heating. When compared to microwave heating, the heating duration using microwave to achieve the same temperature could be dramatically reduced. **Figure 5** shows the heating mechanism for both conventional heating and microwave heating.

Microwave radiation serves as an alternative to conventional heating especially in organic synthesis. The advantages of microwave heating over conventional heating are accelerating chemical reactions and promoting higher conversion and selectivity and low energy consumption as reported by [14] with respect to the same operating conditions under microwave heating. Plenty of studies have been carried out on the application of microwave in organic synthesis, for example, reactions in solvent-free condition, polymer synthesis, homogeneous and heterogeneous catalysis, medicinal and pharmaceutical chemistry, etc. [15].

Bonding and microwave	Energy (eV)
Brownian motion at 37°C	2.7×10^{-3}
Biological compound	13.6
Covalent bond	5
Hydrogen bond	2
Van der Waals intermolecular interactions	<2
Microwave at 0.3 GHz	1.24×10^{-6}
Microwave at 2.45 GHz (domestic oven)	1×10^{-5}
Microwave at 300 GHz	1.24×10^{-3}

Table 1.
Energy of different bonds and microwaves.

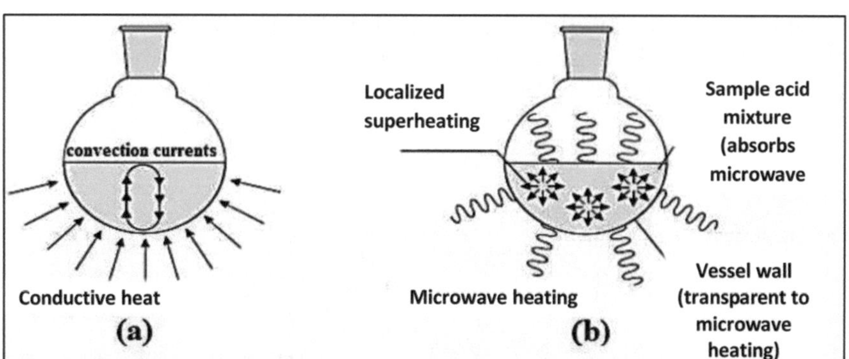

Figure 5.
Heating mechanisms for (a) conventional heating and (b) microwave heating.

Theories that are applied to explain the effect of microwave in enhancing a reaction are based on thermal and nonthermal effects as presented by [14] in **Table 2**.

Most of the studies on microwave effects describe the presence of hot spots or localized heating which enhances the chemical reactions due to direct adsorption of radiation by the polar molecules. Microwave energy is transferred through the relaxation of polar molecules, thus promoting molecular friction and collision [10]. As reported by [15], the hot spot will possess higher energy at the active sites, promoting the activities of the species adsorbed on the catalyst surface, whereas [16] presented that the conversion of hydrogen sulfide is higher than the theoretical equilibrium conversion; however, the results from conventional heating are similar to the theoretical value. They claimed that the reaction temperature at some site in the catalyst bed was much higher than the average temperature measured. However, studies by [14] illustrated that the occurrence of hot spot on catalyst surface is more likely to lower the catalytic activities, hence reducing the productivity. For an exothermic reaction, for example, the desulfurization, both microwave and conventional, provides similar conversion at the same operating conditions. Further increment in temperature in microwave system would cause the conversion to drop (lower than conventional heating), due to the shifting of equilibrium to less favorable reactions [16].

Besides localized heating effects, decreasing in the activation energy and improving the pre-exponential factor are the two main factors in enhancing microwave-assisted organic synthesis. When inducing microwave, a portion of the radiation heats up the system due to microwave thermal effects; another portion directly interacts with chemical or catalytic reaction system to reduce the activation energy, changing the interior energy level of molecules. The increment in the pre-exponential factor is due to the effective collision between molecules assisted by electromagnetic wave which change the movement of the molecules from disordered motion to ordered motion [17].

Even though in most cases microwave heating enhances the reactions, in the reaction with high sensitivity to the temperature change especially in a strong exothermic reaction, conventional heating would be preferred. The study in Fischer glycosidation showed a higher conversion under conventional heating which may be due to the higher overall reaction time, and the reaction temperature is reached smoothly. The reaction under conventional heating does not present localized overheating which can be found in microwave heating. Localized overheating in microwave was observed from the temperature overshoot above the desired temperature, causing glucose decomposition [18].

For the etherification reaction, most of the researches were conducted by conventional heating, and typical reaction times were longer than 8 hours. Microwave radiation is proven to be a more effective heating method in the etherification of glycerol. The required reaction time to produce targeted polyglycerols facilitated by

Thermal effects	Nonthermal effects
Dipolar polarization	$K = A \exp(-Ea/RT)$
Ionic conduction	Enhancement in collision probabilities
Uniform heat distribution	Decrease in activation energy
Specificity for polar molecule	Hot spots
Effect increase with polarity	High localized microscopic temperatures

Table 2.
Microwave heating mechanisms via thermal and nonthermal effects.

the microwave irradiation was much shorter than the etherification carried out by conventional heating which required at least 6 hours, whereas it took only 1 hour for microwave heating.

3. Conclusion

Microwave irradiation-assisted heating found to be a cost-efficient technology to be applied for the biodiesel conversion into polyglycerol. The application of modified heterogenous base catalyst could minimize the waste besides being compatible for the reaction of biodiesel waste into polyglycerol. Besides, it has provided insight on the oligomerization reaction of glycerol to maximize the yield of the valuable di-, tri-, and tetraglycerol oligomers for numerous applications with current industrial demand.

Acknowledgements

The authors would like to thank the Yayasan Universiti Teknologi PETRONAS with research grant FRGS 0153AB-L63 for providing financial support for this study.

Author details

Muhammad Ayoub[1*], Wan Jie Wei[1], Manzoor Ahmad[2], Ranitha Mathialagan[1], Sarah Farrukh[3], Mohammed Danish[4], Sami Ullah[5] and Salman Raza Naqvi[3]

1 Centre for Biofuel and Biochemical Research, Department of Chemical Engineering, Universiti Teknologi PETRONAS, Malaysia

2 Department of Computer and Information Sciences, Universiti Teknologi PETRONAS, Malaysia

3 Department of Chemical Engineering, National University of Science and Technology, Islamabad, Pakistan

4 Bioengineering Technology Section, Universiti Kuala Lumpur Malaysian Institute of Chemical and Bioengineering Technology, Malaysia

5 Department of Chemistry, College of Science, King Khalid University, Abha, Kingdom of Saudi Arabia

*Address all correspondence to: muhammad.ayoub@utp.edu.my

References

[1] Martin A, Richter M. Oligomerization of glycerol—A critical review. European Journal of Lipid Science and Technology. 2011;**113**(1): 100-117. DOI: 10.1002/ejlt.201000386

[2] Mbamalu VC. Glycerin and the Market. Chattanooga, Tennessee: University of Tennessee at Chattanooga; 2013

[3] Anitha M, Kamarudin SK, Kofli NT. The potential of glycerol as a value-added commodity. Chemical Engineering Journal. 2016;**295**:119-130. DOI: 10.1016/j.cej.2016.03.012

[4] Kong PS, Aroua MK, Daud WMAW. Conversion of crude and pure glycerol into derivatives: A feasibility evaluation. Renewable and Sustainable Energy Reviews. 2016;**63**:533-555. DOI: 10.1016/j.rser.2016.05.054

[5] Barros FJS, Moreno-Tost R, Cecilia JA, Ledesma-Muñoz AL, de Oliveira LCC, Luna FMT, et al. Glycerol oligomers production by etherification using calcined eggshell as catalyst. Molecular Catalysis. 2017;**433**:282-290

[6] Sivaiah MV, Robles-Manuel S, Valange S, Barrault J. Recent developments in acid and base-catalyzed etherification of glycerol to polyglycerols. Catalysis Today. 2012; **198**(1):305-313

[7] Ayoub M, Khayoon MS, Abdullah AZ. Synthesis of oxygenated fuel additives via the solventless etherification of glycerol. Bioresource Technology. 2012;**112**:308-312. DOI: 10.1016/j.biortech.2012.02.103

[8] Din NSMNM, Idris Z, Kian YS, Hassan HA. Preparation of polyglycerol from palm-biodiesel crude glycerine. Journal of Oil Palm Research. 2012; **25**(3):289-297

[9] Richter M, Krisnandi Y, Eckelt R, Martin A. Homogeneously catalyzed batch reactor glycerol etherification by CsHCO3. Catalysis Communications. 2008;**9**(11–12):2112-2116. DOI: 10.1016/j.catcom.2008.04.007

[10] Teng WK, Ngoh GC, Yusoff R, Aroua MK, Heng JS. Microwave assisted solvent-free catalytic transesterification of glycerol to glycerol carbonate. International Scholarly and Scientific Research and Innovation. 2015;**9**(9): 1140-1143

[11] Ruppert AM, Meeldijk JD, Kuipers BW, Erne BH, Weckhuysen BM. Glycerol etherification over highly active CaO-based materials: New mechanistic aspects and related colloidal particle formation. Chemistry. 2008;**14**(7): 2016-2024

[12] Barrault J, Clacens JM, Pouilloux Y. Selective oligomerization of glycerol over mesoporous catalysts. Topics in Catalysis. 2004;**27**(1):137-142

[13] Ayoub M, Abdullah AZ. Diglycerol synthesis via solvent-free selective glycerol etherification process over lithium-modified clay catalyst. Chemical Engineering Journal. 2013;**225**: 784-789

[14] Nomanbhay S, Ong MY. A review of microwave-assisted reactions for biodiesel production. Bioengineering (Basel). 2017;**4**(2):57

[15] Das BC, Bhowmik D, Chaudhuri S. Microwave system. The Pharma Innovation. 2012;**1**(6):1-16

[16] Zhang X, Hayward DO, Mingos DMP. Effects of microwave dielectric heating on heterogeneous catalysis. Catalysis Letters. 2003;**8**(1-2): 33-38

[17] Zhou J, Xu W, Wang Z, Luo Y, Gao L, Yin C, et al. A new type of power energy for accelerating chemical reactions: The nature of a microwave-driving force for accelerating chemical reactions. Scientific Reports. 2016;**6**: 25149

[18] Ceron-Camacho R, Aburto JA, Montiel LE, Martinez-Palou R. Microwave-assisted organic synthesis versus conventional heating. A comparative study for Fisher Glycosidation of monosaccharides. Comptes Rendus Chimie. 2013;**16**(5): 427-432

Section 3

Cholesterol

Chapter 8

Human Cholesterol Biosynthesis Defects

Erin Anderson and David Coman

Abstract

Cholesterol plays an essential role in normal embryogenesis and perturbations in its *de novo* synthesis are responsible for organ malformations in the cholesterol biosynthesis defects. Ten distinct inherited disorders have been linked to different enzyme defects in the isoprenoid/cholesterol biosynthetic pathway: mevalonic aciduria, hyperimmunoglobulinemia syndrome, squalene synthase deficiency, lanosterol synthase deficiency, hydrops-ectopic calcification-moth-eaten (Greenberg) skeletal dysplasia, X-linked dominant chondrodysplasia punctata, congenital hemidysplasia with ichthyosiform erythroderma and limb defects syndrome, lathosterolosis, Smith-Lemli-Opitz syndrome and desmosterolosis. These Mendelian disorders are clinically heterogeneous with protean manifestations reflecting the important role of cholesterol, and its intermediary metabolites, in embryogenesis and development. Key clinical features commonly represented by the cholesterol biosynthesis defects include structural brain malformations, axial skeletal developmental anomalies and genital and cardiac malformations. The aetiology of the underlying pathophysiology is unclear and multifactorial but may be due to lowered cholesterol and/or the elevated, teratogenic levels of the intermediate sterol precursors. Herein, we will review clinical, biochemical and molecular aspects of the known human cholesterol biosynthesis defects.

Keywords: cholesterol biosynthesis defects, mevalonate, squalene, skeletal dysplasia, chondrodysplasia, Smith-Lemli-Opitz

1. Introduction

Cholesterol is essential for normal cellular function. All nucleated cells can synthesise cholesterol from acetyl-CoA in the isoprenoid biosynthesis pathway via enzymatic reactions that are localised to the endoplasmic reticulum. Isoprenoids function in a variety of important cellular processes, including cell growth and differentiation, protein glycosylation, as precursors of oxysterols, steroid hormones and bile, in mitochondrial electron transport and signal transduction pathways, especially that of the hedgehog pathway [1–3]. Cholesterol biosynthesis is divided into two major pathways: pre-squalene cholesterol synthesis and post-squalene cholesterol synthesis. Pre-squalene cholesterol synthesis contributes to both sterol and isoprenoid synthesis, whereas post-squalene cholesterol synthesis is a committed pathway to sterol and vitamin D synthesis [3].

Isoprenoid biosynthesis (**Figure 1**) begins with the C2 compound acetyl-CoA, which, via six subsequent enzyme reactions, is converted into isopentenyl-pyrophosphate, the basic C5 isoprene unit used for synthesis of all subsequent

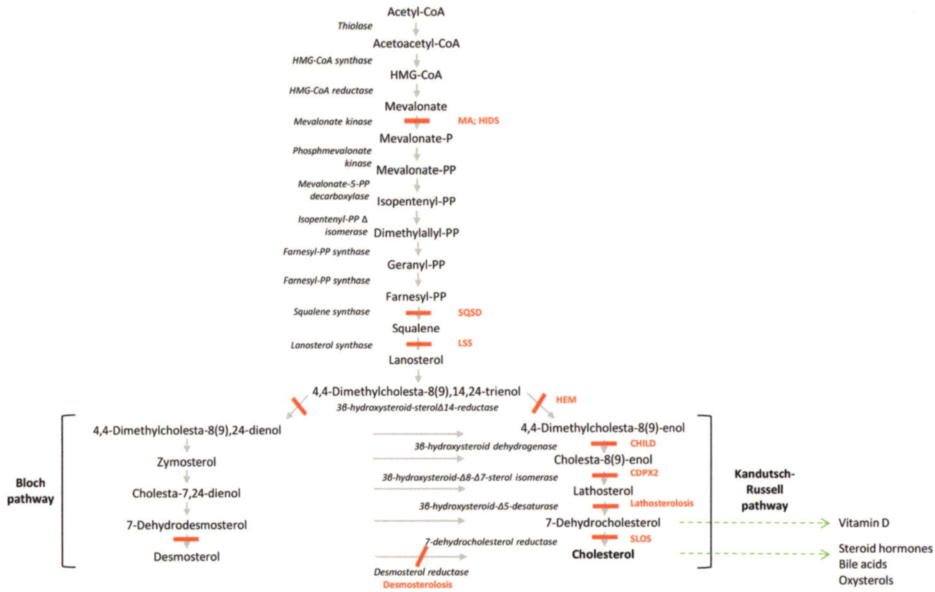

Figure 1.
Schematic representation of the human cholesterol biosynthesis pathway. HMG-CoA, 3-hydroxy-3-methyl-glutaryl-coenzyme A; P, phosphate; PP, pyrophosphate; MA, mevalonic aciduria, HIDS, hyper IgD syndrome; SQSD, squalene synthase deficiency; LSS, lanosterol synthase deficiency; HEM, hydrops-ectopic calcification-moth-eaten; CHILD, congenital hemidysplasia with ichthyosiform erythroderma and limb defects; CDPX2, X-linked chondrodysplasia punctate 2; SLOS, Smith-Lemli-Opitz syndrome.

isoprenoids [3]. The first committed step to the production of sterol isoprenoids is C30 squalene (composed of 6 isoprene units) which, after cyclisation, is converted into C30 lanosterol (4,4,14-α-trimethyl-cholesta-8(9),24-dien-3β-ol) [4]. Following this transformation, cholesterol can be synthesised via one of two independent routes; the Bloch pathway [5] or the Kandutsch-Russell pathway [6]. Both pathways utilise the same enzymes, but in different orders in a tissue-dependent manner, leading to the formation of different intermediates [7]. C27 cholesterol is subsequently produced from lanosterol via a series of at least eight different enzyme reactions, including one demethylation at C14, two demethylations at C4, one isomerisation of the D8 [9] double bond to D7, three reductions of the D24, D14 and D7 double bonds and one desaturation between C-5 and C-6 [3].

Currently, 10 Mendelian disorders of cholesterol biosynthesis have been characterised, all with complex multisystem clinical phenotypes, supporting the importance of cholesterol in embryogenesis and development (see **Figure 1** and **Table 1**). Currently, the only reported defects in the pre-squalene pathway are the mevalonate kinase deficiency allelic conditions of mevalonic aciduria (MA, OMIM 610377) and hyper IgD syndrome (HIDS, OMIM 260960), squalene synthase deficiency (SQSD, OMIM 618156) and lanosterol synthase deficiency (LSS, OMIM 600909). Six Mendelian diseases in the post-squalene pathway have been reported: hydrops-ectopic calcification-moth-eaten skeletal dysplasia (HEM, OMIM 215140), congenital hemidysplasia with ichthyosiform erythroderma and limb defects syndrome (CHILD, OMIM 308050), chondrodysplasia punctate 2 (CDPX2, OMIM 302960), lathosterolosis (OMIM 607330), Smith-Lemli-Opitz syndrome (SLOS, OMIM 270440) and desmosterolosis (OMIM 602398). Improved understanding of molecular mechanisms associated with intracellular trafficking of cholesterol and regulation of key rate limiting steps in cholesterol synthesis (e.g. via the ubiquitin proteasome system) has generated opportunities for identification of other novel Mendelian defects associated with cholesterol homeostasis [8, 9].

DOI: http://dx.doi.org/10.5772/intechopen.87150

Syndrome	OMIM	Chromosome location	Gene	Enzyme	Key features	Inheritance
MA	610377	12q24.11	*MVK*	Mevalonate kinase	Autoinflammatory flares, dysmorphia, DD, psychomotor retardation and hepatosplenomegaly	AR
HIDS	260960	12q24.11	*MVK*	Mevalonate kinase	Recurrent cyclical fevers and abdominal pain	AR
SQSD	618156	8p23.1	*FDFT1*	Squalene synthase	Dysmorphia, DD, male genital malformations, brain malformations, seizures and abnormal urine organic acids	AR
LSS	600909	21q22.3	*LSS*	Lanosterol synthase	Congenital cataracts and hypotrichosis	AR
HEM skeletal dysplasia	215140	1q42.12	*LBR*	3β-hydroxysteroid-sterol Δ^{14}-reductase	Non-immune hydrops fetalis, stippling and erroneous calcification and dwarfism	AR
CHILD	308050	Xq28	*NSDHL*	Sterol C4-demethylase aka 3β-hydroxysteroid dehydrogenase	Unilateral ichthyosis, male-lethal, ipsilateral limb reduction	XLD
CDPX2	302960	Xp11.22–23	*EBP*	3β-hydroxysteroid-Δ^{8}-Δ^{7}-sterol isomerase	Rhizomelia, calcific stippling cataracts	XLD
Lathosterolosis	607330	11q23.3-q24.1	*SC5DL*	3β-hydroxysteroid-Δ^{5}-desaturase	Microcephaly, cataracts, poly and syndactyly, DD, II	AR
SLOS	270400	11q13.4	*DHCR7*	7-dehydrocholesterol reductase	2–3 syndactyly, cleft palate, II, typical craniofacial stigmata	AR
Desmosterolosis	602398	1p32.3	*DHCR24*	24-dehydrocholesterol reductase	SLOS-like dysmorphia, CHD, microcephaly, DD, II	AR

Table 1.
Known human defects of cholesterol biosynthesis.

Modulating flux through the cholesterol biosynthesis pathway has been of interest for many years as a pharmacological treatment option for hypercholesterolemia. The statin family of drugs inhibit HMG-CoA reductase, the rate limiting step in the pre-squalene pathway, and similar efforts have focused on inhibitors of squalene synthase as this enzyme is the first committed step in cholesterol biosynthesis. Animal and human models of squalene synthase inhibitors generated a complex array of farnesol-derived metabolites [10–12], the recognition of which was instrumental in describing SQSD, a newly described pre-squalene cholesterol biosynthesis defect [4]. That pathogenesis of the cholesterol biosynthesis defects is complex, reflective of the multisystem nature of the clinical phenotypes.

2. Disorders of the pre-squalene cholesterol pathway

2.1 Mevalonate kinase deficiency

Mevalonate kinase phosphorylates mevalonate, the product of the reduction of 3-hydroxy-3-methylglutarylcoenzyme A (HMG-CoA), which is important in cholesterol biosynthesis and for farnesylation and isoprenylation of proteins [13]. Mevalonate kinase deficiency (MKD) is a recessively inherited autoinflammatory disorder in the isoprenoid biosynthetic pathway with a spectrum of manifestations, including the well-defined allelic clinical phenotypes of HIDS and MA [14], both of which were identified in the mid-1980s [15, 16].

Mevalonate kinase is essential for the biosynthesis of non-sterol isoprenoids, which mediate protein prenylation. MKD is caused by mutations in the *MKD* gene which encodes mevalonate kinase, with the degree of residual enzyme activity largely determining disease severity. MKD leads to perturbations in the mevalonate pathway of cholesterol synthesis with episodes of hyperinflammation [17]. MKD is now viewed as a phenotypic continuum based on the degree of enzyme deficiency, with MA the most severe phenotype and HIDS the mild end of the spectrum [18].

MKD is characterised by autoinflammatory flares with fever, abdominal pain, mucoid and cutaneous lesions and arthralgias [14]. The more severely affected patients with MA classically have developmental delay, dysmorphism, psychomotor retardation, hepatosplenomegaly and ocular abnormalities [14]. During attacks, patients with MKD have increased levels of acute-phase proteins including C-reactive protein and cytokines such as TNF-α, IL-6 and interferon-γ [19, 20]. The MA phenotype characteristically presents in the first few months of life, with antenatal presentations linked with a high rate of stillbirth [21]. Commonly reported dysmorphic features include frontal bossing, hypertelorism, long eyelashes and triangular-shaped facies, as well as failure to thrive, developmental delay, ataxia, seizures, myopathies and autoinflammatory attacks [21, 22]. MA is a multisystem phenotype with gastrointestinal manifestations including cholestasis and liver dysfunction [23], and ocular findings including recurrent conjunctivitis, cataracts and uveitis [24].

The HIDS phenotype typically presents with recurrent (four-to-six weekly) self-limited bouts of multisystem inflammation characterised by fever, abdominal pain, adenopathy, rash and arthralgia [14]. As these are common symptoms of many childhood infectious illnesses, the diagnosis of HIDS is often delayed for many years. HIDS episodes usually last 3–7 days, occurring in a cyclical fashion or induced by a provocative physiological stress such as illness, injury or vaccination. Acute abdominal pain may be the most marked and debilitating feature of systemic inflammation and can mimic a 'surgical acute abdomen' [24]. A long-term follow-up study of 103 HIDS patients revealed that the frequency of the attacks decreases over time, but 50% of patients greater than 20 years of age still experience six or more attacks per year, impacting on the quality of life [24].

The epidemiology of MKD is largely unknown. At least 300 people are documented worldwide, the majority with HIDS, although this is likely to be underdiagnosed as recurrent fevers in childhood are a common occurrence. The highest documented prevalence is in the Netherlands, with an estimated 1:200,000 affected nationwide, consequent to a high carrier rate which is estimated at 1:65 [24, 25].

Elevations in IgD in MKD are inconsistent and can be normal in up to 20% of cases [24]. Serum amyloidosis is a long-term sequela of prolonged inflammatory activation, with elevations in serum amyloid A noted in approximately 3% of HIDS patients [24]. Urinary excretion of mevalonic acid can persist in MA and maybe present in some HIDS patients during febrile attacks [21]. The diagnosis is confirmed by identifying pathogenic mutations in the *MVK* gene; currently more than 120 sequence variants in this gene have been reported in association with MKD [26], most of which are missense mutations that impair mevalonate kinase stability [27]. Some genotype-phenotype correlations exist: *MVK* variants located in the core of the protein (affecting folding and stability) are highly associated with the more severe MA phenotype [25, 28, 29]. In contrast, other variants such as the C-terminal V377I substitution typically manifest as the HIDS phenotype and are rarely associated with MA [25].

Although the precise pathogenesis of MKD remains unclear, increasing evidence suggests that deficiency in protein prenylation leads to innate immune activation and systemic hyperinflammation, which has assisted in the development of cytokine-directed biologic therapy. Corticosteroids induce a complete response in 24% of HIDS patients [30]. Biologics targeting IL-1, including anakinra and canakinumab, and TNF-α blocking agents such as etanercept and adalimumab, have been used with varying success [30]. Some cases that have failed to respond to anakinra have demonstrated a successful reduction in symptoms with tocilizumab, a monoclonal antibody targeted against the IL-6 receptor [31]. One patient with MKD, treated with alendronate for steroid-induced osteoporosis, subsequently achieved complete remission [32]. Alendronate inhibits farnesol-pyrophosphate synthase. For refractory cases of MA phenotype, the last consideration for therapy includes liver transplantation or haematopoietic stem cell transplantation [14].

Blockade of the mevalonate pathway with the HMG-CoA reductase inhibitors reduces both mevalonic acid levels and residual isoprenoid production and but can trigger disease flares [22]. The inflammatory hyper-responsiveness in MKD appears to be due to lack of isoprenoid products and not accumulation of mevalonic acid. This appears to be due to the need for geranylgeranylation rather than other mevalonate pathway products, such as cholesterol biosynthesis, in mediating the hypersecretion of IL-1β [27]. The use of statins in this disease process has therefore largely been abandoned [30]. $Mvk^{+/-}$ mice do have some features of immune dysfunction, including increased serum IgD and TNF-α levels, as well as increased expression of activation markers on T-lymphocytes and macrophages [33].

2.2 Squalene synthase deficiency

Squalene synthase deficiency (SQSD) is a recently identified pre-squalene defect to have been characterised. In 2018, three patients were reported with this novel cholesterol biosynthesis defect [4]. Salient clinical features include facial dysmorphism, dry skin with photosensitivity, generalised tonic-clonic seizures, structural brain malformations, cortical visual impairment, profound global developmental delay and genital malformations in the two males [4].

Gas chromatography-mass spectrometry (GC-MS) and nuclear magnetic resonance spectroscopy profiles yielded a consistent and complex pattern of abnormal metabolites including accumulation of methylsuccinic acid, mevalonate lactone, mesaconic acid, 3-methyladipic acid, saturated and unsaturated branched-chain dicarboxylic acids and glucuronides derived from farnesol [4]. A similar metabolite profile has previously been observed in the urine of animal models and humans treated with pharmacological inhibitors of squalene synthase, as well as in animals loaded with farnesol [10–12]. This urine metabolic profile is specific to and thus diagnostic of SQSD. Plasma total farnesol levels (the sum of free farnesol and farnesyl-pyrophosphate) in affected individuals were, however, significantly increased (1.5–3.9 mmol/L; reference <0.12) while plasma squalene levels were reduced or normal (0.17–0.93 mmol/L, reference 0.36–1.04).

A range of pathogenic *FDFT1* molecular variants have been described in the three SQSD patients identified thus far (a sibship and an unrelated patient) [4]. The sibship was compound heterozygous for a maternally-inherited 120 kb deletion, resulting in loss of exons 6–10 of *FDFT1* and the entire coding sequence of the neighbouring *CTSB* gene (encoding cathepsin B (OMIM 116810)); and a paternally inherited variant c.88024_88023delinsAG, which created a novel splice acceptor site. The unrelated patient was homozygous for a novel 16-bp intronic deletion. Functional characterisation of the variants demonstrated a partial splicing defect and altered promoter and/or enhancer activity, reflecting essential mechanisms for regulating cholesterol biosynthesis and/or uptake in steady state [4].

Fdft1-null mice demonstrate embryonic lethality at day 12.5 in conjunction with growth restriction and neurodevelopmental disorders [34]. The fact that the *FDFT1* variants in the human SQSD cases are compatible with life may be explained by the fact that all individuals have some form of residual FDFT1 activity, either resulting from the diminished levels of correctly-spliced enzyme or by functional compensation for disrupted regulation [4].

2.3 Lanosterol synthase deficiency

In the cholesterol biosynthesis pathway, lanosterol synthase leads to the cyclisation of (S)-2,3-oxidosqualene into lanosterol. Pathogenic mutations in the *LSS* gene have recently been reported in a spectrum of clinical phenotypes including congenital cataracts in three families [35], hypotrichosis simplex (HS) in three families [36] and a more severe neuroectodermal syndrome formerly named alopecia with mental retardation (APMR) syndrome in six unrelated families [37]. HS (OMIM 618275) is a rare form of hereditary alopecia characterised by childhood onset of diffuse and progressive scalp and body hair loss [36]. APMR syndrome (OMIM 203650) is a rare disorder with autosomal recessive transmission. A recent report identified 11 individuals from seven unrelated families affected with alopecia, male genital abnormalities, variable MRI abnormalities and neurological symptoms [37]. In this cohort, total alopecia was universal with other common dermatological manifestations including ichthyosis and erythroderma. Neurological manifestations included significant developmental delay, microcephaly, epilepsy and hypomyelination [37].

Sterol profiling in lanosterol synthase deficiency cases has not identified any specific abnormalities, thus supporting the previously proposed hypothesis of an alternative cholesterol pathway [36]. *LSS* variants identified to date include truncating, missense and splicing variants. *LSS* has also been associated with congenital cataracts in rat [38]. Mice homozygous for the $Lss^{tm1b(KOMP)Wtsi}$ allele demonstrate variable lethality, from embryonic day 9.5 to postnatal prior to weaning [39].

3. Disorders of the post-squalene cholesterol pathway

3.1 Hydrops-ectopic calcification-moth-eaten skeletal dysplasia

Most proximal in the post-squalene pathway is hydrops-ectopic calcification-moth-eaten (HEM) skeletal dysplasia, or Greenberg dysplasia. This very rare and severe autosomal recessive disorder was first described in 1988 [40] with only 11 examples identified in the literature to date. All but one of these have been lethal *in utero*, with the remaining case dying at 2 days of age [41]. HEM skeletal dysplasia is characterised by significant non-immune hydrops fetalis, erroneous chondro-osseous calcification of vertebrae, ribs, pelvis, larynx and trachea as well as a diagnostic mottled 'moth-eaten' appearance of long bones on radiography [42–44]. Further skeletal abnormalities can include rhizomelic and mesomelic shortening of the limbs, platyspondyly, decreased skull ossification and distal dysmorphisms such as absent phalanges or postaxial polydactyly [42–45]. Non-skeletal congenital malformations include pulmonary hypoplasia, intestinal malrotation, cystic hygroma and excessive extramedullary haematopoiesis [45, 46]. Histology shows significant bone and cartilage disorganisation [43, 45].

HEM skeletal dysplasia was first suggested as an inborn error of cholesterol biosynthesis by Kelley et al. [47] with identification of increased levels of 4,4-dimethylcholesta-8 [9],14-dien-3β-ol and 4,4-dimethylcholesta-8(9),14,24-trien-3β-ol in cultured fibroblasts, indicating a deficiency of sterol Δ^{14}-reductase. This enzyme converts these sterols to 4,4-dimethylcholesta-8(9)-en-3β-ol and 4,4-dimethylcholesta-8(9),24-dien-3β-ol, respectively. This point on the cholesterol biosynthesis pathway is unique with sterol Δ^{14}-reductase activity by both the lamin B receptor (LBR) and a second enzyme DHCR14 (TM7FS2), although functional redundancy is disputed [48, 49]. It was originally thought that the more prominent role in sterol biosynthesis was that of DHCR14 compared to the lamin B receptor. However, it has more recently been demonstrated that it is a deficiency in the lamin B receptor due to mutations in *LBR* at 1q42.12 that is causative for HEM skeletal dysplasia [50, 51] and that it is the LBR, not DHCR14, that is required for cholesterol biosynthesis [48, 52].

The involvement of *LBR* has raised contention as to whether HEM skeletal dysplasia should be classified as a laminopathy rather than as an error of cholesterol synthesis [49]; however, it is appropriate to recognise that mutations in *LBR* can cause different disorders in different contexts. The type of mutation (missense, nonsense or splice-site), the functional location of each mutation in the *LBR* gene and the residual protein activity affect the clinical outcome of this disorder [53, 54]. The LBR protein has both a nuclear domain involved in anchoring chromatin to the nuclear membrane, and a transmembrane domain with sterol Δ^{14}-reductase activity critical for cholesterol synthesis [48], the latter primarily where mutations causing HEM dysplasia are located [50]. Some mutations found in *LBR* in HEM dysplasia patients have been identified in the heterozygous state in the relatively benign autosomal dominant condition of Pelger-Huët anomaly in which granulocytes have bilobed nuclei but patients are otherwise clinically normal. These two conditions may represent different allele patterns of the same disorder for some mutations [53, 55]. The less common homozygous Pelger-Huët is clinically more severe with round or ovoid granulocyte nuclei and some cases with mild skeletal abnormalities [56, 57]. This highlights the role of the lamin B receptor sterol reductase function as essential in prenatal development but also the phenotypic continuum that can occur for various allele combinations of the *LBR* gene.

Species variation with respect to the role of the LBR can make mouse model outcomes difficult to elucidate. Studies of mutations in both *LBR* and *DHCR14/TM7FS2* have been investigated in ichthyosis (*ic*) mice with contrasting conclusions, including those highlighted above and as reviewed by Herman and Kratz [58].

3.2 Congenital hemidysplasia with ichthyosiform erythroderma and limb defects syndrome

Congenital hemidysplasia with ichthyosiform erythroderma and limb defects (CHILD) syndrome is a rare X-linked dominant disorder of cholesterol biosynthesis, with fewer than 100 cases discussed in the literature [58]. The earliest identification of the condition is thought to be in 1903 [59], through the proposal of the syndromic acronym in 1980 [60]. CHILD syndrome is nearly always male-lethal although perhaps two males with this syndrome have been identified, one with a 46, XY karyotype, postulated to have survived due to a postzygotic mutation [61]. The distinguishing hallmark of the condition is that of unilateral skin lesions with ipsilateral limb defects [60, 62, 63]. The characteristic yellow, scaly plaques are usually present at birth or emerge in the first few months of life and while there may be some resolution over time, they often remain for life [60]. These markings may follow the lines of Blaschko, but more commonly, there is a striking delineation at the midline with the lesions showing a unique lateralisation pattern [62]. This has been proposed to be due to interactions between X-inactivation and the organisation of left-right axis symmetry in the developing embryo [60]. The lateralisation of these lesions and their persistence is a distinguishing feature of CHILD syndrome compared to differentials such as CDPX2, a similar but distinct inborn error of cholesterol biosynthesis [64].

CHILD syndrome demonstrates complete limb aplasia, severe phocomelia or severe hypoplasia on the same side of the body as ichthyosiform lesions [60]. Infant radiography may show epiphyseal stippling such as that seen in CDPX2, as well as milder skeletal malformations such as scoliosis, hypo or hemi-plastic vertebrae, distal digit shortening, syndactyly or polydactyly [65]. Non-skeletal manifestations include alopecia, verruciform xanthoma, dystrophic nails and congenital malformations on the affected side that can involve the heart, kidneys and CNS [60, 65]. Intelligence may be normal or slightly reduced. Despite the severity of these symptoms, mild cases of CHILD syndrome with no skeletal and/or cutaneous involvement have been identified through molecular analysis [66].

Molecular investigation has identified various mutations in the *NSDHL* (NADH steroid dehydrogenase-like) gene as causative for CHILD syndrome [67]. *NSHDL* is located at Xq28 and encodes 3β-hydroxysteroid dehydrogenase, part of a three-part enzyme complex. This C4 demethylation complex acts on the sterol ring in the post-squalene pathway, converting 4,4-dimethylcholesta-8(9),24-dien-3β-ol to zymosterol and 4,4-diemthylcholesta-8(9)-en-3β-ol to cholesta-8(9)-en-3β-ol. Mutations are most often loss-of-function [68]. Cholesterol and sterol levels are normal and so a diagnosis requires clinical and molecular assessment. Various *NSDHL* mutations have been studied in the allelic 'bare patches' (*Bpa*) and 'striated' (S*tr*) murine models [69]. These have given insights into facets of CHILD syndrome and cholesterol synthesis disorders in general, for example, the roles of the maternal placenta [70] and Hedgehog signalling pathways [71] in disease presentation.

Treatment options for CHILD syndrome have generally focused on topical management of skin lesions with symptomatic remedies such as emollients or with pathogenesis-based therapies generally involving combinations of cholesterol and a cholesterol synthesis-inhibitor [72, 73], the latter with some efficacy.

3.3 X-linked dominant chondrodysplasia punctata 2

X-linked dominant chondrodysplasia punctata 2 (CDPX2), or Conradi-Hünermann-Happle syndrome, is estimated to have an incidence of 1/400,000 and, similarly to CHILD syndrome, is almost entirely male-lethal. The CDPX2 phenotype, like the other cholesterol biosynthesis disorders, is heavily based on the skeletal and cutaneous domains, and there can be significant variability even within family lines [74–76] with generational anticipation [74]. Severe manifestations can result in neonatal or infant death with considerable skeletal and internal abnormalities, while mild cases may be nearly asymptomatic. This range of phenotypic variability is likely due to the combination of somatic and/or gonadal mosaicism and X-inactivation patterns [76]. Occasional male patients are identified with CDPX2, usually due to somatic mosaicism [77, 78] with one case of 46,XXY [79]. Gonadal mosaicism is possible which is relevant for recurrence risk [80].

Widespread epiphyseal stippling is seen on infant radiographs, often including not just the long bones but the trachea and vertebrae as well [81–83]. Additional skeletal stigmata include short stature and scoliosis (which can be congenital), clubfoot, joint contractures, and postaxial polydactyly [74, 77, 83]. Cutaneous manifestations include skin with patches of scaly hypo or hyper-pigmentation, which usually follows the lines of Blaschko. The initial skin scaling and erythroderma present at birth usually fades over the first few months of life, leaving follicular atrophoderma, pigmentation and alopecia, although ichthyosis can persist [84, 85]. The pattern and then resolution of skin scaling as well as its histological profile is a differentiating diagnostic feature for CDPX2 compared to CHILD syndrome. Diagnosis of CDPX2 in adulthood can be difficult due to the childhood resolution of the characteristic skin lesions and epiphyseal stippling [86]; however, a combination of cutaneous manifestations, asymmetric limb reduction and cataracts (found in 65% of patients) is a good suggestion of this condition for further investigation [86]. CDPX2 presents with characteristic facial features including frontal bossing, midface hypoplasia and flat nasal bridge [74, 81]. The condition is also associated with microphthalmia or microcornea, congenital heart disease, renal abnormalities including hypoplasia and hydronephrosis and sensorineural hearing loss [87]. Cognition is usually normal [87].

CDPX2 is caused by mutations in the *EBP* (emopamil binding protein) gene [88, 89] located at Xp11.23 and encoding a Δ^8-Δ^7-sterol isomerase. This enzyme functions downstream of the C4-demethylation complex affected in CHILD syndrome and converts zymosterol and cholesta-8(9)-en-3β-ol to cholesta-7,24-dien-3β-ol and lathosterol, respectively. There is a phenotypic correlation with enzyme function with lethality of homozygous females and clinically affected heterozygous females; however, there is no clear genotype-phenotype correlation, presumably due to X-inactivation patterns [74, 76]. Surviving males with CDPX2 are almost always due to mosaic postzygotic mutations as a hemizygous male genotype is lethal *in utero*. CDPX2 mutations (including deletions, insertions, nonsense, missense and splice-site) of *EBP* have been identified as both *de novo* and inherited mutations and are found throughout the entire length of the gene [74].

While there is no clear CDPX2 genotype-phenotype correlation, there is a distinct association between genotype and CDPX2 sterol profile [74], and plasma sterol assay is a highly specific indicator for an *EBP* mutation [83]. Plasma shows increased 8-dehydrocholesterol and 8(9)cholesterol levels, with the ratios compared to cholesterol increased 0.71–0.80% [74]. Plasma cholesterol is usually normal. Treatment and surveillance are symptomatic, and studies in these areas have been advanced by the 'tattered' (*Td*) mouse which shares both phenotypic and molecular similarities with human CDPX2 [89].

3.4 Lathosterolosis

Lathosterolosis (OMIM 607330) results from impaired 3-hydroxysteroid-5-desaturase (SC5D) activity [90]. In the Kandutsch-Russel synthetic pathway, SC5D catalyses the conversion of lathosterol to 7-dehydrocholesterol (7DHC) in the enzymatic step immediately preceding the defect in SLOS, whereas in the Bloch pathway of cholesterol synthesis, SC5D catalyses the conversion of cholesta-7,24-dienol to 7-dehydrodesmosterol [90].

To date, deleterious missense mutations of *SC5D* have been reported in six patients from three families [91–95]. The clinical features include microcephaly, facial dysmorphism, bitemporal narrowing, ptosis, cataracts, anteverted nares, micrognathia, postaxial polydactyly, syndactyly, ambiguous genitalia, non-neuronal mucolipidosis, global developmental delay, intellectual impairment, hepatic cirrhosis, and early lethality [91–95]). One surviving patient who developed end-stage hepatic failure and received a liver transplantation had improvement of lathosterolosis symptoms [96]. Another patient had a milder clinical phenotype of microcephaly and learning defects with cataracts [91] highlighting the possible under-diagnosis of the syndrome without plasma sterol analysis.

Plasma cholesterol levels are normal with accumulation of lathosterol in plasma and in cultured fibroblasts, and lamellar inclusions within cellular lysosomes [95]. *Sc5d−/−* pups are stillborn and demonstrate craniofacial malformations including cleft palate and limb defects such as postaxial polydactyly [94].

3.5 Smith-Lemli-Opitz syndrome

Smith-Lemli-Opitz syndrome (SLOS) is the prototypical inborn error of cholesterol biosynthesis first described in 1964 [97]. It is by far the most common disorder in this group, with an incidence of approximately 1/40,000 although this can range from 1/70,000 to 1/10,000 depending on the population in question [98]. The carrier frequency can range from approximately 1:100 in North American Caucasians to 1:50–1:30 in various Central European populations [99]. While these carrier rates would imply a far greater incidence than is clinically observed, there is thought to be a level of misdiagnosis or non-diagnosis in mildly-affected patients, and *in utero* prenatal demise is estimated to affect 42–88% of conceptuses [100], mostly in the first trimester [98, 101].

SLOS has a broad range of phenotypic variabilities: mild cases can comprise minor physical abnormalities and behavioural or learning difficulties through a wide spectrum to a severe phenotype comprising major and life-limiting congenital abnormalities. Cognition can range from near-normal [102] to profound intellectual impairment, and on MRI, up to 96% of SLOS patients have a structural brain abnormality [103]. There is a correlation of atypical sterol profiles with both intellectual impairment and brain malformations, particularly abnormalities of the septum pellucidum and corpus callosum [103]. CNS myelination is normal despite its high proportion of cholesterol content and the mostly *in situ* synthesis of cholesterol in the CNS [104]. As well as intellectual impairment, patients are often diagnosed with language delays or impairment, autistic spectrum disorder and sleep disturbances, and can engage in self-harm. Global developmental delay, hypotonia and failure to thrive are common [105–107]. The most common physical manifestation reported with SLOS is that of 2,3 toe syndactyly, and a combination of this with other structural or cognitive symptoms should suggest a possible SLOS diagnosis for investigation [108]. Limb anomalies are common, including polydactyly, short proximal thumbs and a single palmar crease [105–107]. Other structural malformations that can occur include microcephaly, cleft palate, bifid uvula and characteristic facies with micrognathia, ptosis and broad nasal tip with anteverted nares [105–107]. This facial dysmorphia can be less recognisable in older patients [107]. Congenital abnormalities can also affect the heart and lungs, gastrointestinal tract and genitalia [105–107]. Patients with SLOS often have severe ultraviolet photosensitivity [109].

The final steps of the post-squalene cholesterol biosynthesis pathway are conversion of 7-dehydrodesmosterol to desmosterol and 7-dehydrocholesterol (7-DHC) to cholesterol. The latter is catalysed by the 3β-hydroxysteroid-Δ^7-reductase (or 7-dehydrocholesterol reductase, DHCR7) enzyme, encoded by the *DHCR7* gene at 11q13.4. Increased levels of 7-DHC and decreased levels of cholesterol led to SLOS being identified as a disorder of sterol biosynthesis in 1993 [110, 111]. This altered plasma profile is a useful diagnostic tool for SLOS, and there is evidence of a relationship between serum sterols and disease severity [112, 113].

Over 100 mutations in *DHCR7* have been identified in SLOS [114] with no clear genotype-phenotype correlations [115, 116], although some mutations are associated with more mild phenotypes due to some residual enzyme activity [117]. There is a significant correlation between SLOS patient phenotype and maternal genotype for *ApoE* and *ABCA1* [118, 119]. These correlations are positive for amelioration of SLOS symptomatology and pathogenesis and with the potential for therapeutic mediation [120].

As well as being the precursor to cholesterol, 7-DHC is also the precursor to vitamin D with exposure of cutaneous 7-DHC to ultraviolet B and subsequent synthesis to vitamin D by the liver and kidney. Increased levels of circulating vitamin D are seen in patients with SLOS [121], despite their increased photosensitivity and ensuing limited sun exposure. One of the primary theories for a possible heterozygous advantage of *DHCR7* mutations is that of protection against vitamin D deficiency [105], particularly given the greater carrier rate seen in populations of northern Europe [98, 99].

A prenatal diagnosis can be obtained via molecular or biochemical analysis (e.g. of amniotic fluid sterols [122]); however, non-invasive techniques can also identify pregnancies requiring SLOS investigation. Measurement of a low maternal serum unconjugated estriol (uE3), particularly when combined with abnormal sonography results, can be utilised for prenatal screening although this can yield false positive results and uE3 levels can also be predictive for other disorders [101]. Baseline screening for a SLOS-affected pregnancy is also possible non-invasively via serial measurement of steroids such as pregnanetriol in maternal urine [123, 124]. Abnormal plasma sterol ratios in unaffected heterozygotes [125] mean that carrier status may be determined prior to pregnancy for increased reproductive options.

Current treatment protocols for SLOS usually involve endogenous cholesterol supplementation with or without adjunct therapies such as simvastatin [126]. There is broad anecdotal evidence throughout the literature as to the positive benefit of cholesterol supplementation for patient growth, overall health (including improved photosensitivity and response to infection) and behaviour, as well as measurable changes towards typical plasma sterols [127–129]. These improvements have been reported following initiation of cholesterol treatment in both children and adults [130], although with greater rate of improvement with earlier intervention [131]. Limitations to the efficacy of cholesterol treatment certainly exist, such as cholesterol's inability to cross the blood-brain barrier in any practical quantity (which makes the apparent behavioural improvements reported interesting). The real value of cholesterol supplementation is yet to be definitively determined as trials of increased dietary cholesterol both with and without placebo controls have yielded very mixed results [132–134]. Antioxidant [135, 136] and virus vector [137] therapies have also been explored as an avenue for improved patient outcomes for SLOS and other disorders of cholesterol synthesis. Both mouse and rat models of null and hypomorphic alleles in *DHCR7* have been useful homologues for characterisation and investigation of human SLOS [138, 139].

3.6 Desmosterolosis

Desmosterolosis (OMIM 602398) is currently the final inborn error of cholesterol biosynthesis and is caused by defective enzymatic function of 3-hydroxysterol-delta 24-reductase (DHCR24). This reaction causes the reduction of the C-24 bond in the aliphatic side chain of cholesterol [140]. Reduction of the C-24 bond catalysed by DHCR24 can occur at different times in the cholesterol synthetic pathway: this step occurs early in the Kandutsch-Russel cholesterol synthetic pathway [6] but is the penultimate step in the Bloch pathway of cholesterol synthesis [5].

While first described in 1998, the molecular mechanisms of desmosterolosis were not characterised until 2001 [140, 141]. To date, only nine cases have been reported and clinical features include SLOS-like dysmorphism, thick alveolar ridges, gingival nodules, cleft palate, short limbs, severe congenital heart defect, atherosclerosis, arthrogryposis, ambiguous genitalia, microcephaly, agenesis of the corpus callosum, global developmental delay and intellectual impairment [141–147]. The diagnosis of desmosterolosis is made by demonstrating elevated levels of desmosterol by GC-MS analysis, with serum cholesterol levels usually normal [141, 142]. Reported *DHCR24* pathogenic mutations thus far have all been missense mutations.

A targeted mouse model for desmosterolosis has been generated, and *Dhcr24*$^{-/-}$ mice are viable with some postnatal growth retardation and infertility [148]. Pharmacological inhibitors of DHCR24 have been developed for studies in rat models [135, 149, 150]. Treatment of pregnant rats with these inhibitors of sterol-D24-reductase is teratogenic and produces cataracts, CNS abnormalities, genitourinary and skeletal anomalies [149–151].

4. Cholesterol biosynthesis genes in other Mendelian diseases

Inherited defects in genes encoding cholesterol biosynthetic enzymes or regulators of cholesterol homeostasis create severe clinical phenotypes as discussed above and highlighted in **Table 1**. The central nervous system is highly susceptible to perturbations in cholesterol biosynthesis, with manifestations including structural brain malformations, defects in myelin structures and, in some cases, profound developmental delay. While the cholesterol biosynthesis defects are genetically distinct individual disorders, their characterisation has demonstrated interrelation between human disease processes. This underscores the importance of cholesterol in normal cellular function and opens the possibility of novel therapies for Mendelian disorders associated with cholesterol synthesis, transport and regulation. Lessons learnt from abrogation of the cholesterol biosynthesis pathway, either by deliberate pharmacological manipulation or via inherited Mendelian diseases, serve to provide vital information amongst a raft of seemingly unrelated human disease such as inflammatory bowel disease (IBD), the cholesterol trafficking disorders Niemann-Pick disease type C (NPC, OMIM 257220) and Tangier disease (TD, OMIM 205400) and neurodegenerative diseases such as Alzheimer's disease (OMIM 104300).

NPC is an autosomal recessive lysosomal storage disorder of cholesterol trafficking due to mutations in the *NPC1* and *NPC2* genes [152]. *NPC1* encodes a 13-transmembrane-spanning protein in late endosomes/lysosomes, while *NPC2* encodes a soluble lysosomal cholesterol-binding protein [153]. This is a devastating disease characterised by a relentless neurodegenerative disease course that is usually fatal in the second decade of life, although a subset of patients will die in infancy consequent to hepatic or pulmonary failure [152]. Free cholesterol is stored in the late endosome/lysosome with minimal escape of cholesterol from the acidic compartment to the endoplasmic reticulum. NPC leads to a block in trafficking/fusion essential for the functioning of the endosomal/lysosomal system, causing the secondary storage of cholesterol, glycosphingolipids and sphingomyelin [154]. It is likely that cholesterol accumulation is a secondary storage metabolite in NPC [154].

TD has been reported in approximately 100 patients and is caused by mutations in the gene encoding ABCA1 [155, 156]. Patients have minimal circulating HDL and accumulate cholesterol, leading to the formation of foam cells and the development of cardiovascular disease, orange-coloured tonsils, enlarged spleen, liver and lymph nodes and peripheral neuropathy. The membrane-associated protein ABCA1 regulates cellular cholesterol and phospholipid homeostasis by functioning as a cholesterol efflux pump [157]. Tangier disease patients have structurally abnormal late endocytic vesicles, which are also observed in the cells of patients with NPC disease [158]. There exists a link between ABCA1 expression and function with the NPC pathway [158, 159]. NPC disease is characterised at the cellular level by storage of glycosphingolipids, fatty acids, cholesterol, sphingomyelin and sphingosine. NPC cells also have low levels of calcium in the late endosome/lysosome. These cellular hallmarks were also identified in TD patients, suggesting that the loss of function of ABCA1 inhibits the NPC pathway through an unknown mechanism. A recent serendipitous clinical observation has provided a further link between TD and the NPC pathway: an adult patient thought to have NPC1 was treated with miglustat and demonstrated measurable clinical improvements in neurological and haematological parameters. TD was ultimately diagnosed when the molecular investigations for NPC were negative [160].

SLOS cellular pathophysiology should theoretically be correctable with cholesterol replacement therapy, as this should bypass the enzymatic defect in the conversion of 7DHC to cholesterol. However, when SLOS patient fibroblasts are cultured in a lipid-depleted medium to induce *de novo* cholesterol synthesis, cells exhibit a significant cholesterol trafficking defect leading to the accumulation of unesterified cholesterol in the late endosome/lysosome, which mimics the fate of LDL-derived cholesterol in NPC cells [161]. This proposes a possible mechanistic convergence between these very different inborn errors of metabolism. 7DHC could be interfering with the function of NPC1 and NPC2 by inhibition, akin to the U18666A drug that induces NPC cellular phenotypes [162]. In SLOS patient fibroblasts, accumulation of 7DHC led to the accumulation of metabolic indicators of NPC, that is, the lysosomal storage of cholesterol, sphingomyelin and multiple glycosphingolipids [163]. Elevated sphingosine levels in SLOS patient cerebrospinal fluid have been described. This serendipitous discovery of a link between the NPC pathway, two cholesterol trafficking disorders and the prototypic cholesterol biosynthesis defect SLOS will prove important in delineating the pathogenesis of these diseases and the development of novel therapies. Miglustat is an iminosugar drug that inhibits glucosylceramide synthase, the enzyme that catalyses the first step in glycosphingolipid biosynthesis, and it is in use as a substrate reduction therapy for a number of lysosomal storage defects including NPC [164]. The finding that SLOS and TD involve secondary inhibition of the NPC pathway suggests that miglustat could be a novel therapy for SLOS and TD.

The recently described SQSD exhibits a characteristic sterol pattern dominated by farnesol-derived dicarboxylic acids secondary to accumulation of farnesol-PP proximal to the enzymatic block. The role of these metabolites in the pathogenesis of this rare disease remains to be determined but is of interest as farnesol and its products exhibit a wide variety of biological activities including cell growth inhibition, induction of apoptosis and regulation of bile acid secretion [165]. Evidence is emerging that dysregulation of the mevalonate pathway may be involved in the progression of neurodegeneration in disorders such as Alzheimer's disease [166].

Inflammatory bowel disease (IBD) comprises a spectrum of phenotypes from Crohn's disease to ulcerative colitis. IBD usually occurs in young adults; however, onset in infancy and childhood are described. IBD occurs both in isolation and in monogenic syndromes with early-onset autoinflammation including the *NOD2*, *ATG16L1*, *IL23R*, *IL10R*, *IL10* and *XIAP* genes which have previously been correlated with IBD both in multifactorial and in Mendelian models [167]. MVK mutations may perhaps then synergistically augment the risk of developing IBD, especially as severe neonatal onset colitis responsive to anakinra has been reported as a feature of MVK deficiency [14, 168, 169].

Recent studies have implicated the accumulation of pre-cholesterol sterols and the replacement of cholesterol with some of these sterols in lipid rafts as playing a key role in the underlying pathophysiology of cholesterol synthesis defects [170]. The meiosis-activating sterols were the first group of cholesterol biogenesis intermediates that were found to have important extrahepatic functions in mammals. Mutations in sterol-C4-methyl oxidase-like gene (*SC4MOL*) are causative for a rare autosomal recessive syndrome associated with psoriasiform dermatitis, arthralgias, congenital cataracts, microcephaly and developmental delay [171, 172]. This gene encodes a sterol-C4-methyl oxidase (SMO) which catalyses demethylation of C4-methylsterols in the cholesterol synthesis pathway [172]. C4-methylsterols are meiosis-activating sterols, and further work is required to understand the role of these novel biomolecules in the pathogenesis of the cholesterol biosynthesis defects.

5. Conclusion

Inborn errors of cholesterol metabolism have provided many fundamental insights into normal cholesterol homeostasis and cell biology over several decades. These disorders have been viewed as discrete diseases with their own unique genetic, biochemical and cellular consequences that in turn cause the clinical spectrum of symptoms associated with each disease. There remain specific pre-squalene enzymatic defects to be characterised and many unanswered questions regarding the pathogenesis of the cholesterol biosynthesis defects. What has been surprising is that at least three cholesterol-related disorders (SLOS, NPC and TD) all share a common pathological inhibition of the NPC pathway. The precise mechanism that inhibits this pathway in SLOS and TD remains to be fully elucidated, but these findings are suggestive of novel therapeutic approaches to treating SLOS and TD using drugs that modify the cell biology of NPC such as miglustat. Whether other human diseases also involve NPC pathway dysfunction remains to be determined. Current investigation of this question may pave the way for novel approaches to therapy for diseases that currently lack effective treatments.

MA, mevalonic aciduria; HIDS, hyper IgD syndrome; SQSD, squalene synthase deficiency; LSS, lanosterol synthase deficiency; HEM, hydrops-ectopic calcification-moth-eaten; CHILD, congenital hemidysplasia with ichthyosiform erythroderma and limb defects; CDPX2, X-linked chondrodysplasia punctate 2; SLOS, Smith-Lemli-Opitz syndrome; DD, developmental delay; II, intellectual impairment; CHD, congenital heart defect; AR, autosomal recessive; XLD, X-linked dominant.

Acknowledgements

The authors thank Kevin Milo for proof reading and insightful comments in the manuscript preparation.

Conflict of interest

The authors have no COI to declare.

Author details

Erin Anderson[1] and David Coman[1,2,3]*

1 Virtus Diagnostics, Brisbane, Australia

2 Department of Metabolic Medicine Queensland Children's Hospital, Brisbane, Australia

3 School of Medicine, Griffith University Gold Coast, University of Queensland Brisbane, Australia

*Address all correspondence to: enquiries@drdavidcoman.com.au

References

[1] Cooper MK, Wassif CA, Krakowiak PA, Taipale J, Gong R, Kelley RI, et al. A defective response to Hedgehog signaling in disorders of cholesterol biosynthesis. Nature Genetics. 2003;**33**(4):508-513

[2] Goldstein JL, Brown MS. Regulation of the mevalonate pathway. Nature. 1990;**343**(6257):425

[3] Waterham HR. Defects of cholesterol biosynthesis. FEBS Letters. 2006;**580**(23):5442-5449

[4] Coman D, Vissers LELM, Riley LG, Kwint MP, Hauck R, Koster J, et al. Squalene synthase deficiency: Clinical, biochemical and molecular characterisation of a defect in cholesterol biosynthesis. American Journal of Human Genetics. 2018;**103**(1):125-130

[5] Bloch K. The biological synthesis of cholesterol. Science. 1965;**150**(3962):19-28

[6] Kandutsch AA, Russell AE. Preputial gland tumor sterols. The Journal of Biological Chemistry. 1960;**235**(8):2256-2261

[7] Mitsche MA, McDonald JG, Hobbs HH, Cohen JC. Flux analysis of cholesterol biosynthesis in vivo reveals multiple tissue and cell-type specific pathways. eLife. 2015;**4**:e07999

[8] Ikonen E, Jansen M. Cellular sterol trafficking and metabolism: Spotlight on structure. Current Opinion in Cell Biology. 2008;**20**(4):371-377

[9] Sharpe LJ, Cook ECL, Zelcer N, Brown AJ. The UPS and downs of cholesterol homeostasis. Trends in Biochemical Sciences. 2014;**39**(11):527-535

[10] Bostedor RG, Karkas JD, Arison BH, Bansal VS, Vaidya S, Germershausen JI, et al. Farnesol-derived dicarboxylic acids in the urine of animals treated with zaragozic acid A or with farnesol. The Journal of Biological Chemistry. 1997;**272**(14):9197-9203

[11] Jemal M, Ouyang Z. Gas chromatography-mass spectrometric method for quantitative determination in human urine of dicarboxylic (dioic) acids produced in the body as a consequence of cholesterol biosynthesis inhibition. Journal of Chromatography. B, Biomedical Sciences and Applications. 1998;**709**(2):233-241

[12] Vaidya S, Bostedor R, Kurtz MM, Bergstrom JD, Bansal VS. Massive production of farnesol-derived dicarboxylic acids in mice treated with the squalene synthase Inhibitor zaragozic acid A. Archives of Biochemistry and Biophysics. 1998;**355**(1):84-92

[13] van der Meer JWM, Simon A. The challenge of autoinflammatory syndromes: With an emphasis on hyper-IgD syndrome. Rheumatology. 2016;**55**(suppl 2):ii23-ii29

[14] Favier L, Schulert G. Mevalonate kinase deficiency: Current perspectives. The Application of Clinical Genetics. 2016;**9**:101-110

[15] Berger R, Smit GPA, Schierbeek H, Bijsterveld K, le Coultre R. Mevalonic aciduria: An inborn error of cholesterol biosynthesis? Clinica Chimica Acta. 1985;**152**(1-2):219-222

[16] van der Meer JW, Vossen JM, Radl J, van Nieuwkoop JA, Meyer CJ, Lobatto S, et al. Hyperimmunoglobulinaemia D and periodic fever: A new syndrome. Lancet (London, England). 1984;**1**(8386):1087-1090

[17] Frenkel J, Rijkers GT, Mandey SHL, Buurman SWM, Houten SM, Wanders RJA, et al. Lack of isoprenoid products raises ex vivo interleukin-1? Secretion in hyperimmunoglobulinemia D and periodic fever syndrome. Arthritis and Rheumatism. 2002;**46**(10):2794-2803

[18] Simon A, Kremer HPH, Wevers RA, Scheffer H, de Jong JG, van der Meer JWM, et al. CME Mevalonate kinase deficiency. Neurology. 2004;**62**:994-997

[19] Drenth PH, van Deuren M. Cytokine activation during attacks of the hyperimmunoglobulinemia D and periodic fever syndrome. Blood. 1995;**85**(12):3586-3593

[20] Drenth JPH, Powell RJ, Brown NS, Meer JWMVD. Interferon-γ and urine neopterin in attacks of the hyperimmunoglobulinaemia D and periodic fever syndrome. European Journal of Clinical Investigation. 1995;**25**(9):683-686

[21] Haas D, Hoffmann GF. Mevalonate kinase deficiencies: From mevalonic aciduria to hyperimmunoglobulinemia D syndrome. Orphanet Journal of Rare Diseases. 2006;**1**(1):13

[22] Hoffmann F, Charpentier C, Mayatepek E, Mancini J, Leichsenring M, Gibson KM, et al. Clinical and biochemical phenotype in 11 patients with mevalonic aciduria. Pediatrics. 1993;**91**(5):915-921

[23] Hinson DD, Rogers ZR, Hoffmann GF, Schachtele M, Fingerhut R, Kohlschutter A, et al. Hematological abnormalities and cholestatic liver disease in two patients with mevalonate kinase deficiency. American Journal of Medical Genetics. 1998;**78**:408-412

[24] van der Hilst JCH, Bodar EJ, Barron KS, Frenkel J, Drenth JPH, van der Meer JWM, et al. Long-term follow-up, clinical features, and quality of life in a series of 103 patients with hyperimmunoglobulinemia D syndrome. Medicine (Baltimore). 2008;**87**(6):301-310

[25] Houten SM, van Woerden CS, Wijburg FA, Wanders RJA, Waterham HR. Carrier frequency of the V377I (1129G>A) MVK mutation, associated with hyper-IgD and periodic fever syndrome in the Netherlands. European Journal of Human Genetics. 2003;**11**(2):196-200

[26] Galeotti C, Georgin-Lavialle S, Sarrabay G, Touitou I, Koné-Paut I. Mevalonate kinase deficiency in 2016. La Revue de Médecine Interne. 2018;**39**(4):265-270

[27] Mandey SHL, Schneiders MS, Koster J, Waterham HR. Mutational spectrum and genotype–phenotype correlations in mevalonate kinase deficiency. Human Mutation. 2006;**27**(8):796-802

[28] Cuisset L, Drenth JPH, Simon A, Vincent MF, van der Velde Visser S, van der Meer JWM, et al. Molecular analysis of MVK mutations and enzymatic activity in hyper-IgD and periodic fever syndrome. European Journal of Human Genetics. 2001;**9**(4):260-266

[29] Houten SM, Koster J, Romeijn G-J, Frenkel J, Di Rocco M, Caruso U, et al. Organization of the mevalonate kinase (MVK) gene and identification of novel mutations causing mevalonic aciduria and hyperimmunoglobulinaemia D and periodic fever syndrome. European Journal of Human Genetics. 2001;**9**(4):253-259

[30] ter Haar NM, Jeyaratnam J, Lachmann HJ, Simon A, Brogan PA, Doglio M, et al. The phenotype and genotype of mevalonate kinase deficiency: A series of 114 cases from the Eurofever Registry: Phenotype

and genotype of MKD. Arthritis & Rhematology. 2016;**68**(11):2795-2805

[31] Shendi HM, Devlin LA, Edgar JD. Interleukin 6 blockade for hyperimmunoglobulin D and periodic fever syndrome. JCR: Journal of Clinical Rheumatology. 2014;**20**(2):103-105

[32] Cantarini L, Vitale A, Magnotti F, Lucherini O, Caso F, Frediani B, et al. Weekly oral alendronate in mevalonate kinase deficiency. Orphanet Journal of Rare Diseases. 2013;**8**(1):196

[33] Hager EJ, Tse HM, Piganelli JD, Gupta M, Baetscher M, Tse TE, et al. Deletion of a single mevalonate kinase (Mvk) allele yields a murine model of hyper-IgD syndrome. Journal of Inherited Metabolic Disease. 2007;**30**(6):888-895

[34] Tozawa R, Ishibashi S, Osuga J, Yagyu H, Oka T, Chen Z, et al. Embryonic lethality and defective neural tube closure in mice lacking squalene synthase. The Journal of Biological Chemistry. 1999;**274**(43):30843-30848

[35] Chen X, Liu L. Congenital cataract with LSS gene mutations: A new case report. Journal of Pediatric Endocrinology & Metabolism. 2017;**30**(11):1231-1235

[36] Romano M-T, Tafazzoli A, Mattern M, Sivalingam S, Wolf S, Rupp A, et al. Bi-allelic mutations in LSS, encoding lanosterol synthase, cause autosomal-recessive hypotrichosis simplex. American Journal of Human Genetics. 2018;**103**(5):777-785

[37] Besnard T, Sloboda N, Goldenberg A, Küry S, Cogné B, Breheret F, et al. Biallelic pathogenic variants in the lanosterol synthase gene LSS involved in the cholesterol biosynthesis cause alopecia with intellectual disability, a rare recessive neuroectodermal syndrome. Genetics in Medicine. [published online ahead of print 6 February 2019]. DOI:10.1038/s41436-019-0445-x

[38] Mori M, Li G, Abe I, Nakayama J, Guo Z, Sawashita J, et al. Lanosterol synthase mutations cause cholesterol deficiency-associated cataracts in the Shumiya cataract rat. The Journal of Clinical Investigation. 2006;**116**(2):395-404

[39] The International Mouse Phenotyping Consortium, Dickinson ME, Flenniken AM, Ji X, Teboul L, Wong MD, et al. High-throughput discovery of novel developmental phenotypes. Nature. 2016;**537**(7621):508-514

[40] Greenberg CR, Rimoin DL, Gruber HE, DeSa DJB, Reed M, Lachman RS, et al. A new autosomal recessive lethal chondrodystrophy with congenital hydrops. American Journal of Medical Genetics. 1988;**29**(3):623-632

[41] Lubala TK, Lubala N, Munkana AN, Nyenga AM, Mutombo AM. Greenberg skeletal dysplasia: First reported case in the Democratic Republic of Congo. The Pan African Medical Journal. 2013;**14**:55

[42] Chitayat D, Gruber H, Mullen BJ, Pauzner D, Costa T, Lachman R, et al. Hydrops-ectopic calcification-moth-eaten skeletal dysplasia (Greenberg dysplasia): Prenatal diagnosis and further delineation of a rare genetic disorder. American Journal of Medical Genetics. 1993;**47**(2):272-277

[43] Konstantinidou A, Karadimas C, Waterham HR, Superti-Furga A, Kaminopetros P, Grigoriadou M, et al. Pathologic, radiographic and molecular findings in three fetuses diagnosed with HEM/Greenberg skeletal dysplasia. Prenatal Diagnosis. 2008;**28**(4):309-312

[44] Madazli R, Aksoy F, Ocak V, Atasü T. Detailed ultrasonographic findings in Greenberg dysplasia. Prenatal Diagnosis. 2001;**21**(1):65-67

[45] Trajkovski Z, Vrcakovski M, Saveski J, Gucev ZS. Greenberg dysplasia (hydrops-ectopic calcification-moth-eaten skeletal dysplasia): Prenatal ultrasound diagnosis and review of literature. American Journal of Medical Genetics. 2002;**111**(4):415-419

[46] Horn L-C, Faber R, Meiner A, Piskazeck U, Spranger J. Greenberg dysplasia: First reported case with additional non-skeletal malformations and without consanguinity. Prenatal Diagnosis. 2000;**20**(12):1008-1011

[47] Kelley RI, Kratz LE, Wilcox WG. Abnormal metabolism of 14-dehydrosterols in hydrops-ectopic calcification-moth-eaten skeletal dysplasia: Evidence for new defect of cholesterol biosynthesis. Proceedings of the Greenwood Genetic Center. 2000;**20**:116

[48] Tsai P-L, Zhao C, Turner E, Schlieker C. The Lamin B receptor is essential for cholesterol synthesis and perturbed by disease-causing mutations. eLife. 2016;**5**:e16011

[49] Wassif CA, Brownson KE, Sterner AL, Forlino A, Zerfas PM, Wilson WK, et al. HEM dysplasia and ichthyosis are likely laminopathies and not due to 3b-hydroxysterol D14-reductase deficiency. Human Molecular Genetics. 2007;**16**(10):1176-1187

[50] Turner EM, Schlieker C. Pelger-Huët anomaly and Greenberg skeletal dysplasia: LBR-associated diseases of cholesterol metabolism. Rare Diseases. 2016;**4**(1):e1241363

[51] Waterham HR, Koster J, Mooyer P, van Noort G, Kelley RI, Wilcox WR, et al. Autosomal recessive HEM/Greenberg skeletal dysplasia Is caused by 3β-hydroxysterol Δ14-reductase deficiency due to mutations in the lamin B receptor gene. American Journal of Human Genetics. 2003;**72**:1013-1017

[52] Bennati AM, Schiavoni G, Franken S, Piobbico D, Fazia MAD, Caruso D, et al. Disruption of the gene encoding 3β-hydroxysterol Δ14-reductase (Tm7sf2) in mice does not impair cholesterol biosynthesis. The FEBS Journal. 2008;**275**(20):5034-5047

[53] Clayton P, Fischer B, Mann A, Mansour S, Rossier E, Veen M, et al. Mutations causing Greenberg dysplasia but not Pelger anomaly uncouple enzymatic from structural functions of a nuclear membrane protein. Nucleus. 2010;**1**(4):354-366

[54] Hoffmann K, Dreger CK, Olins AL, Olins DE, Shultz LD, Lucke B, et al. Mutations in the gene encoding the lamin B receptor produce an altered nuclear morphology in granulocytes (Pelger-Huët anomaly). Nature Genetics. 2002;**31**(4):410-414

[55] Oosterwijk JC. Congenital abnormalities reported in Pelger-Huët homozygosity as compared to Greenberg/HEM dysplasia: Highly variable expression of allelic phenotypes. Journal of Medical Genetics. 2003;**40**(12):937-941

[56] Borovik L, Modaff P, Waterham HR, Krentz AD, Pauli RM. Pelger-huet anomaly and a mild skeletal phenotype secondary to mutations in LBR. American Journal of Medical Genetics. Part A. 2013;**161**(8):2066-2073

[57] Thompson E, Abdalla E, Superti-Furga A, McAlister W, Kratz L, Unger S, et al. Lamin B receptor-related disorder is associated with a spectrum of skeletal dysplasia phenotypes. Bone. 2019;**120**:354-363

[58] Herman GE, Kratz L. Disorders of sterol synthesis: Beyond Smith-Lemli-Opitz syndrome. American Journal of Medical Genetics. Part C, Seminars in Medical Genetics. 2012;**160C**(4):301-321

[59] Bittar M, Happle R. CHILD syndrome. Journal of the American Academy of Dermatology. 2004;**50**(2):34-37

[60] Happle R, Koch H, Lenz W. The CHILD syndrome: Congenital hemidysplasia with ichthyosiform erythroderma and limb defects. European Journal of Pediatrics. 1980;**134**(1):27-33

[61] Happle R, Effendy I, Megahed M, Orlow SJ, Kiister W. CHILD syndrome in a boy. American Journal of Medical Genetics. 1996;**62**:192-194

[62] Happle R, Mittag H, Kuster W. The CHILD nevus: A distinct skin disorder. Dermatology. 1995;**191**(3):210-216

[63] Hummel M, Cunningham D, Mullett CJ, Kelley RI, Herman GE. Left-sided CHILD syndrome caused by a nonsense mutation in the NSDHL gene. American Journal of Medical Genetics. 2003;**122A**(3):246-251

[64] Herman G. X-Linked dominant disorders of cholesterol biosynthesis in man and mouse. Biochimica et Biophysica Acta (BBA)—Molecular and Cell Biology of Lipids. 2000;**1529**(1-3):357-373

[65] Avgerinou G, Asvesti A, Katsambas A, Nikolaou V, Christofidou E, Grzeschik K, et al. CHILD syndrome: The NSDHL gene and its role in CHILD syndrome, a rare hereditary disorder: CHILD syndrome. Journal of the European Academy of Dermatology and Venereology. 2009;**24**(6):733-736

[66] Bittar M, Happle R, Grzeschik K-H, Leveleki L, Hertl M, Bornholdt D, et al. CHILD syndrome in 3 generations: The importance of mild or minimal skin lesions. Archives of Dermatology. 2006;**142**(3):348-351

[67] König A, Happle R, Bornholdt D, Engel H, Grzeschik K-H. Mutations in the NSDHL gene, encoding a 3β-hydroxysteroid dehydrogenase, cause CHILD syndrome. American Journal of Medical Genetics. 2000;**90**:339-346

[68] Bornholdt D. Mutational spectrum of NSDHL in CHILD syndrome. Journal of Medical Genetics. 2005;**42**(2):e17-e17

[69] Liu XY, Dangel AW, Kelley RI, Zhao W, Denny P, Botcherby M, et al. The gene mutated in bare patches and striated mice encodes a novel 3β-hydroxysteroid dehydrogenase. Nature Genetics. 1999;**22**(2):182-187

[70] Caldas H, Cunningham D, Wang X, Jiang F, Humphries L, Kelley RI, et al. Placental defects are associated with male lethality in bare patches and striated embryos deficient in the NAD(P)H Steroid Dehydrogenase-like (NSDHL) Enzyme. Molecular Genetics and Metabolism. 2005;**84**(1):48-60

[71] Cunningham D, DeBarber AE, Bir N, Binkley L, Merkens LS, Steiner RD, et al. Analysis of hedgehog signaling in cerebellar granule cell precursors in a conditional Nshdl allele demonstrates an essential role for cholesterol in postnatal CNS development. Human Molecular Genetics. 2015;**24**(10):2808-2825

[72] Bergqvist C, Abdallah B, Hasbani D-J, Abbas O, Kibbi AG, Hamie L, et al. CHILD syndrome: A modified pathogenesis-targeted therapeutic approach. American Journal of Medical Genetics. Part A. 2018;**176**(3):733-738

[73] Paller AS, van Steensel MAM, Rodriguez-Martín M, Sorrell J, Heath C, Crumrine D, et al. Pathogenesis-based therapy reverses cutaneous abnormalities in an inherited disorder of distal cholesterol metabolism. The Journal of Investigative Dermatology. 2011;**131**(11):2242-2248

[74] Cañueto J, Girós M, Ciria S, Pi-Castán G, Artigas M, García-Dorado J, et al. Clinical, molecular

and biochemical characterization of nine Spanish families with Conradi-Hünermann-Happle syndrome: New insights into X-linked dominant chondrodysplasia punctata with a comprehensive review of the literature. The British Journal of Dermatology. 2012;**166**(4):830-838

[75] Pacault M, Vincent M, Besnard T, Kannengiesser C, Bénéteau C, Barbarot S, et al. New splicing pathogenic variant in EBP causing extreme familial variability of Conradi-Hünermann-Happle Syndrome. European Journal of Human Genetics. 2018;**26**(12):1784

[76] Has C, Bruckner-Tuderman L, Traupe H, Seedorf U, Kannenberg F, Folkers E, et al. Gas chromatography-mass spectrometry and molecular genetic studies in families with the Conradi-Hünermann-Happle syndrome. The Journal of Investigative Dermatology. 2002;**118**(5):851-858

[77] Aughton DJ, Kelley RI, Metzenberg A, Pureza V, Pauli RM. X-linked dominant chondrodysplasia punctata (CDPX2) caused by single gene mosaicism in a male. American Journal of Medical Genetics. Part A. 2003;**116A**(3):255-260

[78] Milunsky JM, Maher TA, Metzenberg AB. Molecular, biochemical, and phenotypic analysis of a hemizygous male with a severe atypical phenotype for X-linked dominant Conradi-Hunermann-Happle syndrome and a mutation in EBP. American Journal of Medical Genetics. Part A. 2003;**116A**(3):249-254

[79] Sutphen R, Amar MJ, Kousseff BG, Toomey KE. XXY male with X-linked dominant chondrodysplasia punctata (Happle syndrome). American Journal of Medical Genetics. 1995;**57**(3):489-492

[80] Traupe H, Has C. The Conradi-Hünermann-Happle syndrome is caused by mutations in the gene that encodes a 8-7 sterol isomerase and is biochemically related to the CHILD syndrome. European Journal of Dermatology. 2000;**10**(6):425-428

[81] Hellenbroich Y, Grzeschik K-H, Krapp M, Jarutat T, Lehrmann-Petersen C, Buiting K, et al. Reduced penetrance in a family with X-linked dominant chondrodysplasia punctata. European Journal of Medical Genetics. 2007;**50**(5):392-398

[82] Pazzaglia UE, Zarattini G, Donzelli C, Benetti A, Bondioni MP, Groli C. The nature of cartilage stippling in chrondrodysplasia punctata: Histopathological study of Conradi-Hunermann-Happle syndrome. Fetal and Pediatric Pathology. 2008;**27**(2):71-81

[83] Herman GE, Kelley RI, Pureza V, Smith D, Kopacz K, Pitt J, et al. Characterization of mutations in 22 females with X-linked dominant chondrodysplasia punctata (Happle syndrome). Genetics in Medicine. 2002;**4**(6):434-438

[84] Kelley RI, Herman GE. Inborn errors of sterol biosynthesis. Annual Review of Genomics and Human Genetics. 2001;**2**(1):299-341

[85] Happle R. X-linked dominant chondrodysplasia punctata. Human Genetics. 1979;**53**:65-73

[86] Posey JE, Burrage LC, Campeau PM, Lu JT, Eble TN, Kratz L, et al. Adult presentation of X-linked Conradi-Hünermann-Happle syndrome. American Journal of Medical Genetics. Part A. 2015;**167**(6):1309-1314

[87] Dempsey M, Tan C, Herman GE. Chondrodysplasia punctata 2, X-linked. In: Adam MP, Ardinger HH, Pagon RA, editors. Gene Reviews. Seattle (WA): University of Washington; 2011

[88] Braverman N, Lin P, Moebius FF, Obie C, Moser A, Glossmann H, et al. Mutations in the gene encoding 3 beta-hydroxysteroid-delta 8, delta 7-isomerase cause X-linked dominant Conradi-Hünermann syndrome. Nature Genetics. 1999;**22**(3):291-294

[89] Derry JM, Gormally E, Means GD, Zhao W, Meindl A, Kelley RI, et al. Mutations in a delta-8-delta-7 sterol isomerase in the tattered mouse and X-linked dominant chondrodysplasia punctata. Nature Genetics. 1999;**22**(3):286-290

[90] Porter FD, Herman GE. Malformation syndromes caused by disorders of cholesterol synthesis. Journal of Lipid Research. 2011;**52**(1):6-34

[91] Anderson R, Rust S, Ashworth J, Clayton-Smith J, Taylor RL, Clayton PT, et al. Lathosterolosis: A relatively mild case with cataracts and learning difficulties. In: Morava E, Baumgartner M, Patterson M, Rahman S, Zschocke J, Peters V, editors. JIMD Reports. Berlin, Heidelberg: Springer Berlin Heidelberg; 2018. pp. 79-84

[92] Brunetti-Pierri N, Corso G, Rossi M, Ferrari P, Balli F, Rivasi F, et al. Lathosterolosis, a novel multiple-malformation/mental retardation syndrome due to deficiency of 3β-hydroxysteroid Δ5-desaturase. American Journal of Human Genetics. 2002;**7**:952-958

[93] Ho ACC, Fung CW, Siu TS, Ma OCK, Lam CW, Tam S, et al. Lathosterolosis: A disorder of cholesterol biosynthesis resembling Smith-Lemli-Opitz syndrome. JIMD Reports. 2013;**12**:129-134

[94] Krakowiak PA. Lathosterolosis: An inborn error of human and murine cholesterol synthesis due to lathosterol 5-desaturase deficiency. Human Molecular Genetics. 2003;**12**(13):1631-1641

[95] Rossi M, D'Armiento M, Parisi I, Ferrari P, Hall CM, Cervasio M, et al. Clinical phenotype of lathosterolosis. American Journal of Medical Genetics. Part A. 2007;**143A**(20):2371-2381

[96] Calvo PL, Brunati A, Spada M, Romagnoli R, Corso G, Parenti G, et al. Liver transplantation in defects of cholesterol biosynthesis: The case of lathosterolosis. American Journal of Transplantation. 2014;**14**(4):960-965

[97] Smith DW, Lemli L, Opitz JM. A newly recognized syndrome of multiple congenital anomalies. The Journal of Pediatrics. 1964;**64**(2):210-217

[98] Nowaczyk MJM, Waye JS, Douketis JD. DHCR7 mutation carrier rates and prevalence of the RSH/Smith-Lemli-Opitz syndrome: Where are the patients? American Journal of Medical Genetics. Part A. 2006;**140A**(19):2057-2062

[99] Cross JL, Iben J, Simpson C, Thurm A, Swedo S, Tierney E, et al. Determination of the allelic frequency in Smith-Lemli-Opitz syndrome by analysis of massively parallel sequencing data sets. Clinical Genetics. 2015;**87**(6):570-575

[100] Lazarin GA, Haque IS, Evans EA, Goldberg JD. Smith-Lemli-Opitz syndrome carrier frequency and estimates of in utero mortality rates. Prenatal Diagnosis. 2017;**37**(4):350-355

[101] Schoen E. Maternal serum unconjugated estriol as a predictor for Smith-Lemli-Opitz syndrome and other fetal conditions. Obstetrics and Gynecology. 2003;**102**(1):167-172

[102] Eroglu Y, Nguyen-Driver M, Steiner RD, Merkens L, Merkens M, Roullet J-B, et al. Normal IQ is possible in Smith-Lemli-Opitz syndrome. American Journal of Medical Genetics. Part A. 2017;**173**(8):2097-2100

[103] Lee RWY, Conley SK, Gropman A, Porter FD, Baker EH. Brain magnetic resonance imaging findings in Smith-Lemli-Opitz syndrome. American Journal of Medical Genetics. Part A. 2013;**161**(10):2407-2419

[104] Jurevics H, Morell P. Cholesterol for synthesis of myelin is made locally, not imported into brain. Journal of Neurochemistry. 1995;**64**(2):895-901

[105] Kelley RI, Hennekam RCM. The Smith-Lemli-Opitz syndrome. Journal of Medical Genetics. 2000;**37**(5):321-335

[106] Nowaczyk MJM, Irons MB. Smith-Lemli-Opitz syndrome: Phenotype, natural history, and epidemiology. American Journal of Medical Genetics. Part C, Seminars in Medical Genetics. 2012;**160C**(4):250-262

[107] Ryan AK, Bartlett K, Clayton P, Eaton S, Mills L, Donnai D, et al. Smith-Lemli-Opitz syndrome: A variable clinical and biochemical phenotype. Journal of Medical Genetics. 1998;**35**(7):558-565

[108] Porter FD. Smith-Lemli-Opitz syndrome: Pathogenesis, diagnosis and management. European Journal of Human Genetics. 2008;**16**(5):535-541

[109] Charman R, Tyrrell P, Arlett K, et al. Photosensitivity associated with the Smith-Lemli-Opitz syndrome. The British Journal of Dermatology. 1998;**138**(5):885-888

[110] Irons M, Roy Elias E, Salen G, Tint GS, Batta Ashok K. Defective cholesterol biosynthesis in Smith-Lemli-Opitz syndrome. The Lancet. 1993;**341**(8857):1414

[111] Tint GS, Irons M, Elias ER, Batta AK, Frieden R, Chen TS, et al. Defective cholesterol biosynthesis associated with the Smith-Lemli-Optiz syndrome. The New England Journal of Medicine. 1994;**330**(2):107-113

[112] Donoghue SE, Pitt JJ, Boneh A, White SM. Smith-Lemli-Opitz syndrome: Clinical and biochemical correlates. Journal of Pediatric Endocrinology & Metabolism. 2018;**31**(4):451-459

[113] Oláh AV, Szabó GP, Varga J, Balogh L, Csábi G, Csákváry V, et al. Relation between biomarkers and clinical severity in patients with Smith-Lemli-Opitz syndrome. European Journal of Pediatrics. 2013;**172**(5):623-630

[114] Correa-Cerro LS, Porter FD. 3β-Hydroxysterol Δ7-reductase and the Smith-Lemli-Opitz syndrome. Molecular Genetics and Metabolism. 2005;**84**(2):112-126

[115] Ciara E, Nowaczyk MJM, Witsch-Baumgartner M, Malunowicz E, Popowska E, Jezela-Stanek A, et al. DHCR7 mutations and genotype-phenotype correlation in 37 Polish patients with Smith-Lemli-Opitz syndrome. Clinical Genetics. 2004;**66**(6):517-524

[116] Yu H, Lee M-H, Starck L, Elias ER, Irons M, Salen G, et al. Spectrum of Δ7-dehydrocholesterol reductase mutations in patients with the Smith-Lemli-Opitz (RSH) syndrome. Human Molecular Genetics. 2000;**9**(9):1385-1391

[117] Wassif CA, Krakowiak PA, Wright BS, Gewandter JS, Sterner AL, Javitt N, et al. Residual cholesterol synthesis and simvastatin induction of cholesterol synthesis in Smith-Lemli-Opitz syndrome fibroblasts. Molecular Genetics and Metabolism. 2005;**85**(2):96-107

[118] Lanthaler B, Steichen-Gersdorf E, Kollerits B, Zschocke J, Witsch-Baumgartner M. Maternal ABCA1

genotype is associated with severity of Smith-Lemli-Opitz syndrome and with viability of patients homozygous for null mutations. European Journal of Human Genetics. 2013;**21**(3):286-293

[119] Witsch-Baumgartner M, Gruber M, Kraft HG, Rossi M, Clayton P, Giros M, et al. Maternal apo E genotype is a modifier of the Smith-Lemli-Opitz syndrome. Journal of Medical Genetics. 2004;**41**(8):577-584

[120] Lindegaard ML, Wassif CA, Vaisman B, Amar M, Wasmuth EV, Shamburek R, et al. Characterization of placental cholesterol transport: ABCA1 is a potential target for in utero therapy of Smith-Lemli-Opitz syndrome. Human Molecular Genetics. 2008;**17**(23):3806-3813

[121] Movassaghi M, Bianconi S, Feinn R, Wassif CA, Porter FD. Vitamin D levels in Smith-Lemli-Opitz syndrome. American Journal of Medical Genetics. Part A. 2017;**173**(10):2577-2583

[122] Tint GS, Abuelo D, Till M, Cordier MP, Batta AK, Shefer S, et al. Fetal Smith-Lemli-Opitz syndrome can be detected accurately and reliably by measuring amniotic fluid dehydrocholesterols. Prenatal Diagnosis. 1998;**18**(7):651-658

[123] Jezela-Stanek A, Małunowicz EM, Ciara E, Popowska E, Goryluk-Kozakiewicz B, Spodar K, et al. Maternal urinary steroid profiles in prenatal diagnosis of Smith-Lemli-Opitz syndrome: First patient series comparing biochemical and molecular studies. Clinical Genetics. 2006;**69**(1):77-85

[124] Shackleton CHL, Marcos J, Palomaki GE, Craig WY, Kelley RI, Kratz LE, et al. Dehydrosteroid measurements in maternal urine or serum for the prenatal diagnosis of Smith-Lemli-Opitz syndrome (SLOS). American Journal of Medical Genetics. Part A. 2007;**143A**(18):2129-2136

[125] McGaughran J, Donnai D, Clayton P, Mills K. Diagnosis of Smith-Lemli-Opitz syndrome. The New England Journal of Medicine. 1994;**330**(23):1685-1686

[126] Jira PE, Wevers RA, de Jong J, Rubio-Gozalbo E, Janssen-Zijlstra FSM, van Heyst AFJ, et al. Simvastatin: A new therapeutic approach for Smith-Lemli-Opitz syndrome. Journal of Lipid Research. 2000;**41**(8):1339-1346

[127] Azurdia RM, Anstey AV, Rhodes LE. Cholesterol supplementation objectively reduces photosensitivity in the Smith-Lemli-Opitz syndrome. The British Journal of Dermatology. 2001;**144**(1):143-145

[128] Steiner RD, Linck LM, Flavell DP, Lin DS, Connor WE. Sterol balance in the Smith-Lemli-Opitz syndrome: Reduction in whole body cholesterol synthesis and normal bile acid production. Journal of Lipid Research. 2000;**41**(9):1437-1447

[129] Svoboda MD, Christie JM, Eroglu Y, Freeman KA, Steiner RD. Treatment of Smith-Lemli-Opitz syndrome and other sterol disorders. American Journal of Medical Genetics. Part C, Seminars in Medical Genetics. 2012;**160C**:285-294

[130] Pauli RM, Williams MS, Josephson KD, Tint GS. Smith-Lemli-Opitz syndrome: Thirty-year follow-up of "S" of "RSH" syndrome. American Journal of Medical Genetics. 1997;**68**(3):260-262

[131] Elias ER, Irons MB, Hurley AD, Tint GS, Salen G. Clinical effects of cholesterol supplementation in six patients with the Smith-Lemli-Opitz syndrome (SLOS). American Journal of Medical Genetics. 1997;**68**(3):305-310

[132] Sikora DM, Ruggiero M, Petit-Kekel K, Merkens LS, Connor WE, Steiner RD. Cholesterol supplementation does not improve developmental progress in Smith-Lemli-Opitz syndrome. The Journal of Pediatrics. 2004;**144**(6):783-791

[133] Tierney E, Conley SK, Goodwin H, Porter FD. Analysis of short-term behavioral effects of dietary cholesterol supplementation in Smith-Lemli-Opitz syndrome. American Journal of Medical Genetics. Part A. 2010;**152A**(1):91-95

[134] Wassif CA, Kratz L, Sparks SE, Wheeler C, Bianconi S, Gropman A, et al. A placebo-controlled trial of simvastatin therapy in Smith-Lemli-Opitz syndrome. Genetics in Medicine. 2017;**19**(3):297-305

[135] Fliesler SJ. Antioxidants: The missing key to improved therapeutic intervention in Smith-Lemli-Opitz syndrome? Hereditary Genetics: Current Research. 2013;**2**(2):119

[136] Korade Z, Xu L, Harrison FE, Ahsen R, Hart SE, Folkes OM, et al. Antioxidant supplementation ameliorates molecular deficits in Smith-Lemli-Opitz syndrome. Biological Psychiatry. 2014;**75**(3):215-222

[137] Pasta S, Akhile O, Tabron D, Ting F, Shackleton C, Watson G. Delivery of the 7-dehydrocholesterol reductase gene to the central nervous system using adeno-associated virus vector in a mouse model of Smith-Lemli-Opitz Syndrome. Molecular Genetics and Metabolism Reports. 2015;**4**:92-98

[138] Correa-Cerro LS, Wassif CA, Kratz L, Miller GF, Munasinghe JP, Grinberg A, et al. Development and characterization of a hypomorphic Smith-Lemli-Opitz syndrome mouse model and efficacy of simvastatin therapy. Human Molecular Genetics. 2006;**15**(6):839-851

[139] Dehart DB, Lanoue L, Tint GS, Sulik KK. Pathogenesis of malformations in a rodent model for Smith-Lemli-Opitz syndrome. American Journal of Medical Genetics. 1997;**68**(3):328-337

[140] Waterham HR, Koster J, Romeijn GJ, Hennekam RCM, Vreken P, Andersson HC, et al. Mutations in the 3β-hydroxysterol Δ24-reductase gene cause desmosterolosis, an autosomal recessive disorder of cholesterol biosynthesis. American Journal of Human Genetics. 2001;**69**:685-694

[141] Fitzpatrick DR, Keeling JW, Evans MJ, Kan AE, Bell JE, Porteous MEM, et al. Clinical phenotype of desmosterolosis. American Journal of Medical Genetics. 1998;**75**:145-152

[142] Andersson HC, Kratz L, Kelley R. Desmosterolosis presenting with multiple congenital anomalies and profound developmental delay. American Journal of Medical Genetics. 2002;**113**(4):315-319

[143] Clayton P, Mills K, Keeling J, FitzPatrick D. Desmosterolosis: A new inborn error of cholesterol biosynthesis. The Lancet. 1996;**348**(9024):404

[144] Dias C, Rupps R, Millar B, Choi K, Marra M, Demos M, et al. Desmosterolosis: An illustration of diagnostic ambiguity of cholesterol synthesis disorders. Orphanet Journal of Rare Diseases. 2014;**9**(1):94

[145] Rohanizadegan M, Sacharow S. Desmosterolosis presenting with multiple congenital anomalies. European Journal of Medical Genetics. 2018;**61**(3):152-156

[146] Schaaf CP, Koster J, Katsonis P, Kratz L, Shchelochkov OA, Scaglia F, et al. Desmosterolosis-phenotypic

and molecular characterization of a third case and review of the literature. American Journal of Medical Genetics. Part A. 2011;**155**(7):1597-1604

[147] Zolotushko J, Flusser H, Markus B, Shelef I, Langer Y, Heverin M, et al. The desmosterolosis phenotype: Spasticity, microcephaly and micrognathia with agenesis of corpus callosum and loss of white matter. European Journal of Human Genetics. 2011;**19**(9):942-946

[148] Wechsler A. Generation of viable cholesterol-free mice. Science. 2003;**302**(5653):2087-2087

[149] Cenedella RJ. Cholesterol synthesis inhibitor U18666A and the role of sterol metabolism and trafficking in numerous pathophysiological processes. Lipids. 2009;**44**(6):477-487

[150] Roux C. Teratogenic action of triparanol in animals. Archives Françaises de Pédiatrie. 1964;**21**:451-464

[151] Gofflot F, Hars C, Illien F, Chevy F, Wolf C, Picard JJ, et al. Molecular mechanisms underlying limb anomalies associated with cholesterol deficiency during gestation: Implications of Hedgehog signaling. Human Molecular Genetics. 2003;**12**(10):1187-1198

[152] Vanier MT. Niemann-Pick disease type C. Orphanet Journal of Rare Diseases. 2010;**5**(16):18

[153] Xu Z, Farver W, Kodukula S, Storch J. Regulation of sterol transport between membranes and NPC2. Biochemistry. 2008;**47**(42):11134-11143

[154] Lloyd-Evans E, Morgan AJ, He X, Smith DA, Elliot-Smith E, Sillence DJ, et al. Niemann-Pick disease type C1 is a sphingosine storage disease that causes deregulation of lysosomal calcium. Nature Medicine. 2008;**14**(11):1247-1255

[155] Fredrickson DS. The inheritance of high density lipoprotein deficiency (Tangier disease). The Journal of Clinical Investigation. 1964;**43**(2):228-236

[156] Puntoni M, Sbrana F, Bigazzi F, Sampietro T. Tangier disease: Epidemiology, pathophysiology and management. American Journal of Cardiovascular Drugs. 2012;**12**(5):303-311

[157] Sahoo D, Trischuk TC, Chan T, Drover VAB, Ho S, Chimini G, et al. ABCA1-dependent lipid efflux to apolipoprotein A-I mediates HDL particle formation and decreases VLDL secretion from murine hepatocytes. Journal of Lipid Research. 2004;**45**(6):1122-1131

[158] Boadu E, Francis GA. The role of vesicular transport in ABCA1-dependent lipid efflux and its connection with NPC pathways. Journal of Molecular Medicine. 2006;**84**(4):266-275

[159] Boadu E, Nelson RC, Francis GA. ABCA1-dependent mobilization of lysosomal cholesterol requires functional Niemann-Pick C2 but not Niemann-Pick C1 protein. Biochimica et Biophysica Acta (BBA)—Molecular and Cell Biology. 2012;**1821**(3):396-404

[160] Sechi A, Dardis A, Zampieri S, Rabacchi C, Zanoni P, Calandra S, et al. Effects of miglustat treatment in a patient affected by an atypical form of Tangier disease. Orphanet Journal of Rare Diseases. 2014;**9**(1):143

[161] Wassif CA, Vied D, Tsokos M, Connor WE, Steiner RD, Porter FD. Cholesterol storage defect in RSH/Smith-Lemli-Opitz syndrome fibroblasts. Molecular Genetics and Metabolism. 2002;**75**(4):325-334

[162] Liscum L, Ruggiero RM, Faust JR. The intracellular transport of low density lipoprotein-derived cholesterol

is defective in Niemann-Pick type C fibroblasts. The Journal of Cell Biology. 1989;**108**(5):1625-1636

[163] Platt FM, Wassif C, Colaco A, Dardis A, Lloyd-Evans E, Bembi B, et al. Disorders of cholesterol metabolism and their unanticipated convergent mechanisms of disease. Annual Review of Genomics and Human Genetics. 2014;**15**(1):173-194

[164] Patterson MC, Vecchio D, Prady H, Abel L, Wraith JE. Miglustat for treatment of Niemann-Pick C disease: A randomised controlled study. Lancet Neurology. 2007;**6**(9):765-772

[165] Joo JH, Jetten AM. Molecular mechanisms involved in farnesol-induced apoptosis. Cancer Letters. 2010;**287**(2):123-135

[166] Hottman DA, Li L. Protein prenylation and synaptic plasticity: Implications for Alzheimer's disease. Molecular Neurobiology. 2014;**50**(1):177-185

[167] Bianco AM. Genetics of inflammatory bowel disease from multifactorial to monogenic forms. World Journal of Gastroenterology. 2015;**21**(43):12296

[168] Bianco AM, Girardelli M, Vozzi D, Crovella S, Kleiner G, Marcuzzi A. Mevalonate kinase deficiency and IBD: Shared genetic background. Gut. 2014;**63**(8):1367-1368

[169] Levy M, Arion A, Berrebi D, Cuisset L, Jeanne-Pasquier C, Bader-Meunier B, et al. Severe early-onset colitis revealing mevalonate kinase deficiency. Pediatrics. 2013;**132**(3):e779-e783

[170] Rakheja D, Boriack RL. Precholesterol sterols accumulate in lipid rafts of patients with Smith-Lemli-Opitz syndrome and X-linked dominant chondrodysplasia punctata. Pediatric and Developmental Pathology. 2008;**11**(2):128-132

[171] Frisso G, Gelzo M, Procopio E, Sica C, Lenza MP, Dello Russo A, et al. A rare case of sterol-C4-methyl oxidase deficiency in a young Italian male: Biochemical and molecular characterization. Molecular Genetics and Metabolism. 2017;**121**(4):329-335

[172] He M, Kratz LE, Michel JJ, Vallejo AN, Ferris L, Kelley RI, et al. Mutations in the human SC4MOL gene encoding a methyl sterol oxidase cause psoriasiform dermatitis, microcephaly, and developmental delay. The Journal of Clinical Investigation. 2011;**121**(3):976-984

Chapter 9

Low-Density Lipoprotein: Biochemical and Metabolic Characteristics and Its Pathogenic Mechanism

Jie Lin

Abstract

This chapter mainly introduces the physicochemical properties, physiological function, and metabolic pathway of low-density lipoprotein (LDL), with emphasis on the mechanism of atherosclerosis caused by LDL and the existing treatment methods. The content of this paper is detailed and comprehensive, including the latest research results in recent years. Different from the published articles, this paper adopts an innovative "tree" structure, with the disease development process as the main line and the molecular mechanism as the branch, showing the clinical significance of the molecular mechanism. The main purpose of the author of this paper is to take LDL as an example, establish a multidimensional knowledge system, and strive to make the complex molecular mechanism vivid.

Keywords: low-density lipoprotein, metabolism, atherosclerosis, familial hypercholesterolemia, treatment

1. Introduction

In recent years, the research on abnormal metabolism of blood lipids has gradually increased, and the importance of blood lipid management in primary prevention and secondary prevention of cardiovascular diseases has been clarified. This is because dyslipidemia, as a major cardiovascular risk factor, is the common pathway of multiple pathogenic factors, mainly manifested as the qualitative and quantitative changes in low-density lipoprotein (LDL), attacking the endothelium, initiating an inflammatory response, promoting the occurrence and development of atherosclerosis, and ultimately producing cardiovascular events. Thus, the understanding of the quantitative and qualitative changes in LDL is the basis of atherosclerosis treatment and prevention.

2. Physicochemical characteristics and metabolic process of LDL

LDL is a spherical particle, core lipids of LDL particle are composed of cholesterol ester (CE) and triglyceride (TG), an outer monolayer is composed of free cholesterol (FC) and phospholipid (PL) including phosphatidylcholine (PC), and one molecule of apolipoprotein B100 surrounds the LDL particle [1]. Apolipoprotein

IntechOpen

B-100 (apoB-100) is composed of the amphoteric α-helical domain and β-sheet domain alternately (NH2-βα1-β1-α2-β2-α3-coOH), in which β-sheet structure accounts for 40%, which is related to the stability of LDL particles, and α-helix accounts for 25%, which is related to the amphiphilicity of LDL particles and the potential of self-repair [2].

LDL is the plasma metabolite of very low-density lipoprotein (VLDL). Plasma lipoprotein lipase or liver lipase catalyzes the hydrolysis of triglyceride (TG) in VLDL particles. At the same time, under the action of cholesterol ester transfer proteins (CETPs), cholesterol ester (CE) of HDL is transferred to VLDL, and phospholipids, apolipoprotein C (ApoC), and cholesterol are transferred to high-density lipoprotein (HDL) on the surface of VLDL. This process continues. In VLDL, TG decreases continuously, CE increases gradually, particles become smaller, and density increases gradually. First intermediate density lipoprotein (IDL) is formed, and then LDL is formed [3]. According to the formation process of LDL, it is easy to see that LDL is not a kind of particle but a class of particles with different sizes, densities, chemical composition, or different charges. In recent years, LDL particles have been divided into two phenotypes: type A (large and light LDL), LDL particle diameter ≥25.5 nm, and type B (small and dense LDL, sd-LDL), LDL particle diameter <25.5 nm. Compared with type A LDL, sd-LDL has a stronger ability to cause atherosclerosis and has been identified for a long time as a new risk factor of cardiovascular disease by the American Cholesterol Education Program and Adult Treatment Program III (NCEPIII) [4].

Natural LDL (nLDL) is in charge of the transport of endogenous cholesterol, and its metabolic process is the transport process of endogenous cholesterol. Among them, two-third were metabolized through the LDL receptor pathway [5]. LDL receptor (LDLR) is widely distributed in the whole body, especially on the cell membrane surface of the liver, adrenocortical, ovarian, testicular, and arterial wall, and specifically binds to ApoB100 on the surface of LDL. Internalization causes the membrane at the junction to sink in for endocytosis. Under the action of the proton pump (H+ -ATPase), the pH of endocytotic vesicle contents decreased, and LDL is separated from the receptor and fused with the lysosome. ApoB100 is decomposed into amino acids by a lysosomal proteolytic enzyme, and CE is hydrolyzed into free cholesterol and fatty acids by cholesterol esterase for cell utilization.

The remaining one-third is cleared by the mononuclear macrophage pathway. As a major member of innate immunity, macrophages are endowed with an advanced arsenal of sensors, composed of various pattern-recognition receptors (PRR). It is able to identify and bind foreign substances or altered substances to inactivate and degrade them. Therefore, the monocyte–macrophage clearance pathway is mainly aimed at LDL, which has changed its structure for a variety of reasons. It can also be called modified LDL (mLDL). mLDL specifically binds to the scavenger receptors (SRs) on the surface of macrophages and is subsequently removed. The half-life of plasma LDL in normal people is 2–4 days.

The cells prepare their cholesterol needs via two pathways: an exogenous pathway mediated with the LDLR and an endogenous pathway activated with the substrates of mevalonate and HMG-CoA reductase [6]. When the intracellular cholesterol level is too high, sterol regulatory element-binding protein (SREBP), a nuclear transcription factor, is activated, which inhibits the expression of LDL receptor gene from the transcription level, inhibits the synthesis of the receptor protein, and reduces the further uptake of LDL by the cell [7]. It is suggested that the simple increase of the nLDL-C level cannot fully explain the occurrence of atherosclerotic disease. More significantly, contrary to LDLR, SR expression is not inhibited by elevated intracellular cholesterol levels [1]. Then, when the number of modified LDL absorbed by macrophages through SRs far exceeds their scavenging capacity, significant cholesterol accumulation will occur in macrophages, and then macrophages will be transformed into foam cells, which is a key link in the occurrence of atherosclerosis [8, 9].

3. LDL modification: endowing LDL with special biological activity

3.1 Changes in lipid composition of LDL particles and primary structure of apoB-100

At present, more research is placed on the effects of oxidation and glycosylation on LDL.

LDL oxidation: when circulating LDL is out of extremely high levels, oxidized LDL (ox-LDL) is rarely found in circulation due to the presence of plasma antioxidants and vitamin C. In this case, the oxidation of LDL occurs mainly in the arterial wall. Vascular wall cells (endothelial cells, smooth muscle cells, and macrophages), stimulated by the attack factors, produce and release a large number of reactive oxygen species (ROS) and reactive nitrogen species (RNS). Free radicals rapidly oxidize the polyunsaturated fatty acids (PUFAs) on the surface of LDL into fatty acid fragments [10, 11]. This modification is produced in the surface of the LDL particles, so the physical and chemical properties of LDL change little. We call this LDL the "minimum modified LDL," which retains the affinity for the LDL receptor. Then the "minimum modified LDL" activates the endothelial anti-apoptosis signaling pathway, induces endothelial cells to express tissue factors and chemokines, promotes the aggregation of inflammatory cells, triggers an inflammatory reaction, generates a large number of free radicals, and leads to the continuous oxidation of LDL. Continuous oxidation further converts fatty acid fragments into aldehydes, and aldehydes interact with the lysine residues of apoB-100 to form new antigenic determinants, inducing the formation of autoantibodies [12, 13]. After complete oxidation, ox-LDL completely loses its affinity to LDLR and binds specifically to the scavenger receptors (SRs) [14].

With an extremely high level of circulating LDL, the antioxidants in the body are insufficient to maintain the antioxidant protection of nLDL. And nLDL oxidizes rapidly even without strong attack factors. Meanwhile, high levels of LDL, in turn, promote the binding of NO with hydrogen peroxide to produce peroxynitrite (ONOO-), which is a strong oxidant and constantly attacks endothelial cells, resulting in endothelial dysfunction [15, 16].

LDL glycosylation: LDL glycosylation is a nonenzymatic reaction, and the reaction rate depends on the level of glucose and the duration of exposure [17]. The 67th amino acid region of the N-terminus of apoB-100 is the main site of glycosylation and also the attachment site of LDLR. Glycosylation at this site reduces the affinity of LDL to LDLR, promotes the uptake of LDL by SRs on the surface of macrophages, and induces the formation of foam cells. Glycosylation of LDL enhances the susceptibility to further oxidation of LDL. In addition, the glycosylation process is accompanied by the production of free radicals, which often result in the simultaneous existence of LDL glycosylation and LDL oxidation [18]. This is among the causes of peripheral vascular disease in patients with advanced diabetes.

3.2 Changes in lipid content of LDL particles and secondary structure of apoB-100

The conformation of apoB-100 on the surface of LDL is more dependent on the physical and chemical state of the lipid core, which is linked to the shape and size of the LDL particles. Related experiments have found that the conformation of ApoB-100 is associated with the coverage of the interfacial area effect existing at the LDL particle-water solution interface. The coverage of the interfacial area effect increases with the decrease of LDL particle size, which can be regarded as an adaptive conformation change. This change is to avoid or minimize the contact between hydrophobic protein residues and aqueous solutions largely. It also ensures the stable existence of cholesterol in a hydrophilic environment and leads to successfully complete the targeted transport. However, this conformation regulation also causes more β-sheet domains to be parallel to the phospholipid monolayer of LDL, making it more vulnerable to adverse factors [19]. This may be one reason why sd-LDL with higher atherosclerosis is more easily oxidized.

The effect of oxidative modification on the secondary structure of LDL is mainly characterized by the destruction of β-sheet domain in the initial stage, resulting in the generation of disordered rings and corners. This change reduces the proportion of β-sheet in ApoB-100 and destroys the stability of LDL, but the alpha-helix ratio increases, triggering the self-repairing potential of apoB-100. At this point, the physical and chemical properties of LDL particles changed very little. When the oxidant continuously penetrates into the core, the physical state and accumulation mode of the lipid inside the hydrophobic core change, leading to the loss of the secondary structure of apoB-100, and the physical and chemical properties of LDL completely change [19].

In summary, compared to nLDL, mLDL has the following characteristics:

- the negative charge on the surface of particles is increased,
- the particle density is increased,
- due to the oxidation of lipids, polyunsaturated fatty acids decreased, and the contents of hemolytic lecithin, cholesterol oxide, and lipid hydroperoxide increased,
- ApoB100 is fragmented,
- reduced affinity with LDLR and specifically binds to scavenger receptors (SRs),
- acquired immunogenicity.

4. Modified LDL (mLDL) is involved in the formation of atherosclerotic lesions

Combined with the current known research results, the concept that the LDL-C level is critical to the occurrence and development of atherosclerotic lesions is too broad. It is better to say that the increase of the mLDL level caused by LDL qualitative change and quantitative change is the "turning point" of the entire event. Now, let us reexamine the role of mLDL in this process by combining the macro with the micro.

The physiological process of life can be divided into two levels: the basic physiological process and the complex regulatory network. Abnormal physiological processes are at the root of all diseases. According to the level of abnormality, we divide the disease into functional disorder and congenital defect.

4.1 Functional disorder: abnormal regulation of the network and damage to the ability to retain stability

The main purpose of the regulation network is to check and balance the basic physiological process and maintain the local and overall steady state. According to the understanding of atherosclerotic lesions, we regard the homeostasis maintenance here as the balance between injury factors and protective factors, and the balance can be broken for the following reasons:

4.1.1 Excessive damage factors

Poor lifestyle such as smoking and alcohol abuse as well as basic diseases such as hyperglycemia, insulin resistance, and hypertension damage vascular endothelial cells, causing oxidative stress on the blood vessel wall and producing a large number of free radicals (ROS and RNS) [11, 20]. There is a close interaction between LDL and vascular endothelial cells, followed by oxidative modification of LDL to form ox-LDL. Ox-LDL is most likely to remain in the intima-media than nLDL. And ox-LDL, which is stuck in the middle layer of the intima, affects the activity of endothelial cells (ECs) by changing the level of microRNAs and other epigenetic factors [21, 22]. It can induce the inflammatory activation of endothelial cells mediated by $\alpha5\beta1$ integrity and promote the production of vascular cell adhesion molecule-1 (VCAM-1). These cytokines attract monocytes to adhere to the designated location, promote the transendothelial transport of monocytes, enable them to accumulate in the endometrium, and differentiate into macrophages [23]. This process upregulates the expression of SRs on the surface of macrophages and promotes the uptake of ox-LDL by macrophages. At the same time, ox-LDL acts as an antigen to form an immune complex with autoantibodies, and then the immune complex is recognized and internalized by the Fcγ receptor on the surface of macrophages [24, 25]. These two mechanisms are associated with the formation of foam cells. LPC, the main phospholipid component of ox-LDL, plays a pro-inflammatory role by regulating the function of a series of immune cells (monocytes, macrophages, T cells, dendritic cells) and vascular cells (endothelial cells, vascular smooth muscle cells). It also acts as a ligand for specific G protein-coupled receptors, activating several atherogenic signal transduction pathways [26].

4.1.2 Abnormal protective factors

At present, the research is mainly about the protective effect of nitric oxide (NO)-based antioxidants and the innate immunity of mononuclear macrophages.

NO, as a signal transduction molecule with free radical properties, can maintain vascular diastolic tension and regulate lipid peroxidation reaction, which subsequently changes the expression of pro-inflammatory genes in endothelial cells. When NO synthesis is blocked, the balance between reactive oxygen species (ROS) released by stressed tissues and antioxidants is broken, and ox-LDL is produced in large quantity. Stress also induces endothelial dysfunction and permeability of endothelial changes. The main consequence of this series of changes is the accumulation of ox-LDL in the subendothelial area [11, 15, 16].

As the principal carrier of circulating cholesterol, LDL contains a certain amount of polyunsaturated fatty acid (PUFAs), which is very sensitive to oxidative modification and is prone to structural changes [10]. In the normal state, innate immunity is activated to clear the ox-LDL produced in the physiological process. When the immune system is extremely responsive, macrophages play an antigen-presenting role and stimulate the proliferation of helper T lymphocytes, and the helper T lymphocytes produce specific cytokines, realizing cross-dialog between different immune cell groups and initiating adaptive immunity. All the mechanisms work together to launch a sharp attack on ox-LDL, forming a chronic inflammatory process unique to atherosclerotic diseases [27].

The persistent inflammatory response stimulates smooth muscle cells to migrate to the intima-media, proliferate, and secrete large amounts of collagen, and the lesion enters the fibrous plaque stage. In this stage, intima thickening and remodeling first lead to increasing arterial wall stiffness and increased interfacial pressure in the environment where LDL is located, which promotes secondary structural changes of apoB-100 [28]. Secondly, the ability of regional oxygen diffusion was weakened, and the intermediate area of atherosclerotic lesions and intima-media showed ischemia and hypoxia. For the purpose of compensation, glycolysis becomes the primary mode of cell productivity in this region. This metabolic process produces lactic acid, which leads to extracellular space acidification. The state of regional low PH value is beneficial to the activation of macrophages and the upregulation of ox-LDL receptor (mainly LOX-1) expression in macrophages. The upregulated LOX-1 increases the lipid absorption of macrophages and promotes the formation of foam cells [29]. In addition, upregulated LOX-1 increases endothelial permeability and promotes the migration of ox-LDL to subendothelial space by reducing the expression of desmoglein-1 (DSG-1) and desmocollin-2 (DSC-2) [30, 31]. Under the stimulation of ox-LDL, the above process is repeated, the fiber cap becomes thickened, the declining foam cells form a necrotic lipid core, and the lesion enters the atherosclerotic plaque stage. When the compensatory arterial dilatation is unable to compensate for the stenosis caused by plaques protruding into the lumen, blood flow changes. The clinical manifestation is stable coronary syndrome. After that, if the disease continues to evolve, under the action of mechanical stimulation or inflammatory mediators, the fibrous cap becomes thinner, and local macrophages are activated, followed by focal necrosis, plaque rupture, content flow out, and thrombosis. The initiation of this process is the main cause of acute cardiovascular events.

Most of the atherosclerotic lesions caused by functional disorders progress slowly, and the clinical symptoms appear relatively late, which is mainly due to the existence of a series of compensation mechanisms, from the molecular level to the organ level, forming the natural defense line of the body. The breakthrough of the line of defense is characterized by micro decompensation, which corresponds to the progress of the disease at the macro level. The occurrence of cardiovascular events represents the loss of the ability to maintain stability, and treatment such as stenting can quickly relieve symptoms. However, this does not solve problems fundamentally, and it is necessary to effectively control the concentration of LDL particles having atherogenic properties. Combined with clinical experience, controlling the level of circulating LDL-C is the only effective method at present. Fortunately, the basic physiological processes of these patients are complete, and their regulatory network still exists. The focus of treatment is on using drugs to help the body to build a new balance. On the basis of a healthy lifestyle, adequate statin administration is effective in controlling LDL-C levels. For patients with moderate- or high-dose statin intolerance, the use of statins can be reduced by combining with ezetimibe or proprotein convertase subtilisin/kexin type 9 (PCSK9) inhibitors, which can delay or even prevent the progression of the disease and largely avoid cardiovascular endpoint events [32].

4.2 Congenital defect: familial hypercholesterolemia (FH)

Familial hypercholesterolemia is an autosomal dominant genetic disease characterized by high plasma levels of low-density lipoprotein cholesterol (LDL-C) and premature coronary heart disease, mainly caused by mutations in low-density lipoprotein receptor (LDLR), apolipoprotein B (APOB), and proprotein convertase subtilisin/kexin type 9 genes. The first half of this article has introduced how extremely high levels of LDL-C can initiate LDL modification and attack the endothelium, which will not be described here. The high level of LDL-C in patients with FH since childhood is rooted in the damage to the basic physiological processes, which are genetically determined and inherent defects. Moreover, since the expression products of related genes are widely distributed under normal conditions, gene abnormalities also lead to regulatory network defects with compensation function damage, and the disease progresses rapidly in a short time. The traditional lipid-lowering method based on statins acts on the intracellular cholesterol synthesis process, which depends on the integrity of the basic physiological process, so it cannot effectively control the LDL-C level of FH patients. The key point of FH treatment is to make up for the defects in the basic physiological process and ensure survival. The discovery of PCSK9 inhibitors is a boon to such patients. PCSK9 combines with LDLR to promote LDLR degradation and interrupt the recycling of LDLR. Combined with statins, it can effectively control the LDL-C level of most FH patients. The reason why most but not all FH patients benefit is that this precision also leads to the poor treatment effect of PCSK9 inhibitor on LDLR-deficient FH homozygous [32–36].

5. Conclusions

LDL participates in the whole process of the formation and development of atherosclerotic lesions in the form of mLDL until plaque rupture, thrombosis, and cardiovascular events occur. The understanding of the characteristics of atherosclerosis caused by LDL should be built on the understanding of individual context. At present, a single LDL-C level has been impossible to accurately predict the progression of atherosclerotic lesions. The application value of mLDL, sd-LDL, etc. in the clinical practice needs to be further explored in the future.

Conflict of interest

To the best of our knowledge, the named authors have no conflict of interest, financially or otherwise.

Author details

Jie Lin[1,2]

1 Beijing Institute of Heart, Lung and Blood Vessel Diseases, Beijing Anzhen Hospital, Capital Medical University, Beijing, China

2 Department of Atherosclerosis, Beijing Anzhen Hospital, Capital Medical University, Beijing, China

*Address all correspondence to: jielinaz@ccmu.edu.cn

References

[1] Steinberg D. Atherogenesis in perspective: Hypercholesterolemia and inflammation as partners in crime. Nature Medicine. 2002;**8**:1211-1217. DOI: 10.1038/nm1102-1211

[2] Yabg C-Y, Gu Z-W, Weng S-A, Kim TW, Chen S-H, Pownall HJ, et al. Structure of apolipoprotein B-100 of human low-density lipoproteins. Arteriosclerosis. 1989;**9**:96-108. DOI: 10.1161/01.ATV.9.1.96

[3] Shelness GS, Sellers JA. Very-low-density lipoprotein assembly and secretion. Current Opinion in Lipidology. 2001;**12**:151-157

[4] Hirayama S, Miida T. Small dense LDL: An emerging risk factor for cardiovascular disease. Clinica Chimica Acta. 2012;**414**:215-224. DOI: 10.1016/j.cca.2012.09.010

[5] Goedeke L, Fernández-Hernando C. Regulation of cholesterol homeostasis. Cellular and Molecular Life Sciences. 2012;**69**:915-930. DOI: 10.1007/s00018-011-0857-5

[6] Jasiñska M, Owczarek J, Orszulak-Michalak D. Statins: A new insight into their mechanisms of action and consequent pleiotropic effects. Pharmacological Reports. 2007;**59**:483

[7] Sakakura Y, Shimano H, Sone H, Takahashi A, Inoue K, Toyoshima H, et al. Sterol regulatory element-binding proteins induce an entire pathway of cholesterol synthesis. Biochemical and Biophysical Research Communications. 2001;**286**:176-183. DOI: 10.1006/bbrc.2001.5375

[8] Schulz C, Massberg S. Atherosclerosis—Multiple pathways to lesional macrophages. Science Translational Medicine. 2014;**6**:239ps2. DOI: 10.1126/scitranslmed.3008922

[9] Kruth HS. Receptor-independent fluid-phase pinocytosis mechanisms for induction of foam cell formation with native LDL particles. Current Opinion in Lipidology. 2011;**22**:386. DOI: 10.1097/MOL.0b013e32834adadb

[10] Yoshida H, Kisugi R. Mechanisms of LDL oxidation. Clinica Chimica Acta. 2010;**411**:1875-1882. DOI: 10.1016/j.cca.2010.08.038

[11] Schleicher E, Friess U. Oxidative stress, AGE, and atherosclerosis. Kidney International Supplement. 2007;**72**:S17-S26. DOI: 10.1038/sj.ki.5002382

[12] Tsimikas S, Brilakis ES, Miller ER, McConnell JP, Lennon RJ, Kornman KS, et al. Oxidized phospholipids, Lp (a) lipoprotein, and coronary artery disease. The New England Journal of Medicine. 2005;**353**:46-57. DOI: 10.1056/NEJMoa043175

[13] Boullier A, Li Y, Quehenberger O, et al. Minimally oxidized LDL offsets the apoptotic effects of extensively oxidized LDL and free cholesterol in macrophages. Arteriosclerosis, Thrombosis, and Vascular Biology. 2006;**26**:1169-1176. DOI: 10.1161/01.ATV.0000210279.97308.9a

[14] Syväranta S, Alanne-Kinnunen M, Öörni K, Oksjoki R, Kupari M, Kovanen PT, et al. Potential pathological roles for oxidized low-density lipoprotein and scavenger receptors SR-AI, CD36, and LOX-1 in aortic valve stenosis. Atherosclerosis. 2014;**235**:398-407. DOI: 10.1016/j.atherosclerosis.2014.05.933

[15] Seo H, Oh H, Park H, Park M, Jang Y, Lee M. Contribution of dietary intakes of antioxidants to homocysteine-induced low density lipoprotein (LDL) oxidation in atherosclerotic patients. Yonsei Medical Journal. 2010;**51**:526-533. DOI: 10.3349/ymj.2010.51.4.526

[16] Forstermann U, Sessa WC. Nitric oxide synthases: Regulation and function. European Heart Journal. 2012;**33**:829-837. DOI: 10.1093/eurheartj/ehr304

[17] Sasaki J, Cottam GL. Glycosylation of LDL decreases its ability to interact with high-affinity receptors of human fibroblasts in vitro and decreases its clearance from rabbit plasma in vivo. Biochimica et Biophysica Acta. 1982;**713**:199-207

[18] Soran H, Durrington PN. Susceptibility of LDL and its subfractions to glycation. Current Opinion in Lipidology. 2011;**22**:254-261. DOI: 10.1097/MOL.0b013e328348a43f

[19] B Bore'n J, Williams KJ. The central role of arterial retention of cholesterol-rich apolipoprotein-B-containing lipoproteins in the pathogenesis of atherosclerosis: A triumph of simplicity. Current Opinion in Lipidology. 2016;**27**:473-483. DOI: 10.1097/MOL.0000000000000330

[20] Alique M, Luna C, Carracedo J, Ramírez R. LDL biochemical modifications: A link between atherosclerosis and aging. Food & Nutrition Research. 2015;**59**:29240. DOI: 10.3402/fnr.v59.29240

[21] Itabe H, Obama T, Kato R. The dynamics of oxidized LDL during atherogenesis. J Lipid. 2011;**2011**:1-9. DOI: 10.1155/2011/418313

[22] Zhang E, Wu Y. MicroRNAs: Important modulators of oxLDL-mediated signaling in atherosclerosis. Journal of Atherosclerosis and Thrombosis. 2013;**20**:215. DOI: 10.5551/jat.15180

[23] Yurdagul A Jr, Green J, Albert P, McInnis MC, Mazar AP, Orr AW. α5β1 integrin signaling mediates oxidized low-density lipoprotein-induced inflammation and early atherosclerosis. Arteriosclerosis, Thrombosis, and Vascular Biology. 2014;**34**:1362-1373. DOI: 10.1161/ATVBAHA.114.303863

[24] Orekhov AN, Bobryshev YV, Sobenin IA, Melnichenko AA, Chistiakov DA. Modified low density lipoprotein and lipoprotein-containing circulating immune complexes as diagnostic and prognostic biomarkers of atherosclerosis and type 1 diabetes macrovascular disease. International Journal of Molecular Sciences. 2014;**15**:12807-12841. DOI: 10.3390/ijms150712807

[25] Bobryshev YV, Ivanova EA, Chistiakov DA, Nikiforov NG, Orekhov AN. Macrophages and their role in atherosclerosis: Pathophysiology and transcriptome analysis. BioMed Research International. 2016;**2016**:9582430. DOI: 10.1155/2016/9582430

[26] Gistera A, Hansson GK. The immunology of atherosclerosis. Nature Reviews Nephrology. 2017;**13**(6): 368-380. DOI: 10.1038/nrneph.2017.51

[27] Conti P, Shaik-Dasthagirisaeb Y. Atherosclerosis: A chronic inflammatory disease mediated by mast cells. Central European Journal of Immunology. 2015;**40**:380. DOI: 10.5114/ceji.2015.54603

[28] Ollila S, Lamberg A, Lehtivaara M, Koivuniemi A, Sysi-Aho Vattulainen I. Interfacial tension and surface pressure of high-density lipoprotein, low-density lipoprotein, and related lipid droplets. Biophysical Journal. 2012;**103**:1236-1244. DOI: 10.1016/j.bpj.2012.08.023

[29] Chistiakov DA, Orekhov AN, Bobryshev YV. LOX-1-mediated effects on vascular cells in atherosclerosis. Cellular Physiology and Biochemistry. 2016;**38**:1851-1859. DOI: 10.1159/000443123

[30] Li YB, Zhang QH, Chen Z, He ZJ, Yi GH. Oxidized low-density

lipoprotein attenuated desmoglein 1 and desmocollin 2 expression via LOX-1/Ca(2+)/PKC-beta signal in human umbilical vein endothelial cells. Biochemical and Biophysical Research Communications. 2015;**468**:380-386. DOI: 10.1016/j.bbrc.2015.10.079

[31] Yang HY, Bian YF, Zhang HP, Gao F, Xiao CS, Liang B, et al. LOX-1 is implicated in oxidized low-density lipoprotein-induced oxidative stress of macrophages in atherosclerosis. Molecular Medicine Reports. 2015;**12**:5335-5341. DOI: 10.3892/mmr.2015.4066

[32] Grundy SM, Stone NJ, Bailey AL, Beam C, Birtcher KK, Blumenthal RS, et al. AHA/ACC/AACVPR/AAPA/ABC/ACPM/ADA/AGS/APhA/ASPC/NLA/PCNA guideline on the management of blood cholesterol: A report of the american college of cardiology/american heart association task force on clinical practice guidelines. Journal of the American College of Cardiology. 2018. DOI: 10.1016/j.jacc.2018.11.004

[33] Defesche JC, Gidding SS, Harada-Shiba M, Hegele RA, Santos RD, Wierzbicki AS. Familial hypercholesterolaemia. Nature Reviews Disease Primers. 2017;**3**:17093. DOI: 10.1038/nrdp.2017.93

[34] McKenney JM. Understanding PCSK9 and anti-PCSK9 therapies. Journal of Clinical Lipidology. 2015;**9**:170-186. DOI: 10.1016/j.jacl.2015.01.001

[35] Sabatine MS, Giugliano RP, Keech AC, Honarpour N, Wiviott SD, Murphy SA, et al. Evolocumab and clinical outcomes in patients with cardiovascular disease. The New England Journal of Medicine, 1722. 2017;**376**:1713. DOI: 10.1056/NEJMoa1615664

[36] Catapano AL, Graham I, De Backer G, Wiklund O, Chapman MJ, Drexel H, et al. ESC/EAS guidelines for the management of dyslipidaemias. European Heart Journal. 2016;**37**:2999-3058. DOI: 10.1093/eurheartj/ehw272

Section 4

Functional Properties and Therapeutic Effects

Chapter 10

Therapeutic Properties and Anti-Lipidemic Activity of *Cordyceps sinensis*

Kanchana Samarasinghe and Viduranga Y. Waisundara

Abstract

Cordyceps sinensis is an entomophagous medicinal mushroom, which is mainly endemic to the Tibetan Plateau including the adjoining high altitude areas. The fungus attacks many lepidopteran larvae caterpillars and mummifies it. The larvae along with the mummified insect are highly valued for their medicinal property. *Cordyceps sinensis* is one of the most efficient and expensive medicinal mushrooms in traditional medicinal systems such as those in China and Tibet, having multiple medicinal and pharmacological properties. It has been used to treat respiratory and immune disorders; pulmonary diseases; renal, liver, and cardiovascular diseases; hyposexuality; and hyperlipidemia. The extract of this mushroom and its bioactive compounds are noteworthy for their ability to regulate lipid metabolism and thereby exhibit anti-lipidemic activity. Cordycepin in particular, which is a bioactive compound existing in *Cordyceps sinensis*, has been identified as one of the primary compounds of interest in this aspect. Despite the global and scientific interest exerted toward *Cordyceps sinensis*, it appears to be of utmost importance that the price and other market factors owing to the rarity of this herb are managed through artificial means of synthesis.

Keywords: caterpillar, cholesterol, *Cordyceps sinensis*, ethnopharmacology, triglycerides

1. Introduction

Edible mushrooms have been widely used around the world for their high nutritional and therapeutic value as well as in the form of a functional food. Additionally, they have been highly appreciated for their medicinal and therapeutic applications from a traditional perspective as well. This is owing to the availability of a vast variety of bioactive compounds such as polysaccharides, proteoglycans, terpenoids, phenolic compounds, steroids, and lectins, which are available in copious amounts in all its components [1]. *Cordyceps sinensis* is a highly valued fungus in this aspect which thrives at very high altitudes, especially in meadows on the Himalayan Plateau [2]. As a medicinal fungus of a long and illustrious history, *Cordyceps sinensis* is an ascomycetes fungus [3]. An image of this fungus is shown in **Figure 1**. Although it is not actually a mushroom in the taxonomic sense, it has been regarded as a medicinal mushroom throughout history. The name *Cordyceps* comes from the Latin words "cord" and "ceps," meaning "club" and "head," respectively [3]. The term "*Cordyceps*"

Figure 1.
Cordyceps sinensis *growing on a caterpillar.*

usually refer to the specific species *Cordyceps sinensis*, but there are also many other species that come under the genus *Cordyceps*.

The medicinal value of this fungus was first recognized more than 2000 years ago in China and the recognition passed onto the rest of Asia [3]. Early records of *Cordyceps sinensis* as a medicine is as old as the Qing Dynasty in China and this information has been mentioned in *Ben Cao Cong Xin* (New Compilation of Materia Medica) written by Wu Yiluo in around 1757 [4]. However, knowledge of this fungus reached Western scientific audiences only in 1726, when it was introduced at a scientific meeting in Paris. The first specimens were carried back to France by a Jesuit priest, who chronicled his experiences with *Cordyceps sinensis* during his stay at the Chinese Emperor's court [4].

Various names have been used to identify *Cordyceps sinensis* in various languages and cultures, where the most common forms of reference are shown in **Table 1**. In historical and general usage, the term "*Cordyceps*" refers specifically to the species *Cordyceps sinensis* (Berk.) Sacc (syn *Ophiocordyceps sinensis* (Berk.) G.H. Sung, J.M. Sung, Hywel-Jones & Spatafora.), which is the most widely used *Cordyceps* species. Berkeley, the British mycologist, first described this fungus in 1843 as *Sphaeria sinensis* Berk., while later on in 1878, Saccardo renamed it as *Cordyceps sinensis* [5]. The currently accepted scientific name is *Cordyceps sinensis* (Berk.) Sacc.

The taxonomic classification of *Cordyceps sinensis* is shown in **Table 2**.

Given the difficulties involved in harvesting, *Cordyceps sinensis* has been highly priced. It is basically the costliest fungus or medicinal mushroom throughout the world, hence known as "fungal gold" or "soft gold" [6]. The price of *Cordyceps sinensis* has been shown to grow approximately 20% annually and to date is probably 1000% the price which existed 10 years ago [7]. Pricing is achieved by inspecting the size and firmness of the larval host, which is often tested by squeezing it between two fingers; the stiffer it is, the higher the price [8]. The color is also observed when pricing, where a saturated yellowish-brown color is preferred than the pale ones [9]. Other physical characteristics taken into account when assessing quality are size, weight, smell, taste, and robustness [9]. The texture of good-quality specimens is relatively fleshy while the taste is bitter. Despite its cost and rarity, the unprecedented medicinal applications of *Cordyceps sinensis* have made it a highly valued indispensable component of the traditional Chinese and Tibetan medicine.

Language	Vernacular name
Chinese	Dong chong xia cao
Tibetan	Yartsa gunbu
Hindi	Keeda jadi
Nepali language	Yarcha gumba

Table 1.
Vernacular names used for Cordyceps sinensis.

Classification	Category
Kingdom	Fungi
Division	Ascomycota
Class	Sordariomycetes
Order	Hypocreales
Family	Ophiocordycipitaceae
Genus	*Cordyceps*
Species	*sinensis*

Table 2.
Taxonomic classification of Cordyceps sinensis.

Although *Cordyceps sinensis* has been used for a wide range of medicinal applications, it is only apt that its anti-lipidemic properties are given due attention. For this purpose, it is important that other aspects of this fungus are compiled herein prior to highlighting its effects against lipid-related disorders. Thus, this chapter has been structured in a manner which provides a holistic overview of *Cordyceps sinensis*, where any reader can refer to this write-up and obtain an overall understanding of this medicine prior to delving into its anti-lipidemic effects.

2. The life cycle of *Cordyceps sinensis*

Cordyceps sinensis fungus begins off its wondrous life cycle with the combination of an underground caterpillar (insect)—mostly Lepidoptera larvae, which generally lives about 15 cm underground. One of the special factors, which affect the initial point to begin the life of *Cordyceps sinensis* and which enable its identification from other fungi, is the low temperature that is required for its life cycle. This particular temperature is believed to be below approximately 21°C [6]. The life of the caterpillar fungus which begins under these conditions carries out a slow growing procedure thereafter.

The underground caterpillars which survive by feeding on the underground roots start to shed off their skin in the later part of summer and autumn seasons. Therefore, their skin becomes more vulnerable to fungal spores. Fungal spores, which begin to interact with the chemicals on the skin of these caterpillars, release a fungal mycelium, which then starts infecting the caterpillar to its death. With its death, a formation of a stalk-like structure of the mushroom grows out of the upper part of the head of the dead caterpillar and that particular structure emerges above to the ground surface in the season of spring. The infected larvae after its death become a rigid structure due to the production of the fungal sclerotia, which could pause the germination for a while to produce spores. Right after the end of the winter season, the formation of the mushroom body continues, and along with that, the perithecial stroma grows upward in order to emerge above the ground soil, thereby concluding its life cycle.

3. Ethnopharmacological uses of *Cordyceps sinensis*

Cordyceps sinensis is presently regarded as a highly priced fungus at present due to its precious medicinal value especially in the field of traditional Chinese medicine (TCM). This is primarily for the treatment of kidney disease, fatigue, sexual dysfunctions, diabetes, and cancers, which are highly prevalent nowadays. *Cordyceps sinensis* is also regarded as a Bhutanese indigenous medicine other than TCM [10]. It is said that the *Cordyceps* organism was discovered by yak herders in the Himalayas of ancient Tibet and Nepal who, recognizing the ardent behavior of their animals after grazing on *Cordyceps* at high altitudes in the spring, sought the causal agent for their own usage and medical applications [3]. The capless mushroom was then incorporated into TCM ever since starting with kidney, lung, and heart ailments, male and female sexual dysfunction, fatigue, cancer, hiccups, and serious injury, to relieve pain, and the symptoms of tuberculosis and hemorrhoids, to restore general health and appetite and to promote longevity [3].

A tabulated description of the health benefits of *Cordyceps sinensis* in their traditional applications against disease conditions and scientific studies, which have verified its potency, is listed in **Table 3**.

In modern times, *Cordyceps sinensis* appears to be actively used by the elderly people and athletes as well to boost their energy [16]. Experiments show that although *Cordyceps sinensis* could be used as an energy booster, only a small portion of extra energy could be achieved due to the increase of cellular adenosine triphosphate (ATP) when it is taken as an herbal supplement. The energy output and its oxygen capacity help with treating cold intolerance and decreasing fatigue. It is also noteworthy that compared with other medicinal herbs, *Cordyceps sinensis* has

Disease	Description and effect	References
Cancer	*Cordyceps sinensis* has been able to demonstrate the inhibition of colorectal cancer cells	[11]
Liver fibrosis	Experiments have shown that ergosterol in cultured mycelium *Cordyceps sinensis* has an anti-inflammatory as well as an antifibrotic effect *Cordyceps sinensis* culture had helped in the inhibition of hepatocytes and hepatic stellate cell activation	[12]
Respiratory	*Cordyceps sinensis* is used to ease various kinds of diseases associated with the lungs and the respiratory system such as phlegm, cough, and various other diseases associated with the bronchial and asthma	[13]
Kidney	Studies showed that consumption of *Cordyceps sinensis* stem helps in increasing 17-ketosteroid and 17-hydroxycorticosteroids levels in the human body resulting in the prevention of kidney disease. In the meantime, the consumption of *Cordyceps sinensis* by patients with renal failure has helped them decrease their blood pressure by 15%	[14]
AIDS/HIV	*Cordyceps sinensis* has been used to treat HIV infections in West Africa	[15]

Table 3.
Disease conditions which have ethnopharmacologically utilized Cordyceps sinensis *and studies elucidating their scientific verifications.*

a very low toxicity and side effects over its benefits. Studies show that health issues like nausea, diarrhea, and diseases such as dry mouth have not been experienced by the patients who have taken *Cordyceps sinensis* [5].

4. Anti-lipidemic effects of *Cordyceps sinensis*

Age-related diseases are rising as a common issue in the present era due to the evolutionary changes that take place in the dietary patterns and lifestyle changes of humans. Studies which are being carried out to mitigate these rising health issues show methodological approaches to prevent and mitigate them by utilizing the medicinal values of traditional medicines across the globe.

Complications related to lipid metabolism have been identified as an age-related disease condition [17]. As a detrimental health effect related to lipid metabolism, the incidence of obesity has also been increasing steadily in the developed and developing countries worldwide. Analysis of the global burden of obesity revealed that there were 396 million adults with obesity in 2005 and that the expected number is projected to be 573 million individuals in 2030, without the application of an adjustment for secular trends [18]. Excessive fat accumulation that increases the risk of adverse health effects is one of the definitions of obesity [19]. It is a condition that is implicated as a risk factor for various diseases such as hypertension, coronary heart disease, and type II diabetes [20].

Hyperlipidemia is another age-related disease condition. This occurs due to the presence of too much low-density lipoprotein (LDL) in the blood, while it threatens the health of the circulatory system, risking the blockage of arteries with the deposition of fats and lipids. In addition, hyperlipidemia acts as the root cause of diabetes and functional depression in organs such as the liver, heart, and kidney [3, 4].

Cordyceps sinensis has been used in modern times for the treatment of lipid-related disorders. A summary of the anti-lipidemic effects of *Cordyceps sinensis* is shown in **Figure 2**. The fruiting body part of *Cordyceps sinensis* contains a composition of nucleosides, exopolysaccharides (EPS), proteins, and sterols. These bioactive components which play a significant role in treating diseases are also

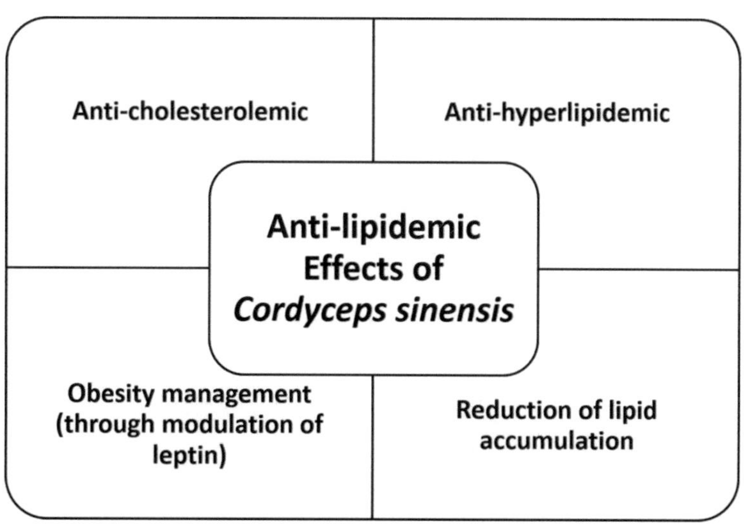

Figure 2.
Schematic depicting the anti-lipidemic effects of Cordyceps sinensis.

used in treating diseases like hyperlipidemia due to their anti-lipidemic and anti-cholesterolemic ability which have demonstrated to lower cholesterol. Treatment of mice with the extract of *Cordyceps sinensis* mycelia has demonstrated to significantly reduce cholesterol and triglyceride concentrations and the synthesis of very-low-density lipoprotein (VLDL, an LDL precursor). Therefore, given these functions, a useful mechanism of action present in *Cordyceps sinensis* for obesity prevention could be considered as the inhibition of preadipocyte differentiation.

In human as well as animal studies, administration of *Cordyceps sinensis* has been associated with a reduction in cholesterol and triglyceride and an increase in the ratio of high-density lipoprotein (HDL) to LDL cholesterol [17]. Whether the causative mechanism for this lipid-balancing effect is through blood sugar stabilization, enhancing liver function, or any other as hitherto unknown cause remains to be seen.

Interestingly, in addition to their mycelium effects obtained from solid-state fermentation, *Cordyceps sinensis* biomass obtained from submerged fermentation has been recognized as an effective agent in lipid metabolism [21]. This hypothesis was supported by some evidence which surfaced about glucan isolated from the *Cordyceps sinensis*. Current interest in the effect of glucans on lipid metabolism mainly is centered on the possibility that the glucans could entrap bile acids in the intestine and thus increase bile acid exclusion in the feces. The results by Freire Dos Santos et al. [22] suggest that *Cordyceps sinensis* biomass supplementation in high-fat diet (HFD)-fed rats for 4 months normalizes the blood lipid and the low testosterone levels induced by HFD. Despite the outcomes of the study, it has to be borne in mind nevertheless that *Cordyceps sinensis* biomass supplementation cannot replace the use of currently available drug regiments for lipid reduction but can easily complement them. It may also enable the use of lower doses of therapeutic drugs, thereby decreasing the risk of dose-related side effects. Furthermore, Freire Dos Santos et al. demonstrated that a long-term intake of HFD caused a significant liver damage which has been reverted by *Cordyceps sinensis* biomass supplementation and, this in turn, normalized decreasing testosterone levels observed in HFD-fed rats.

Leptin, a newly discovered hormonal product of the obesity (ob) gene, is expressed by white adipose tissue. It has been implicated in the regulation of body weight, glucose metabolism, and fertility. Leptin deficiency produces severe obesity, insulin resistance, and impaired glucose tolerance in ob/ob mice, and also, congenital leptin deficiency in humans leads to hyperphagia and marked obesity. Choi suggested that *Cordyceps sinensis* originated EPS are excellent candidates to develop as functional food additives or therapeutic agents for anti-obesity and antidiabetic purposes through improvement in glucose homeostasis and preservation of insulin reserves [23].

Tiamyom et al. [17] investigated the effects of *Cordyceps sinensis* extract and *Gymnema inodorum* extract, used alone and combined, on antiadipogenesis in 3T3-L1 cells. Results from this study indicated that both herbs and their combination had inhibitory effects on lipid accumulation in the adipocytes. The pancreatic lipase assay results indicated that *Cordyceps sinensis* extract inhibited the pancreatic lipase activity in a dose-dependent manner. These results suggested that the use of *Cordyceps sinensis* extract alone and in combination with *Gymnema inodorum* extract may be efficacious as a complementary therapy for hyperlipidemia and obesity management.

5. Bioactive compounds of interest

Cordyceps sinensis appears to contain all of the essential amino acids, vitamins E and K, as well as the water-soluble vitamins B1, B2, and B12. In addition, it contains many sugars, including mono-, di-, and oligosaccharides, and many complex polysaccharides, proteins, sterols, nucleosides, and trace elements. Characterization of cordycepin (**Figure 2**) and 2′-deoxyadenosine was reported by nuclear magnetic resonance (NMR) and infrared (IR) spectroscopy in an extract of *Cordyceps sinensis* [24, 25]. Cordycepin has been identified as one of the compounds involved in the anti-lipidemic effects of *Cordyceps sinensis*, and thus, within the context of this chapter needs to be highlighted as the primary bioactive compound involved in lipid metabolism and regulation. Cordycepin inhibits adipocyte differentiation and accumulation of lipid in mature adipocytes [26]. As cordycepin blocks both adipocyte differentiation and lipid accumulation, it has the potential to be an effective therapeutic agent for obesity and obesity-related disorders [26].

Nucleosides have been found in all species of *Cordyceps*, including uridine. A number of polysaccharides and other sugar derivatives, such as cordycepic acid (D-mannitol), have been also identified in *Cordyceps sinensis* [17].

In recent years, much interest has been focused on EPS produced by *Cordyceps sinensis* and its other varieties due to their various biological and pharmacological

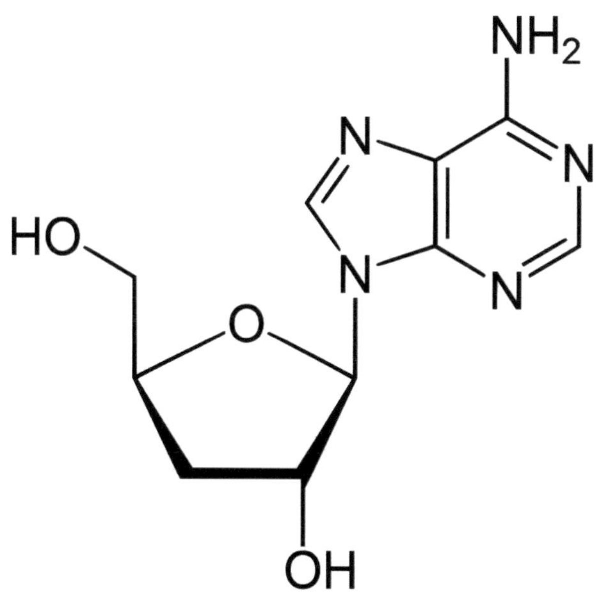

Figure 3.
Chemical structure of cordycepin, which has been isolated from Cordyceps sinensis.

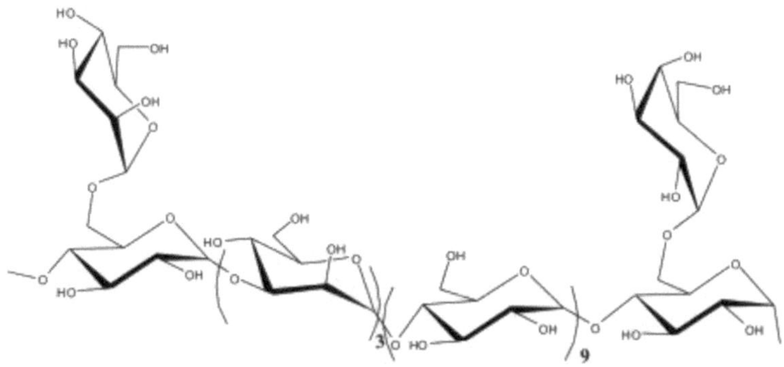

Figure 4.
Chemical structure of CPS-2, which was isolated by Wang et al. [27]. The structure was found to be mostly of α-(1 → 4)-d-glucose and α-(1 → 3)-D-mannose, branched with α-(1 → 4,6)-d-glucose every 12 residues on average.

activities. These polysaccharides are effective in regulating blood sugar and appear to have antimetastatic and antitumor effects [27]. *Cordyceps sinensis* contains proteins, peptides, polyamines, and all essential amino acids as well [28]. Sterol-type compounds have also been found in *Cordyceps sinensis*, including ergosterol, delta-3 ergosterol, ergosterol peroxide, 3-sitosterol, daucosterol, and campesterol [27, 28]. Twenty-eight saturated and unsaturated fatty acids and their derivatives have been isolated from *Cordyceps sinensis* [14]. Polar compounds of *Cordyceps sinensis* extracts include many compounds of alcohols and aldehydes; polycyclic aromatic hydrocarbons produced by *Cordyceps sinensis* as secondary metabolites were also reported. A water-soluble polysaccharide (CPS-2), isolated from the cultured *Cordyceps sinensis*, was obtained by hot water extraction, anion exchange, and gel permeation chromatography by Wang et al. [14]. The changes in blood urea nitrogen and serum creatinine in this particular study revealed that CPS-2 could significantly relieve renal failure caused by fulgerizing kidney. An image of this polysaccharide is shown in **Figure 3**. It is believed that there are several other polysaccharides as such with therapeutic potential in terms of lipid regulation in *Cordyceps sinensis* (**Figure 4**).

6. Other health benefits of *Cordyceps sinensis*

Cordyceps sinensis appear to have a variety of other health benefits which do not necessarily stem from its traditional medicinal applications. Some of them are shown in **Table 4**.

Disease condition	Part of fungus being used	Effect	References
Diabetes	Fruiting body and mycelia	Antiproliferative activity toward breast cancer cells (MCF-7)	[29]
Oxidative damage	Fruiting bodies	Antioxidant	[30]
Immune injury	Fruiting bodies and mycelia	Immune modulating	[31]
Melanogenesis	Mycelia	Anti-tyrosinase activity	[32]

Table 4.
Scientifically verified beneficial properties of Cordyceps sinensis, *which are not part of its ethnopharmacological uses.*

7. Conclusions

Although there is a higher and more favorable medicinal value in traditional medicines such as *Cordyceps sinensis*, the limitations should be considered as well in order to reveal its undiscovered potential of treating lipid-related diseases. However, as the availability of *Cordyceps sinensis* is gradually being decreased in the West, care should be taken when meeting the consumer demands, especially when it comes to the cultivation and growth of the herb. Moreover, as the natural *Cordyceps sinensis* is becoming rare, methods are in the process of being developed for the artificial production of this fungus with the same chemical constituents or even better. The world is now awakening to the importance of this fungus, and the price factor appears to play an important role in the marketplace. A lot of research is still pending on this fungus especially from the medical side. With the present research nevertheless, it is noteworthy that *Cordyceps sinensis* has been identified

DOI: http://dx.doi.org/10.5772/intechopen.92616

as a traditional medicine which is able to regulate lipid levels and demonstrated anti-lipidemic activity. Cordycepin appears to be an important bioactive compound in this aspect, and it is without a doubt that further research will be able to elucidate more bioactive compounds existing in *Cordyceps sinensis* which are able to collectively involve in lipid metabolism and regulation. The potency of *Cordyceps sinensis* would highly depend on its chemical constituents, viz., the cordycepin and polysaccharides that make up the fruiting body, mycelium, or spores. However, further experimental, epidemiological, and clinical studies are needed to identify other molecular targets, resolve the relationships between *Cordyceps sinensis* intake and anti-lipidemic activity, and explore the optimum dosing, efficacy, and safety alone and in combination with existing anti-cholesterolemic and anti-triglyceridemic pharmaceutical therapies.

Acknowledgements

The authors wish to thank the Australian College of Business & Technology, Kandy Campus, for the facilities provided to put this write-up together.

Conflict of interest

The authors declare no conflicts of interest, financial, or otherwise.

Author details

Kanchana Samarasinghe and Viduranga Y. Waisundara*
Australian College of Business and Technology—Kandy Campus, Kandy, Sri Lanka

*Address all correspondence to: viduranga@gmail.com

References

[1] Mehra A, Zaidi KU, Mani A, Thawani V. The health benefits of *Cordyceps militaris*—A review. KAVAKA. 2017;**48**(1):27-32

[2] Chang ST, Miles PG. Mushrooms: Cultivation, Nutritional Value, Medicinal Effect and Environmental Impact. 1st ed. Boca Raton, FL, USA: CRC Press; 2004

[3] Joshi RK. Phytochemical and medicinal aspect of *Cordyceps sinensis* (Berk.): A review. Journal of Medicinal Plants Studies. 2016;**4**(1):65-67

[4] Hobbs CH. Medicinal Mushrooms: An Exploration of Tradition, Healing, and Culture. Vol. 251. Santa Cruz, CA: Botanica Press; 1995

[5] Devkota S. Yarsagumba [*Cordyceps sinensis* (Berk.) Sacc.]: Traditional utilization in Dolpa District, Western Nepal. Our Nature. 2006;**4**:48-52

[6] Panicker S. *Cordyceps* the fungal gold—A review. Advances in Research. 2017;**11**(3):1-16

[7] Zhou XW, Li LJ, Tian EW. Advances in research of the artificial cultivation of *Ophiocordyceps sinensis* in China. Critical Reviews in Biotechnology. 2014;**34**:233-243

[8] Jordan JL, Sullivan AM, Lee TDG. Immune activation by a sterile aqueous extract of *Cordyceps sinensis* mechanism of action. Immunopharmacology and Immunotoxicology. 2008;**30**:53-70

[9] Singh RP, Pachauri V, Verma RC, Mishra KK. Caterpillar fungus (*Cordyceps sinensis*)—A review. Journal of Eco-friendly Agriculture. 2008;**3**:1-15

[10] Panda AK, Swain KC. Traditional uses and medicinal potential of *Cordyceps sinensis* of Sikkim. Journal of Ayurveda and Integrative Medicine. 2011. DOI: 10.4103/0975-9476.78183

[11] Rao YK, Fang SH, Tzeng YM. Evaluation of the anti-inflammatory and anti-proliferation tumoral cells activities of *Antrodia camphorata*, *Cordyceps Sinensis*, and *Cinnamomum osmophloeum* bark extracts. Journal of Ethnopharmacology. 2007;**114**(1):78-85

[12] Yamaguchi Y, Kagota S, Nakamura K, Shinozuka K, Kunitomo M. Antioxidant activity of the extracts from fruiting bodies of cultured *Cordyceps sinensis*. Phytotherapy Research. 2000;**14**(8):647-649

[13] Wang N, Li J, Huang X, Chen W, Chen Y. Herbal medicine *Cordyceps sinensis* improves health-related quality of life in moderate-to-severe asthma. Evidence-Based Complementary and Alternative Medicine. 2016. DOI: 10.1155/2016/6134593

[14] Wang Y, Yin H, Lv X, Wang Y, Gao H, Wang M. Protection of chronic renal failure by a polysaccharide from *Cordyceps sinensis*. Fitoterapia. 2010;**81**(5):397-402

[15] Adotey G, Quarcoo A, Holiday J, Saaka B. Effect of Immunomodulating and antiviral agent of medicinal mushrooms (immune assist 24/7TM) on CD4+ T-lymphocyte counts of HIV-infected patients. International Journal of Medicinal Mushrooms. 2011;**13**(2):109-113

[16] Zhu JS, Rippe J. CordyMax enhances aerobic capability, endurance performance, and exercise metabolism in healthy, mid-age to elderly sedentary humans. In: Proceedings of the American Physiological Society's (APS) Annual Scientific Conference, Experimental Biology. Washington, DC: Convention Center; 2004. pp. 28-31

[17] Tiamyom K, Sirichaiwetchakoon K, Hengpratom T, Kupittayanant S, Srisawat R, Thaeomor A, et al. The effects of *Cordyceps sinensis* (Berk.) Sacc. and *Gymnema inodorum* (Lour.) Decne. Extracts on adipogenesis and lipase activity *in vitro*. Evidence-based Complementary and Alternative Medicine. 2019. DOI: 10.1155/2019/5370473

[18] Kelly T, Yang W, Chen CS, Reynolds K, He J. Global burden of obesity in 2005 and projections to 2030. International Journal of Obesity. 2008;**32**:1431-1437

[19] Kahn SE, Hull RE, Utzschneider KM. Mechanisms linking obesity to insulin resistance and type 2 diabetes. Nature. 2006;**444**(7121):840-846

[20] Guilherme A, Virbasius JV, Puri V, Czech MP. Adipocyte dysfunctions linking obesity to insulin resistance and type 2 diabetes. Nature Reviews. Molecular Cell Biology. 2008;**9**(5):367-377

[21] Chimilovski JS, Habu S, Bosqui Teixeira RF, Thomaz-Soccol V, Noseda MD, Pedroni Medeiros AB, et al. Antitumour activity of *Grifola frondosa* exopolysaccharides produced by submerged fermentation using sugar cane and soy molasses as carbon sources. Food Technology and Biotechnology. 2011;**49**:359-363

[22] Freire Dos Santos L, Rubel R, Bonatto SJ, et al. *Cordyceps sinensis* biomass produced by submerged fermentation in high-fat diet feed rats normalizes the blood lipid and the low testosterone induced by diet. EXCLI Journal. 2012;**11**:767-775

[23] Choi JW. Anti-diabetic effect of the exopolysaccharides (EPS) produced from *Cordyceps sinensis* on *ob/ob* mice. Korean Society for Biotechnology and Bioengineering Journal. 2011;**26**:33-40

[24] Chen SZ, Chu JZ. NMR and IR studies on the characterization of cordycepin and 2′-deoxyadenosine. Zhongguo Kangshengsu Zazhi. 1996;**21**:9-12

[25] Segelken R. Cyclosporin mold's sexual state found in New York Forest Cornell Students' discovery could target additional sources of nature-based pharmaceuticals. Cornell University Science News. 2002

[26] Takahashi S, Tamai M, Nakajima S, et al. Blockade of adipocyte differentiation by cordycepin. British Journal of Pharmacology. 2012;**167**(3):561-575

[27] Wu JW, Chen X, Siu KC. Isolation and structure characterization of an antioxidative glycopeptide from mycelial culture broth of a medicinal fungus. International Journal of Molecular Sciences. 2014;**15**:17318-17332

[28] Vasiljevic JD, Zivkovic LP, Cabarkapa AM, Bajic VP, Djelic NJ, Potparevic BM. *Cordyceps sinensis*: Genotoxic potential in human peripheral blood cells and antigenotoxic properties against hydrogen peroxide by comet assay. Alternative Therapies. 2016;**22**:S213-S221

[29] Ma L, Zhang S, Du M. Cordycepin from *Cordyceps militaris* prevents hyperglycemia in alloxan-induced diabetic mice. Nutrition Research. 2015;**35**(5):431-439

[30] Zhang G, Huang Y, Bian Y, Wong JH, Ng TB, Wang H. Hypoglycemic activity of the fungi *Cordyceps militaris*, *Cordyceps sinensis*, *Tricholoma mongolicum* and *Omphalia lapidescens* in streptozotocin-induced diabetic rats. Applied Microbiology and Biotechnology. 2006;**72**:11521-11526

[31] Nathan C. Points of control in inflammation. Nature. 2002;**420**(6917):846-852

[32] Aramwit P, Bang N, Ratanavaraporn J, Nakp T, Srichana T. An anti-cancer cordycepin produced by *Cordyceps militaris* growing on the dead larva of *Bombyx moris* silkworm. Journal of Agricultural Science. 2014;**4**(4):42-53

Chapter 11

Coenzyme Q10: Regulators of Mitochondria and beyond

Gopi Marappan

Abstract

The role of coenzyme Q10 (CoQ10) was relatively unknown except its involvement in the oxidative phosphorylation at electron transport chain. Recent researches revealed its association in conditions like maintenance of cardiac and pulmonary functions, regulation of cell proliferation to cancer prevention, etc. CoQ10, a potent lipophilic antioxidant, prevents the cellular biomolecules *viz.*, DNA, RNA, lipid bilayers, etc. Endogenous *de novo* synthesis will be sufficient to maintain the daily body needs; however, synthesis showed age-dependent reduction. Commercial preparations are available for oral consumption; there are even food-grade preparations for cattle, swine and poultry. A major concern with oral intake of CoQ10 was bioavailability due to its lipophilic nature. CoQ10 has been recommended for patients under continuous statin therapy as these drugs inhibit the pathway of CoQ10 biosynthesis. The use of CoQ10 in various cardiac and tumor conditions indicates that its activity is not only due to its antioxidant activity but also due to its apoptosis property. Apart from human uses, CoQ10 is now used in food animals especially broilers as they were fed with high energy dense diet there will be leakage of electrons at electron transport chain level which adversely affects the bird's performance and also used in treatment of ascites mortality.

Keywords: coenzyme Q10, antioxidant, mitochondrial regulators, reduced, cholesterol synthesis, cancer

1. Introduction

Coenzymes are the cofactors in the body, which are essential for numerous enzymatic reactions at various levels. One such enzyme is coenzyme Q10 (CoQ10) and also widely known as ubiquinone. As the name ubiquinone suggests, this coenzyme is ubiquitous in nature. However, the identification of coenzyme Q10 (CoQ10) was accidental when Crane and co-authors in 1957 [1] were involved in the investigation of the mitochondrial electron transport system, they first identified and isolated this enzyme from the beef heart. The fundamental role of CoQ10 is in the mitochondrial respiratory chain and in oxidative phosphorylation [2], for which he was awarded the Nobel Prize in Chemistry in 1978 [3]. The CoQ10 is an endogenously synthesized lipophilic compound present in all living cells (ubiquitous in nature); hence, it is also designated as ubiquinones [4]. Coenzyme Q10 (CoQ10) is a lipid-soluble compound involved in mitochondrial adenosine triphosphate (ATP) synthesis (bioenergetics) and reduces the pulmonary hypertension syndrome and ascites mortality [5]. CoQ10 does various roles along with its three important functions in the body, namely as an electron carrier in respiratory chain, antioxidant [6] and cell signaling and gene expression [7]. These functions have practical applications in clinical practice and its use as food/feed supplementation [8]. Supplementing coenzyme Q10 is known to provide health benefits, much like nutraceuticals even in healthy individuals [9] and individuals with metabolic disorders like oxidative phosphorylation disorder [10]. CoQ10 also maintains membrane fluidity [11] and protects membranous phospholipid against peroxidation [12] and in plant photosynthesis [13]. Normal respiratory rate requires the maintenance of a high CoQ10 concentration, and even a small decrease is deleterious [14].

2. Coenzyme Q10

2.1 Is coenzyme Q10 a vitamin?

CoQ10 is similar to vitamin K in its chemical structure, but it is not considered a vitamin because it is synthesized in the body [15, 16]. All the fat-soluble vitamins (A, D, E and K) possess isoprene units in their structures. Likewise, coenzyme Q also has an isoprenoid (seen as A, D, E and K), a quinone structure (as in vitamin K) and cyclized chromanol (vitamin E). A definition to which a molecule is considered as a vitamin is as follows: an organic compound with small molecular weight, not to be synthesized in the body and supplemented through the diet; the absence of this will lead to a deficiency syndrome; converted to an active coenzyme form required for metabolic activity. Day-to-day findings make CoQ10 nearly fit into the typical definition for a vitamin. Being endogenously synthesized by all animal tissues might rule them out for a vitamin status. But vitamin D3 and vitamin C are endogenously synthesized from cholesterol and glucose, respectively, and are still given the vitamin status; hence, CoQ10 might be termed as vitamin Q as expressed by folkers. Supplementing coenzyme Q10 provides health benefits to the likes of nutraceuticals [9].

2.2 Chemistry of coenzyme Q10

CoQ10 is a 2,3-dimethoxy-5-methyl-6-decaprenyl-1,4-benzoquinone [16]. It contains 10 isoprene units and is the predominant form in both mammals and birds, whereas CoQ9 (9 isoprene units) is predominant in rodents [14]. Due to its lipophilic nature and higher molecular weight (863 Da), the oral bioavailability of CoQ10 is low [8]. Following absorption, it is taken up by the liver for incorporation into very low density lipoprotein (VLDL) particles before being released into circulation [6, 15]. An increase (about 160%) in CoQ10 levels in the VLDL and LDL fractions following its dietary intake. To counteract the problem of low bioavailability, currently different types of carriers like lipid emulsion of solid triglyceride, tocopherol succinate and phospholipids (Ultrasome®) [17], different cyclodextrins [18] and gel form (UbiQGel®) [19], are being tried with great success.

2.3 Biosynthesis of ubiquinone

CoQ10 is endogenously synthesized in all human and animal cells [20]. Two pathways are involved in CoQ10 biosynthesis in the body. The biosynthesis of polyprenyl side chain occurs through the mevalonate pathway. This reaction starts with acetyl-coenzyme A and ends up with farnesyl pyrophosphate (FPP). This FPP also acts as a substrate for the biosynthesis of isoprenylated proteins, dolichol and cholesterol. However, the quinone head is synthesized from either the amino acid tyrosine or the phenylalanine [21].

Major findings of coenzyme Q10 [22] are as follows:

- Coenzyme Q is distributed throughout all cell components.

- Unlike vitamins K and E, exogenous CoQ is absorbed into liver and not in other tissues.

- All the tissues in the body have the capacity to independently synthesis CoQ, but this capacity is less during developing early embryonic tissues.

- The mevalonate pathway is used by animals, plants and fungi for the synthesis of CoQ but not used by some bacteria and also for synthesis of vitamin K in mycobacteria.

- In liver, accumulation of CoQ occurs due to lower catabolism or enhanced synthesis under conditions like deficiency of vitamin A, cold stress exposure and excess thyroid secretion.

- Excesses of CoQ in liver either by endogenous synthesis or by absorption has a negative feedback mechanism to inhibit its own synthesis, which also leads to low serum cholesterol content as *de novo* synthesis of cholesterol shares the same biosynthetic pathway.

Presently, coenzyme Q10 is produced by chemical synthesis, semi-chemical synthesis or microbial conversion and is commonly available. Humans or animals fed with non-vegetarian diet will have higher CoQ10 intake and its absorption varies with the amount and uptake increases with increase in fat content. The absorption of reduced form is more than that of the oxidized CoQ10, and with its large molecular weight, about 60% of intake is excreted through the feces [23]. The yeast fermentation technique, which involves with inclusion of B vitamins in their culture, is the major form of industrial CoQ10 synthesis. Recently, CoQ10 is available as feed grade in powder form for swine and poultry but in gel form for human preparations [24, 25].

3. Therapeutic indications

3.1 Cancer therapy

CoQ10 a lipophilic antioxidant exhibits different biological activities like immune boosting, free radical scavenging and DNA protection. Studies on administration of CoQ10 have revealed promising results in prevention and/or treating cancers. Positive effect with breast cancer in patients consuming CoQ10 has been reported in recent publications [26–29].

3.2 Insufficiency of CoQ10 levels are considered as one of the risk factors

A significant lower level of CoQ10 is observed in cancerous tissues when compared to the normal tissues. CoQ10 is known for its counteraction of ROS in cellular and DNA integrity [30]. A case-control study [31, 32] revealed an inverse association between CoQ10 levels and incidence of breast cancer. An *in vitro* study with MCF-7 breast cancer cell lines where the cells were co-incubated with CoQ10 showed results with significant decrease in intracellular peroxide formation and matrix metalloproteinase 2 activity and the effects were in a dose-dependent manner [33].

Results further showed that CoQ10 had no inhibitory effect on apoptotic, anti-growth and anti-colonization effects of doxorubicin at any doses [34]. An animal study with mammary carcinoma model revealed that administration of CoQ10 at 40 mg/kg body weight restored the normal antioxidant level [35]. Reports suggest that increasing the dose of CoQ10 to 390 mg from 300 mg for more periods revealed resolution of tumor residue without any metastases [36–38] and increased the survivability [39]. Daily intake of a combination of 100 mg CoQ10, 10 mg riboflavin and 50 mg niacin reduced the circulating tumor markers [40–46]. Consuming a combination of CoQ10 along with lipotropic factor L-carnitine reduced the tumor-related fatigue in subjects [47–50].

3.3 Coenzyme Q10 on ascites heart index (AHI) and ascites mortality

In fast-growing broilers, the impact of ascites mortality is very high (after 5 weeks of age) as the farmers are not only losing the birds but also are incurring the feeding and rearing cost by the time. Feed restriction or skip-a-day feeding is followed in broiler during finisher phase to avoid the problem of ascites, which results in poor body weight and feed efficiency. Few researchers are suggesting that ascites might be due to the bird's inability to endogenously synthesize the CoQ10 demand. To counteract this, CoQ10 was used, and in fact, the importance of CoQ10 was felt with a reduction in ascites mortality in broilers when fed with CoQ10 [51]. Then, the term ascites heart index (AHI) comes into prominence, which gives more information about the susceptibility of the birds to ascites. Ascites heart index (AHI), a sensitive index of pulmonary hypertension, is based on the relative ratio of the right ventricle to the total ventricle [52]. The AHI was further made into a more useful tool where broilers with AHI value of less than 0.27 without any fluid accumulation in abdomen are normal and those birds having AHI value more than 0.30 with fluid accumulation are pulmonary hypertensioned and prone to ascites mortality [53]. The relative heart weight of birds receiving CoQ10 at 20 mg kgG1 of diet was higher [54–58]. A reduction in AHI ratio in broilers fed with CoQ10 at 40 mg/kg of feed was noticed [59]. However, there was a lower heart weight with respect to percentage of body weight when broilers fed with 20 and 40 mg of CoQ10 [60]. Ascites mortality in broilers was reduced around 75% by CoQ10 supplementation at both 20 and 40 mg/kg of diet. But at 40 mg/kg of diet supplementation, the incidence of leg problem was high. This reduction in ascites mortality (around 40%) was observed when broilers were fed CoQ9. These studies imply that CoQ, either 9 or 10 isoprene units, is able to reduce the broiler's mortality due to ascites.

3.4 Coenzyme Q10 on lipid metabolism

Clinical human and animal studies suggested that dietary CoQ10 supplementation improved the cholesterol metabolism in mammals. Nearly 10% lower cholesterol concentration was found in heart tissue of broilers when supplemented with CoQ10 [8]. CoQ10 supplementation decreased plasma total cholesterol concentration in humans [61] and rats [62]. CoQ10 was reportedly able to suppress the hepatic cholesterogenesis in rats [63] and in hens [64]. In an experiment in layer, chicks revealed reduced hepatic total cholesterol, plasma cholesterol and very low density lipoprotein (VLDL) cholesterol concentration by supplementation of CoQ10 at 400 and 800 mg/kg feed [65]. However, the plasma HDL, LDL cholesterol and total bile acids were not influenced by CoQ10 supplementation. The reduction in cholesterol level was due to decreased enzymatic activity of

3-hydroxy-3-methylglutaryl coenzyme A reductase (HMGR) in the liver, but it had no influence on the enzymatic activity of 3-hydroxy-3-methylglutaryl coenzyme A synthetase (HMGS). Dietary CoQ10 supplementation suppressed hepatic cholesterogenesis in laying hens [64] and observed a decrease in egg yolk cholesterol concentration by 7–10% on CoQ10 supplementation.

In a long-term CoQ10 feeding trial, reduced cholesterol synthesis with suppression in cholesterol catabolism was observed resulting in return of hepatic cholesterol to normal level [65]. However, long-term (0–42 days) supplementation of CoQ10 at 20 and 40 mg kgG1 reduced the levels of serum total cholesterol and serum LDL cholesterol [58, 66]. The reduction in serum LDL cholesterol due to CoQ10 supplementation was attributed to the action of reduced form of CoQ10(H2), which induces characteristic gene expression patterns, which are translated into reduced LDL cholesterol level in human subjects. However, there were no reports of increase in the HDL cholesterol levels [58, 65]. CoQ10 reduced cholesterol metabolism in the plasma of patients with myocardial infarction [67] and in diabetic rats [62]. CoQ9, a major coenzyme Q in rats, decreases plasma total cholesterol concentration and suppresses hepatic cholesterogenesis [68].

3.5 Coenzyme Q10 antioxidant properties

Under the present intensive system of poultry production especially in tropics, stresses due to environment, metabolic, managemental, etc. are inevitable, resulting in lower productivity, less nutrient retention, decreased serum and tissue vitamin level, humoral immunity (HI) and molecular changes like protein, nucleic acid denaturation and lipid peroxidation. Increased reactive oxygen species (ROS) metabolites compromise cell membrane integrity [53], which results in drip loss in muscles [60] affect keeping quality of muscles. Different nutrients and additives (like the use of synthetic amino acids, low heat increment nutrients, vitamins C, E and minerals such as selenium, zinc and magnesium or additive such as genistein and melatonin, and essential oils) are tried with varied success to counteract these stresses [69]. Aside from its role in mitochondrial bioenergetics, ubiquinone also affects membrane fluidity [11] and protects membrane phospholipids against peroxidation [12]. CoQ10 in its reduced form possesses free radical scavenging and increases total antioxidant capacity [70, 71]. CoQ10 is preferred over α-tocopherol [72] as CoQ10 enhances the activity of other enzymatic and non-enzymatic antioxidants. The serum vitamin E level was increased by CoQ10 at 20 mg/kg [7, 58]. CoQ10 shows the property of regenerating the oxidized (inactive) α-tocopherol to reduce (an active form of vitamin E) [73]. Serum or liver malondialdehyde (MDA) is a product of lipid peroxidation and serves as a biomarker for oxidative damage in lipids. This suggested the protective action on lipid peroxidation in liver mitochondria by CoQ10.

Superoxide dismutase (SOD) activity was increased in accordance with CoQ10 supplementation in broilers and in rats [74]. An increase in hepatic SOD and anti-ROS capacity in broilers was observed by CoQ10 supplementation [55]. The supplementation of CoQ10 increases the SOD activity by antagonizing nitric oxide (NO) inactivation, thereby making more NO availability for the biological function that leads to extracellular SOD gene expression. The reduced glutathione and glutathione peroxidase activity was also increased by CoQ10 at 20 mg/kg. This synergistic action of CoQ10 is possible as it acts as a primary regenerating antioxidant [75]. However, supplementation at 40 mg kgG1 of diet resulted in no effect on serum vitamin E and SOD levels. This ineffectiveness of CoQ10 at 40 mg kgG1 of diet is due to the auto-oxidation of CoQ10 resulting in higher production of mitochondrial reactive oxygen species (ROS), which leads to oxidative stress in the body. The development of auto-oxidation was observed in birds fed higher level of CoQ10 for prolonged duration.

4. Absorption and distribution

4.1 Coenzyme Q10 absorption among body tissues

The content of CoQ10 in different body tissues is well studied in human subjects, but there are not enough studies in farm animals or birds. The highest concentration of CoQ10 was found in the most active organs like heart, kidney and liver. The CoQ10 concentration depends on a balance between inputs and outputs. Inputs are the level of CoQ10, which is endogenously synthesized, plus dietary supply and the outputs are the usage by oxidative stress and cellular metabolism. An adult human body has approximately 2 g of CoQ10, where a daily replacement of 0.5 g should be done by both endogenous synthesis and dietary means. Therefore, an average body CoQ10 content turnover rate was around 4 days and dietary supply becomes essential with impairment in endogenous synthesis. The body content of CoQ10 decreased rapidly after the age of 40 years in humans with reduced biosynthesis. CoQ10 supplementation reversed the reduced circulating CoQ10 concentrations in statin-treated subjects as statin inhibits the pathways involved in both cholesterol and CoQ10 supplementation. Various authors recommended daily intake of CoQ10 of about 30–100 mg for healthy people over 40 years and 60–1200 mg for those undergoing an adjunctive therapy for some medical conditions.

4.2 Distribution in body tissues

The CoQ10 level in human tissues varies with inappropriate nutrition, smoking and different medical conditions such as cardiomyopathy, diabetes and neurological disorder conditions [76]. Similarly in broiler chicken, the concentration of CoQ10 among different body tissues was recorded (**Table 1**) [77].

Among the organelles, larger amount of CoQ10 is found in mitochondria of heart cells (92.3–282.0 mg/kg), followed by liver (22.7–132.2 mg/kg) of cattle, swine and chicken. Being lipophilic, vegetable oils especially rape seed and peanut oils have a high content (63.5–77.0 mg/kg) of CoQ. This again proved that CoQ10 is required more by tissues that are very active.

CoQ10 to cholesterol index (QCI) is increasingly used as a measure for assessment of meat quality. QCI was used as a reliable indicator of oxidative status, and the possible oxidative stresses induced by different food ingredients and consider them as oxidant foods [8]. In simple terms, muscles with higher oxidative stress

Tissues	Concentration (mg/kg)
Heart	92.3–192
Liver	116.2–132.2
Thigh	24.2–25
Breast	7.8–17.1
Wing	11.0
Whole chicken	14–21

Table 1.
Concentration of CoQ10 in various body tissues [77].

due to either metabolic activity or food would have a reduction of QCI value. The QCI value was higher in 20 mg/kg supplemented group suggestive of low oxidative stress. Auto-oxidation of CoQ10 at 40 mg of CoQ10/kg increased muscle metabolic activity leading to reduced QCI value [78]. Due to its antioxidant property, CoQ10 supplementation will be helpful in reducing drip loss during meat storage. Supplementation with CoQ10 at 40 mg kgG1 diet improved breast muscle yield and reduced the drip loss in broilers [60]. The reduction in muscle drip loss was attributed to the reduced reactive oxygen metabolites, thereby improving the cell membrane integrity and improved water retention.

5. Conclusion

The role of coenzyme Q10 is widely being studied under various health conditions including cancer and cardiac hypertrophy. Its importance in normal healthy life is quite evident and physicians are prescribing it for oral intake for persons who continuously smoke as well as for those under statin drug therapy. Recently, CoQ10 was widely used in food animals especially broilers, which are highly susceptible to mortality due to ascites/sudden death syndrome as a result of its rapid growth rate.

Conflict of interest

The author declares no conflict of interest.

Author details

Gopi Marappan
ICAR-Central Avian Research Institute, Izatnagar, India

*Address all correspondence to: gopsgopi72@gmail.com

References

[1] Crane FL, Hatefi Y, Lester RL, Widmer C. Isolation of a quinone from beef heart mitochondria. Biochemica et Biophysica Acta. 1957;**25**:220-221

[2] Mitchell P. Protonmotive redox mechanism of the cytochrome *b-c*1 complex in the respiratory chain: Protonmotive ubiquinone cycle. FEBS Letters. 1975;**56**:1-6

[3] Bliznakov EG, Chopra RK, Bhagavan HN. Coenzyme Q10 and neoplasia: Overview of experimental and clinical evidence. In: Bagchi D, Preuss HG, editors. Phytopharmaceuticals in Cancer Chemoprevention. Boca Raton: CRC Press; 2004. pp. 599-622

[4] Lenaz G, Esposti D. Physical properties of ubiquinones in model systems and membranes. In: Lenaz G, editor. Coenzyme Q. Biochemistry, Bioenergetics and Clinical Applications of Ubiquinone. New Jersey, United States: John Wiley and Sons; 1985. pp. 83-105

[5] Geng AL, Guo YM, Yuan J. Effects of dietary L-carnitine and coenzyme Q10 supplementation on performance and ascites mortality of broilers. Archives of Animal Nutrition. 2004;**58**:473-482

[6] Kaikkonen J, Nyyssonen K, Porkkala-Sarataho E, Poulsen HE, Metsa-Ketela T. Effect of oral coenzyme Q10 supplementation on the oxidation resistance of human VLDL+LDL fraction: Absorption and antioxidative properties of oil and granule-based preparations. Free Radical Biological Medicine. 1997;**22**:1195-1202

[7] Ernster L, Dallner G. Biochemical, physiological and medical aspects of ubiquinone function. Biochimica et Biophysica Acta: Molecular Basis of Disease. 1995;**1271**:195-204

[8] Krizman PJ, Prosek M, Smidovnik A, Wondra AG, Glaser R, Zelenko B, et al. Poultry products with increased content of CoQ10 prepared from chickens fed with supplemental CoQ10. In: Eissa AHA, editor. Trends in Vital Food and Control Engineering. Rijeka, Croatia. ISBN-13: 9789535104490: IntechOpen; 2012. pp. 165-186

[9] Ramasarma T. A touch of history and a peep into the future of the lipidquinone known as coenzyme Q and ubiquinone. Current Science. 2012;**102**:1459-1471

[10] Marriage BJ, Clandinin MT, Macdonald IM, Glerum DM. Cofactor treatment improves ATP synthetic capacity in patients with oxidative phosphorylation disorders. Molecular Genetics and Metabolism. 2004;**81**:263-272

[11] Fato R, Bertoli E, Castelli GP, Lenaz G. Fluidizing effect of endogenous ubiquinone in bovine heart mitochondrial membranes. FEBS Letters. 1984;**172**:6-10

[12] Takayanagi R, Takeshige K, Minakami S. NADH and NADPH dependent lipid peroxidation in bovine heart submitochondrial particles. Dependence on the rate of electron flow in the respiratory chain and an antioxidant role of ubiquinol. Journal of Biochemistry. 1980;**192**:853-860

[13] Redfearn ER. Mode of action of ubiquinones (coenzymes Q) in electron transport systems. Vitamins Hormones. 1966;**24**:465-488

[14] Battino M, Fato R, Parenti-Castelli G, Lenaz G. Coenzyme Q can control the efficiency of oxidative phosphorylation. International Journal of Tissue Reactions. 1990;**12**:137-144

[15] Bhagavan HN, Chopra RK. Coenzyme Q10: Absorption, tissue

uptake, metabolism and pharmacokinetics. Free Radical Research. 2006;**40**:445-453

[16] Wang Y, Oxer D, Hekimi S. Mitochondrial function and lifespan of mice with controlled ubiquinone biosynthesis. Nature Communications. 2015. DOI: 10.1038/ncomms7393

[17] Amselem S. Solid lipid compositions of lipophilic compounds for enhanced oral bioavailability. Patent No. 5. 1999. 989,583. Available from: http://www.pharmcast.com/Patents/112399OG/5989583_bioavail112399.htm

[18] Fir MM, Milivojevic L, Prosek M, Smidovnik A. Property studies of coenzyme Q10-cyclodextrins complexes. Acta Chimica Slovenica. 2009;**56**:885-891

[19] Natural Medicines Comprehensive Database: Coenzymes Q10 Monograph. 2003. Available from: http://www.naturaldatabase.com

[20] Elmberger PG, Kalen A, Appelkvist EL, Dallner G. *In vitro* and *in vivo* synthesis of dolichol and other main mevalonate products in various organs of the rat. European Journal of Biochemistry. 1987;**168**:1-11

[21] Turunen M, Olsson J, Dallner G. Metabolism and function of coenzyme Q. Biochimica et Biophysica Acta (BBA)-Biomembranes. 2004;**1660**:171-199

[22] Ramasarma T. Studies on ubiquinone. Journal of Scientific and Industrial Research;**27**:147-164

[23] Zlatohlavek L, Vrablik B, Grauova EM, Ceska R. The effect of coenzyme Q10 in statin myopathy. Neuroendocrinology Letters. 2012;**33**:98-101

[24] Ioana VS, Lasio V, Uivarosan D. Stimulation of biosynthesis of coenzyme Q10 by *Sacharomyces cerevisiae* under the influence of vitamin B1. Analele Universitatii Din Oradea. 2009;**2**:693-700

[25] Lambrechts P, Siebrecht S. Coenzyme Q10 and ubiquinol as adjunctive therapy for heart failure. Agro Food Industry Hi Tech. 2013;**24**:60-62

[26] Tafazoli A. Coenzyme Q10 in breast cancer care. Future Oncology. 2017;**13**(11):1035-1041. DOI: 10/2217/fon-2016-0547

[27] Hill GJ, Shriver BJ, Arnett DB. Examining intentions to use CoQ10 amongst breast cancer patients. American Journal of Health Behavior. 2006;**30**(3):313-321

[28] Complementary treatments highlighted at recent meeting. Oncology (Williston Park, N.Y.). 1999;**13**(2):166

[29] Sivak LA, Askol'skii AV, Lial'kin SA, Klimanov M, Maidanevich NN, Kasap NV. Cardiotoxicity of conservative treatment of solid tumors. Likars'ka Sprava. 2011;**3-4**:51-59

[30] Portakal O, Ozkaya O, Erden Inal M, Bozan B, Kosan M, Sayek I. Coenzyme Q10 concentrations and antioxidant status in tissues of breast cancer patients. Clinical Biochemistry. 2000;**33**(4):279-284

[31] Sinatra ST. Care, cancer and coenzyme Q10. Journal of the American College of Cardiology. 1999;**33**(3):897-899

[32] Cooney RV, Dai Q, Gao YT. Low plasma coenzyme Q(10) levels and breast cancer risk in Chinese women. Cancer Epidemiology, Biomarkers & Prevention. 2011;**20**(6):1124-1130

[33] Bahar M, Khaghani S, Pasalar P. Exogenous coenzyme Q10 modulates

MMP-2 activity in MCF-7 cell line as a breast cancer cellular model. Nutrition Journal. 2010;**9**:62

[34] Greenlee H, Shaw J, Lau YK, Naini A, Maurer M. Lack of effect of coenzyme q10 on doxorubicin cytotoxicity in breast cancer cell cultures. Integrative Cancer Therapies. 2012;**11**(3):243-250

[35] Perumal SS, Shanthi P, Sachdanandam P. Augmented efficacy of tamoxifen in rat breast tumorigenesis when gavaged along with riboflavin, niacin and CoQ10: Effects on lipid peroxidation and antioxidants in mitochondria. Chemico-biological Interactions. 2005;**152**(1):49-58

[36] Lockwood K, Moesgaard S, Folkers K. Partial and complete regression of breast cancer in patients in relation to dosage of coenzyme Q10. Biochemical and Biophysical Research Communications. 1994;**199**(3):1504-1508

[37] Lockwood K, Moesgaard S, Hanioka T, Folkers K. Apparent partial remission of breast cancer in 'high risk' patients supplemented with nutritional antioxidants, essential fatty acids and coenzyme Q10. Molecular Aspects of Medicine. 1994;**15**:s231-s240

[38] Lockwood K, Moesgaard S, Yamamoto T, Folkers K. Progress on therapy of breast cancer with vitamin Q10 and the regression of metastases. Biochemical and Biophysical Research Communications. 1995;**212**(1):172-177

[39] Bjorklund G. The adjuvant nutritional intervention in cancer (ANICA) trial. Nutrition and Cancer. 2015;**67**(8):1355-1358

[40] Premkumar VG, Yuvaraj S, Vijayasarathy K, Gangadaran SG, Sachdanandam P. Effect of coenzyme Q10, riboflavin and niacin on serum CEA and CA 15-3 levels in breast cancer patients undergoing tamoxifen therapy. Biological and Pharmaceutical Bulletin. 2007a;**30**(2):367-370

[41] Premkumar VG, Yuvaraj S, Shanthi P, Sachdanandam P. Co-enzyme Q10, riboflavin and niacin supplementation on alteration of DNA repair enzyme and DNA methylation in breast cancer patients undergoing tamoxifen therapy. The British Journal of Nutrition. 2008;**100**(6):1179-1182

[42] Premkumar VG, Yuvaraj S, Sathish S, Shanthi P, Sachdanandam P. Anti-angiogenic potential of Coenzyme Q10, riboflavin and niacin in breast cancer patients undergoing tamoxifen therapy. Vascular Pharmacology. 2008b;**48**(4-6):191-201

[43] Premkumar VG, Yuvaraj S, Vijayasarathy K, Gangadaran SG, Sachdanandam P. Serum cytokine levels of interleukin-1beta, -6, -8, tumour necrosis factor-alpha and vascular endothelial growth factor in breast cancer patients treated with tamoxifen and supplemented with co-enzyme Q(10), riboflavin and niacin. Basic and Clinical Pharmacology and Toxicology. 2007b;**100**(6):387-391

[44] Yuvaraj S, Premkumar VG, Vijayasarathy K, Gangadaran SG, Sachdanandam P. Augmented antioxidant status in tamoxifen treated postmenopausal women with breast cancer on coadministration with coenzyme Q10, niacin and riboflavin. Cancer Chemotherapy and Pharmacology. 2008;**61**(6):933-941

[45] Yuvaraj S, Premkumar VG, Shanthi P, Vijayasarathy K, Gangadaran SG, Sachdanandam P. Effect of coenzyme Q (10), riboflavin and niacin on tamoxifen treated postmenopausal breast cancer women with special reference to blood chemistry profiles. Breast

Cancer Research and Treatment. 2009;**114**(2):377-384

[46] Yuvaraj S, Premkumar VG, Vijayasarathy K, Gangadaran SG, Sachdanandam P. Ameliorating effect of coenzyme Q10, riboflavin and niacin in tamoxifen-treated postmenopausal breast cancer patients with special reference to lipids and lipoproteins. Clinical Biochemistry. 2007;**40**(9-10):623-628

[47] Iwase S, Kawaguchi T, Yotsumoto D. Efficacy and safety of an amino acid jelly containing coenzyme Q10 and L-carnitine in controlling fatigue in breast cancer patients receiving chemotherapy: A multi-institutional, randomized, exploratory trial (JORTC-CAM01). Supportive Care in Cancer. 2016;**24**(2):637-646

[48] Lesser GJ, Case D, Stark N. A randomized, double-blind, placebo-controlled study of oral coenzyme Q10 to relieve self-reported treatment-related fatigue in newly diagnosed patients with breast cancer. The Journal of Supportive Oncology. 2013;**11**(1):31-42

[49] Chai W, Cooney RV, Franke AA. Plasma coenzyme Q10 levels and postmenopausal breast cancer risk: The multiethnic cohort study. Cancer Epidemiology, Biomarkers & Prevention. 2010;**19**(9):2351-2356

[50] Sachdanandam P. Antiangiogenic and hypolipidemic activity of coenzyme Q10 supplementation to breast cancer patients undergoing Tamoxifen therapy. BioFactors (Oxford, England). 2008;**32**(1-4):151-159

[51] Nakamura K, Noguchi K, Aoyama T, Nakajima T, Tanimura N. Protective effect of ubiquinone (coenzyme Q9) on ascites in broiler chickens. British Poultry Science. 2006;**37**:189-195

[52] Burton RR, Besch EL, Smith AH. Effect of chronic hypoxia on the pulmonary arterial blood pressure of the chicken. American Journal of Physiology. 1968;**214**:1438-1442

[53] Cawthon D, Beers K, Bottje WG. Electron transport chain defect and inefficient respiration may underlie pulmonary hypertension syndrome (ascites)-associated mitochondrial dysfunction in broilers. Poultry Science. 2001;**80**:474-484

[54] Azuma J, Harada H, Sawamura A, Ohta H, Awata N. Beneficial effect of coenzyme Q on myocardial slow action potentials in hearts subjected to decreased perfusion pressure hypoxia-substrate-free perfusion. Basic Research in Cardiology. 1985;**80**:147-155

[55] Geng AL, Guo YM, Yuan J. Effects of dietary L-carnitine and coenzyme Q10 supplementation on performance and ascites mortality of broilers. Archives of Animal Nutrition. 2004;**58**:473-482

[56] Geng AL, Guo YM, Yang Y. Reduction of ascites mortality in broilers by coenzyme Q10. Poultry Science. 2004b;**83**:1587-1593

[57] Geng A, Li B, Guo Y. Effects of dietary L-carnitine and coenzyme Q10 at different supplemental ages on growth performance and some immune response in ascites-susceptible broilers. Archives of Animal Nutrition. 2007;**61**:50-60

[58] Gopi M. Bioenergetics role of coenzyme Q10 supplementation on broiler performance [M.V.Sc. thesis]. Chennai, India: Tamil Nadu Veterinary and Animal Sciences University; 2013

[59] Fathi M. Effects of coenzyme Q10 supplementation on growth performance, some hematological parameters, plasma enzymes activities in broilers with pulmonary

hypertension syndrome (PHS). Iranian Journal of Applied Animal Science. 2015;**5**:147-153

[60] Huang B, Guo Y, Hu X, Song Y. Effects of coenzyme Q10 on growth performance and heart mitochondrial function of broilers under high altitude induced hypoxia. Journal of Poultry Science. 2011;**48**:40-46

[61] Cicero AFG, Derosa G, Miconi A, Laghi L, Nascetti S, Gaddi A. Possible role of ubiquinone in the treatment of massive hypertriglyceridemia resistant to PUFA and fibrates. Biomedicine and Pharmacotheraphy. 2005;**59**:312-317

[62] Modi K, Santani DD, Goyal RK, Bhatt PA. Effect of coenzyme Q10 on catalase activity and other antioxidant parameters in streptozotocin-induced diabetic rats. Biological Trace Elements Research. 2006;**109**:25-34

[63] Omkumar RV, Gaikwad AS, Ramasarma T. Feedback-type inhibition of activity of 3-hydroxy-3-methylglutaryl coenzyme a reductase by ubiquinone. Biochemica Biophysics Research Communications. 1992;**184**:1280-1287

[64] Kamisoyama H, Honda K, Kitaguchi K, Hasegawa S. Transfer of dietary coenzyme Q10 into the egg yolk of laying hens. Journal of Poultry Science. 2010;**47**:28-33

[65] Honda K, Kamisoyama H, Motoori T, Saneyasu T, Hasegawa S. Effect of dietary coenzyme Q10 on cholesterol metabolism in growing chickens. Journal of Poultry Science. 2010;**47**:41-47

[66] Schmelzer C, Niklowitz P, Okun JG, Haas D, Menke T, Doring F. Ubiquinol-induced gene expression signatures are translated into altered parameters of erythropoiesis and reduced low density lipoprotein cholesterol levels in humans. IUBMB Life. 2011;**63**:42-48

[67] Singh RB, Neki NS, Kartikey K, Pella D, Kumar A, Niaz MA, et al. Effect of coenzyme Q10 on risk of atherosclerosis in patients with recent myocardial infarction. Molecular and Cellular Biochemistry. 2003;**246**:75-82.67

[68] Krishnaiah KV, Ramasarma T. Regulation of hepatic cholesterolgenesis by ubiquinone. Biochimica et Biophysica Acta. 1970;**202**:332-342

[69] Gopi M, Dhinesh Kumar R, Elaiyaraja G, Karthik K, Manjunatha HV, Gautham K, et al. Dietary essentiality I: Coenzyme Q10 conditionally essential—Review. Asian Journal of Animal and Veterinary Advances. 2015;**10**(9):461-475

[70] Forsmark-Andree P, Lee CP, Dallner G, Ernster L. Lipid peroxidation and changes in the ubiquinone content and the respiratory chain enzymes of submitochondrial particles. Free Radical Biology and Medicine. 1997;**22**:391-400

[71] Armanfar M, Jafari A, Dehghan GR. Effect of coenzyme Q10 supplementation on exercise-induced response of oxidative stress and muscle damage indicators in male runners. Zahedan Journal of Research in Medicinal Science. 2015;**17**:1-5

[72] Tang PH, Miles MV, DeGrauw A, Hershey A, Pesce A. HPLC analysis of reduced and oxidized coenzyme Q(10) in human plasma. Clinical Chemistry. 2001;**47**:256-265

[73] Constantinescu A, Maguire JJ, Packer L. Interactions between ubiquinones and vitamins in membranes and cells. Molecular Aspects in Medicine. 1994;**15**:s57-s65

[74] Lakomkin VL, Konovalova GG, Kalenikova EI, Zabbarova IV, Kaminnyi AI. Changes in antioxidant status of myocardium during oxidative stress under the influence of coenzyme Q10. Biochemistry. 2005;**70**:79-84

[75] Quiles JL, Ochoa JJ, Huertas JR, Mataix J. Coenzyme Q supplementation protects from age-related DNA double-strand breaks and increases lifespan in rats fed on a PUFA-rich diet. Experimental Gerontology. 2004;**39**:189-194

[76] Madmani ME, Solaiman AY, Agha KT, Madmani Y, Madmani Y, Essali A, et al. Coenzyme Q10 for heart failure. Cochrane Database Systemic Review. 2014. DOI: 10.1002/14651858. CD008684.pub2

[77] Pravst I, Zmitek K, Zmitek J. Coenzyme Q10 contents in foods and fortification strategies. Critical Review in Food Science Nutrition. 2010;**50**:269-280

[78] Turrens JF, Alexandre A, Lehninger AL. Ubisemiquinone is the electron donor for superoxide formation by complex III of heart mitochondria. Archives in Biochemistry and Biophysics. 1985;**237**:408-414

Chapter 12

Modulatory Potentials of n-3 Polyunsaturated Fatty Acids in Inflammatory Diseases

Vinayak Uppin, Pooja Acharya
and Ramaprasad Ravichandra Talahalli

Abstract

Inflammatory diseases (bronchitis, irritable bowel diseases, psoriasis, chronic obstructive pulmonary disease, rheumatoid arthritis, conjunctivitis, hepatitis, rhinitis, etc.) are increasingly becoming the cause of health concern across the world. For both developed and developing countries, the public health cost attached to the management of these complications is increasing each year. Control of diet is one of the critical strategies to improve the therapeutic potentials of clinically proven drugs. Among the dietary factors, bioactive lipids, particularly polyunsaturated fatty acids, elicit essential effects of modulating signaling pathways that could provide a beneficial effect in individuals suffering from various inflammatory complications. Among the polyunsaturated fatty acids, both n-3 and n-6 fatty acids exhibit differential effects, and their dietary ratio plays a significant role in the overall impact in an individual. This is most evident with the fact that the decrease in the n-3 intake in recent times has significantly contributed to the severity of chronic inflammation. The mechanism by which these fatty acids and bioactive unsaponifiable embedded in the dietary oils modulate the critical genes, thereby alter the pathology of inflammatory complications is under study for many decades. In this chapter, the role of polyunsaturated fatty acids and their modulatory effects on some of the significant inflammatory complications are discussed.

Keywords: arthritis, chronic obstructive pulmonary disease, inflammation, irritable bowel diseases, polyunsaturated fatty acids, psoriasis

1. Introduction

Inflammation is a natural phenomenon for healing that occurs through the body's immune response. The cellular reactions remove the threat that may be in the form of a pathogen, damaged cell, irritant, and other foreign particles. Often, they are self-limiting; however, under certain circumstances, they can cause permanent damages to healthy tissues and eventually to the entire organ system. The phenomenon of inflammation is caused by specific lipid-derived and small molecular weight proteins referred to as eicosanoids and cytokines, respectively. The cellular and systemic level of eicosanoids and cytokines, and the relative presence of pro- vs. anti-inflammatory mediators determines overall inflammatory milieu in an individual. Uncontrolled and chronic inflammation triggers

IntechOpen

many common conditions and diseases. Bronchitis, conjunctivitis, chronic obstructive pulmonary disease (COPD), hepatitis, irritable bowel disease (IBD), psoriasis, rheumatoid arthritis (RA), and rhinitis are known to have a critical link with eicosanoids productions that eventually leads inflammation [1–8]. Even though effective pharmaceutical interventions are available to treat as well as manage these complications, adjunct therapies involving dietary management is mostly recommended [9, 10]. Nutraceutical intervention in the management of these inflammatory complications is gaining importance due to convenience, safety, and the low-cost attached in it Investigations have pointed out that n-3 polyunsaturated fatty acids (n-3 PUFA) positively alter inflammatory pathways. Their actions aid in reduced levels of eicosanoids (20 carbon lipid-derived molecules) and other inflammatory biomarkers, such as interleukin-1b (IL-1b), IL-6, and tumor necrosis factor-α (TNF- α) in plasma and urine, by serving as secondary messengers or modifying the transcriptional regulation of particular genes involved in inflammation [11]. In contrast, a high intake of dietary n-6 polyunsaturated fatty acids (n-6 PUFA) is associated with increased synthesis of proinflammatory eicosanoids synthesized from arachidonic acid metabolism, which include leukotrienes, lipoxins, thromboxanes, prostaglandins, and hydroxy fatty acids, and suppresses the synthesis of anti-inflammatory eicosanoids derived from EPA and DHA [12].

The modern (mostly western) diet is a poor source of n-3 PUFA, with an abysmal n-3:n-6 fatty acid ratio of 1:15–20, which is in sharp contrast to that found in the diet followed by our ancestors-1:1 [13, 14]. Both n-3 and n-6 PUFA consumed in the diet compete for the same enzymes and also regulate many transcription factors which impact cellular and tissue metabolism, and, a skewed ratio of n-3:n-6 PUFA result in the altered equilibrium in composition and fluidity of cell membrane as well as the functionality of organs [15, 16]. Evidence is accumulating that these physiological changes have led to interactions between constituents of the diet and genes, culminating in the rise of local and systemic inflammation. It is coupled with the disruption of immune/defense mechanism leading to rheumatoid arthritis, chronic obstructive pulmonary disease, irritable bowel diseases, psoriasis, bronchitis, rhinitis, hepatitis, conjunctivitis, etc.

2. Therapeutic potential of polyunsaturated fatty acids in inflammatory diseases

The intervention of cellular functions by bioactive lipids from the diet offers an attractive tool to modify or to prevent many pathophysiological processes of inflammatory diseases including bronchitis, conjunctivitis, COPD, hepatitis, IBD, psoriasis, RA and rhinitis. Dietary n-3 PUFA act via several mechanisms, which they primarily conduct by substituting phospholipid pools containing arachidonic acid (AA) with EPA and DHA. The enhanced physiological levels of EPA and DHA, suppress cyclooxygenase and lipoxygenase and lower synthesis of the inflammatory eicosanoids, thromboxane (TXA2) and leukotriene B4 (LTB4) from 20:4 in platelets and macrophages. Anti-inflammatory effects of n-3 PUFA can also be apart from eicosanoid pathway by affecting the expression pattern of cell surface adhesion molecules which participate in interactions between leukocytes and endothelial cells, thereby facilitating the infiltration of leukocytes into sites of inflammation [17]. Understanding the influence of dietary n-3 PUFA on the modulation of inflammatory complication is hence crucial if they need to be exploited as an adjunct therapy in treating many of these diseases.

2.1 Bronchitis

Chronic bronchitis is associated with high morbidity and mortality worldwide and has a diverse range of associated factors, including occupational, economic, and educational status [18]. The inflammation observed in chronic bronchitis is marked by airway eosinophilia, activated T-lymphocytes, rise in the amount of TNF-α-positive cells and neutrophils in the bronchial mucosa [1]. The advanced stage of the disease also shows similar signs, accompanied by hypertrophy of goblet cells, purulent bronchitis, and dilatation (diffuse or localized). The rise in eosinophils in bronchitis (caused by allergens, chemicals or drugs, etc.) is characterized by asthma and chronic cough and currently treated with anti-inflammatory drugs like inhaled corticosteroids. Although a shortcoming of this treatment is that it is most helpful only in those respiratory illnesses where eosinophils are elevated, and not in other types of bronchitis [19].

Moreover, acute bronchitis which is generally treated with antibiotics has several reported undesired side-effects associated with their use- nausea, vomiting or diarrhea, headaches, skin rash, and vaginitis, apart from population-level harm from antibiotic resistance [20]. Concerning the chronic form of the disease as seen in smokers, etc., rampant use of prophylactic antibiotics does not seem to have a considerable effect [21]. In such a scenario, having an effective treatment option for bronchitis, that eliminates any adverse effects and can always be effective despite widespread and prolonged use, is welcome. Some essential nutrients have proven benefits in this regard. A randomized self-controlled study was done using n-3 PUFA, vitamin C, and Zn supplementation in asthmatic children showed that nutritional intervention could lower the severity of inflammatory diseases. This was confirmed by testing the pulmonary function and inflammatory markers present in the sputum, where a remarkable improvement was seen in results of those on a diet supplemented with n-3 PUFA, vitamin C, and Zn [22]. When n-3 PUFA are considered individually, it was established from a study that daily ingestion of 3 g of n-3 PUFA for a month by atopic patients, showed reduced bronchial hyperactivity and improved responsiveness towards the ultrasonically nebulized distilled water [23]. To validate the connection between dietary fats and bronchitis, the pathophysiology of this condition needs to be addressed. The inflammation is related to altered level in membrane fatty acid composition of erythrocytes and an imbalance in the precursor of a pro to anti-inflammatory eicosanoid ratio [24]. A comparative assessment of the impact of n-3 and n-6 PUFA on respiratory function was done, by supplementation of the diet of asthmatic patients with perilla oil (n-3) or corn oil (n-6). The study revealed that, leucocytes from the corn oil supplemented group generated a greater amount of LTB4 and LTC4 than the perilla oil supplemented group, and the anti-inflammatory benefits in the latter were also reflected through their improved respiratory function parameters of peak expiratory flow (PEF) in the mornings, forced expiratory volume (FEV) and forced vital capacity (FVC) [25]. On deeper investigation of the mechanisms of n-3 fatty acid-mediated actions, it was found that mice exposed to cigarette smoke to mimic lung injury caused by smoke and pulmonary toxicants, on treatment with the resolvin RvD1, exhibited potent anti-inflammatory and resolution of inflammation in the lungs when supplied after the final smoke exposure. The same study also made use of blood monocytes, small airway epithelial cells, and primary human lung fibroblasts, which were treated with cytokine IL-1β or an extract of cigarette smoke along with RvD1 in in vitro. RvD1 inhibits the release of pro-inflammatory mediators by primary human cells in a dose-dependent manner, attenuates neutrophilic lung inflammation and release of pro-inflammatory cytokines, along with the upregulation of the anti-inflammatory cytokine IL-10 [26]. Lipoxins and resolvins serve as natural agonists in resolution of pulmonary inflammation. Briefly, lipoxins are arachidonic acid derivatives generated via multistep enzymatic reactions mediated by lipoxygenases to produce trihydroxy-tetraene-containing eicosanoids. On the other hand, resolvins are endogenous autacoids, derived from EPA and DHA as E-series (RvE) and D-series (RvD) resolvins, respectively. The intracellular signaling pathways resolvins adopt to bring about their actions, involve NF-Ƙβ and kinases. Resolvins can control the NF-Ƙβ signaling pathway, for e.g. RvE1 is an agonist in the signal transduction mechanism to inhibit TNF-α induced activation of NF-Ƙβ in a concentration-dependent manner. It can also suppress pro-inflammatory LTB4-mediated BLT1 signaling by lowering activation of NF-Ƙβ. The mechanism of NF-Ƙβ repression is relevant for resolution of inflammation in pulmonary diseases, as it induces granulocyte apoptosis. With respect to kinases, the pro-inflammatory reactions involved in chronic inflammatory lung diseases, are mainly regulated by phosphatidylinositol 3-kinase (PI3-K) and the MAPK family member extracellular signal regulated kinase (ERK). *In vitro*, RvE1 increases phosphorylation of kinases, which has a protective effect of controlling apoptotic programs in cells. RvE1 also mitigates attenuates inflammatory pain by indirectly inhibiting the ERK signaling pathway [27].

2.2 Irritable bowel disease (IBD)

Inflammatory/irritable bowel disease is a broader term used to describe the disorders that involve the inflammation of the gastrointestinal tract. Crohn's disease (CD) and ulcerative colitis (UC) are both related to IBD and result from interactions between gene, environment, and gut microbiota. Crohn's disease is characterized by chronic inflammation which relapses and remits, ultimately affecting the entire gastrointestinal tract, whereas ulcerative colitis inflammation is mostly confined to the colon-rectum junction. The pathogenesis of the disease mostly involves predisposing genetic, environmental, and immunologic factors [28, 29]. More than 200 gene regions have been identified that confer risk for Crohn's disease or ulcerative colitis in the European population [30]. The currently available drugs to treat IBD work by targeting receptors of T-cell activation (e.g., visilizumab, abatacept), prebiotics, probiotics targeting the intestinal flora antibiotics, adhesion molecule blockers (e.g., MLN-02, alicaforsen, natalizumab), cytokines (e.g., interleukin 10) [31]. Despite the advent of new therapeutic agents in recent years, current treatments are modestly successful with notable side effects [32]. Also, epidemiologic studies reported that increased animal fat/n-6 PUFA intake with the prevalence of both CD [33, 34] and UC [35]. Hence it is necessary to understand the role of n-3 PUFA and its metabolites in disease prevention of IBD. The potential role of n-3 PUFA in inflammation has garnered interest in fatty acid profiling in various metabolic and inflammatory diseases. Like other diseases, the inflammatory state in IBD is a result of the eicosanoid pathway with elevated levels of LTB4, LTC4, and thromboxanes (A1 and A2) in the inflamed intestinal mucosa [4]. Many clinical studies conducted with fish oil or EPA supplementation reported the inhibitory effect on inflammatory molecules production [36–38]. Animal model studies are in line with results of clinical studies concluded that n-3 PUFA ameliorated the intestinal inflammation [39–41]. The immunomodulatory and anti-inflammatory effects of n-3 PUFA are reported either directly [42, 43] or indirectly by the generation of PUFA metabolites containing hydroxyl groups i.e. resolvins, hydroxy fatty acids etc. [44]. Although there is a reasonable amount of data available on anti-inflammatory effect of n-3 PUFA, limited reports available in their use on IBD patients; hence deeper research in the aspects of dose, and mechanistic understanding is needed to replace or enhance the effect of currently available drugs along with nutritional intervention.

2.3 Psoriasis

Psoriasis is an inflammatory skin disease, affecting about 2% of the general population, characterized by hyperproliferation and poor differentiation of keratinocytes, increased vascularization of the skin, leukocyte infiltration and fibroblast activation [45]. It is a skin disease considered to be resulting from T-cell-mediated inflammation, wherein the inflammatory reaction is controlled by diet, lifestyle, and environmental cues such as infections and stress. Although the etiology remains unknown, accumulating evidence indicates that cytokines, mainly tumor necrosis factor α (TNF-α), IL-17 and IL-23, play essential roles in the development [6]. In the psoriatic lesions, inflammatory dermal dendritic cells (DCs) produce IL-23 that enhances the production of T cytotoxic 17 cells (Tc17) or IL-17 from T helper 17 cells (Th17) in the skin, leading to epithelial hyperplasia and inflammatory cell infiltration [46]. A documented and conclusive cure for psoriasis does not exist, some of the commonly used treatments to reduce the severity of disease and lessen the impact are phototherapy, topical and systemic therapies in moderate-to-severe psoriasis and drugs are corticosteroids, methotrexate, tar, cyclosporine, and emollients. Nevertheless, almost all the mentioned treatments are accompanied by major adverse effects [47]. Among nutritional components regulating the pathophysiology of psoriasis, PUFAs proved to be promising in the treatment of many skin diseases as safe, including psoriasis [48]. Calorie restriction or weight loss may also influence the severity of the disease. However randomized control trial study conducted by Guida et al. concluded that calorie restriction (20 kcal/kg/day) along with n-3 PUFA (2.6 g/day) enriched diet improved the clinical response of immune-modulating drugs in obese psoriatic patients [49]. Many clinical studies have demonstrated that the administration of fish oil or n-3 PUFA ethyl esters in psoriatic animal model/patients ameliorated the degree of skin inflammation [44–48, 50]. However, the reports on the exact mechanism are limited. A recent study conducted by Sawada et al. using metabolite produced by n-3 PUFA such as resolvinE1 (RvE1) in imiquimod (IMQ)-induced murine psoriasis mouse model concluded that RvE1 remarkably suppressed the epidermal hyperplasia and inflammatory cell infiltration in the psoriatic skin. These suppressive effects are due to the antagonistic effect of RvE1 on BLT1 production, a receptor of LTB4 and decrease in mRNA expression of cytokines (IL-17 and IL-23) indicating the novel mechanism of psoriatic pathogenesis and potential of RvE1 as an adjuvant in the psoriatic therapy [51].

2.4 Chronic obstructive pulmonary disease (COPD)

COPD is a chronic inflammatory lung disease triggered by exposure to toxic particles and gases, especially, cigarette smoke. Some of the predominant characteristics of COPD include parenchymal destruction, poorly reversible, persistent airflow limitation, and chronic bronchitis. A COPD patient shows some visible symptoms like cough, chest tightness dyspnoea, excess sputum production, and wheezing [52]. COPD is coordinated by a complex network of inflammatory mediators, such as inflammatory enzymes, lipid mediators (eicosanoids, namely prostanoids and leukotrienes, LTs), interleukins (ILs), and adhesion molecules, which may moderate airway inflammation through chemotactic and autocrine or paracrine effects [3]. Eicosanoids are produced by cells resident in the lung and airways, such as fibroblasts, epithelial cells, myofibroblasts, and smooth muscle cells, also by inflammatory cells that have migrated from the circulation to the airways, such as macrophages, mast cells, neutrophils, platelets, eosinophils, and T-lymphocytes [53]. Activated macrophages release cytokines such as IL-8, IL-6, IL-10, TNF-α, and LTB4, which can attract and activate various inflammatory cells.

Smooth muscle cells, fibroblasts, and myofibroblasts play a key regulatory role in airway remodeling by generating different structural components, such as collagen and proteoglycans [54]. Although the mediators involved in structural remodeling of the bronchial wall remains unknown, chemokines, cytokines, and growth factors are thought to have a crucial role. Observational clinical studies suggest that n-3 PUFA levels in COPD are inversely related to systemic inflammation and directly related to clinical outcomes [55–57]. Interventional studies using n-3 PUFAs alone in COPD do not exist; however, many trials are currently underway, which may generate valuable data in the coming years [58–60]. An animal study in pneumonia model of mice, revealed RvE1 metabolite of n-3 PUFA decreased levels of several pro-inflammatory chemokines and cytokines in the lung, and improved survival [61]. However, currently, there is a paucity of data regarding the potential of n-3 PUFAs to be used therapeutically in COPD. Recent studies, conducted using omega-3 PUFAs in COPD has used nutrient combinations, in a manner that the exact effects of omega-3 PUFAs cannot be elucidated.

2.5 Rheumatoid arthritis

Rheumatoid arthritis (RA) is a low grade chronic inflammatory autoimmune disease affecting the joints and bones. RA and systemic lupus erythematosus (SLE) are multisystem autoimmune diseases featuring the production of a variety of autoantibodies, and resulting in higher immune-mediated inflammation. Prevalence of RA is known to affect 17.6 million people worldwide [7] with the huge economic burden to mankind. Pathogenesis of RA in early phase focused on auto-antibodies and immune complexes [62], however, the nature of T-cell mediated antigen-specific responses, T-cell-independent cytokine species, and aggressive tumor-like behavior of rheumatoid synovium have also been reported. Immune complex theory strongly suggests increased neutrophils accumulation in synovial fluid results in engulfment of immune complex and release proteolytic enzymes [62]. A better understanding of the intracellular targets that regulate cytokines in RA can potentially lead to new therapeutic interventions. For instance, studies reported activation of NF-κB in the synovium of RA patients [63] and mitogen-activated protein (MAP) kinases are identified as vital regulators of cytokine and metalloproteinase production. The mechanistic approach to currently available drugs is partially understood and mostly targeting the eicosanoid pathway, and reports are scarce about gene targets on CD28, AP-1, and MAP kinases. Investigators have examined the effect of dietary fatty acid supplementation in different autoimmune diseases, and the effects of both n-6 and n-3 PUFA on RA have been reported [64]. Resolvins derived from EPA and DHA are anti-inflammatory and resolve inflammation, serving as important mediators in regulating various homeostatic functions, including gastric mucosal integrity. Kremer and co-workers reported that a dose of 90 mg/kg/day EPA/DHA (3:2 ratio) showed a shorter period to respond than 45 mg/kg/day [65]. Clinical studies have ambiguity in results, few observed reductions in both tender joints by 36% and swollen joint to 38% in patients, whereas the placebo group showed no improvement in these parameters [66]. Fish oil supplementation in women patients showing a decrease in the production of serum nuclear factor-kappa B (Nf-kB) ligand/osteoprotegerin ratio [67]. Cell culture-based investigations have reported that EPA and DHA prevent the proliferation of human T cells and their generation of IL-2 [68]. Animal studies also reported a beneficial effect of marine n-3 PUFA in RA [69, 70]. Finally, supplementation with dietary n-3 PUFA and its ability to inhibit TNF-α and IL-1b synthesis is rational. Considering the emphasis on recommendations to increase dietary n-3 PUFA intake for health benefits, the possible therapeutic potential of fish oil/n-3 PUFA on autoimmune diseases need to be clearly defined.

2.6 Conjunctivitis

Conjunctivitis is an ocular inflammatory condition of the membrane lining the eyelid, which provides a shelter to the open surface of the sclera [71]. It is said to be the most common cause of "red eye," and the infection, either acute or chronic can originate from three different sources, based on which it is commonly classified: viral, allergic, and bacterial conjunctivitis. The type of conjunctivitis may be determined by a diagnostic investigation which takes into account, the patient's age, time of the year, and physical examination findings [72]. Identification of the pathogen which elicits the inflammatory response is necessary to decide on the required treatment module, since some bacterial conjunctivitis forms are self-limiting, while those caused by *Chlamydia trachomatis* or *Neisseria gonorrhoeae* demand aggressive antibiotic therapy, and primarily good eyelid hygiene [71]. The visual symptoms of inflammation at the ocular surface, include, lid and conjunctival edema–redness, tearing, extreme itching, and photophobia during the acute phase cause immense discomfort, thereby necessitating treatment [73]. Effective pharmacological cure comprises antihistamines, mast cell stabilizers, and non-steroidal anti-inflammatory drugs [74]. The excess wateriness experienced during conjunctivitis and also early dry eye disease is because of enhanced goblet cell mucin secretion into tears, which otherwise produce a regulated amount of mucins only to protect the eye. This phenomenon is due to the action of inflammatory mediators, cysteinyl leukotrienes LTB4, LTC4, LTD4, and LTE4, and prostaglandin PGD2, which stimulate conjunctival goblet cell mucous secretion. Thus, mitigation of this LTD4 effect can aid in reducing mucin secretion, and this has been achieved previously by the action of Resolvins D1 (RvD1) and E1 (RvE1) [2]. The finding of Dartt and co-workers [2] indicates that lipids including fatty acids or lipid mediators play a role in suppression of systemic inflammation as well as local, such as in the case of inflammatory conjunctivitis. The connection between the quality of fat consumed and conjunctivitis was drawn from the results of a Japanese study, which states that meat intake is directly proportional to the prevalence of rhinoconjunctivitis in young adult Japanese women, while no correlation was seen between fish intake and rhinoconjunctivitis [75]. Though, there are not many reports from similar controlled clinical trials that specifically correlate meat intake with incidence of conjunctivitis, to ascertain such claims. The study does raise questions on the role of fats present in meat that possibly trigger the pathways in inflammation. The true potential of n-3 PUFA as an anti-inflammatory in managing conjunctivitis was elucidated through other detailed investigations of their actions and mechanisms at molecular levels. A primary mode of action is that of the resolvin mediated responses, as mentioned earlier. The process of resolution of inflammation is an active process facilitated by pro-resolution lipid mediators. Not only D-series resolvins RvD1, which are produced in the cornea but also aspirin-triggered RvD1 (ATRvD1) have exhibited regulation of inflammatory responses to histamine in allergic conjunctivitis. The effect is implemented by a cross-talk between two types of G protein-coupled receptors (GPCRs). When RvD1 interacts with its receptor GPR32, it prevents the histamine-stimulated H1 receptor-mediated rise in intracellular Ca^{2+}, thereby blocking H1 receptor-mediated responses. Consequentially, this activates extracellularly regulated–protein kinase (ERK) 1/2 [76]. In addition to resolvins, another class of lipid mediators that can mimic these anti-inflammatory actions in conjunctivitis includes lipoxins, especially, lipoxin A4 that has benefits similar to that of ATRvDI and can activate ALX/FPR2 receptor to regulate conjunctival goblet cell secretion, which is particularly useful in maintaining ocular surface homeostasis, and managing the dry eye syndrome [77]. Another evidence of polyunsaturated fats aiding in the treatment of conjunctivitis emerged on evaluating the impact of the feeding of omega-3 and omega-6 PUFA on human leukocyte antigen-DR (HLA-DR) marker of conjunctival inflammation in dry eye patients; it was observed that increased intake of both these fatty acids suppressed the expression of HLA-DR [78]. With this, it is evident that dietary supplementation of n-3 PUFA has a modulatory effect on conjunctivitis. Their efficacy has been tested in mammals such as dogs affected with keratoconjunctivitis sicca, where EPA + DHA + antioxidants in defined proportion was used as an adjuvant [79]. Furthermore, their use of topical treatment has also been suggested in the literature. Based on data from animal preliminary human studies, protectins (NPD1) and DHA turn out to be a safe, effective treatment for dry eye through a topical application [80]. Ultimately, investigators have also proposed formulations for active use in therapy, one such patented formulation by Aleo et al. [81], clearly provided the n-3 and n-6 PUFA proportion for creating an ophthalmic composition suitable for ocular inflammation.

2.7 Hepatitis

Hepatitis is the inflammation of the liver with symptoms that generally include right upper abdominal pain, headache, fatigue and malaise, myalgia, altered sense of smell or taste, nausea and vomiting, coryza, photophobia, diarrhea (may have pale stools and dark urine) [5]. Hepatitis treatment can be managed better without any inconveniences, with the help of dietary interventions, which not only limit the progression of hepatitis but can also be used as a novel therapy. Considering the abundant evidence of n-3 PUFA and their anti-inflammatory effect in most liver maladies, including fatty liver disease, cirrhosis, and acute liver failure, their role in immune-mediated liver diseases is not fully exploited. Fatty acids belonging to the n-3 family have been shown to suppress hepatitis via an autophagy-dependent mechanism and are particularly helpful for therapy in autoimmune hepatitis [82]. In this regard, the n-3 PUFA mediators resolvin D1 and E1 can be particularly useful, as they inhibit concanavalin A-induced liver injury and restricting the progression of hepatitis to liver cancer in mice through suppression of NF-kβ/AP-1 activity [83]. The actions of n-3 PUFA were evaluated in a macrophage-dependent acute D-galactosamine/lipopolysaccharide (D-GalN/LPS) hepatitis model in the transgenic fat-1 mice. The findings clearly stated that the n-3 PUFA supplementation dampened the severity of the inflammatory liver injury and histological liver damage in fat-1 mice. The balance in n-6/n-3 PUFA ratio was improved, and levels of serum alanine aminotransferase and TNF-α and IL-1β cytokine production were lowered. The effect was confirmed through the lowered hepatic gene expression of all the pro-inflammatory cytokines, thus proving their anti-inflammatory benefits in the framework of liver inflammation [84]. The most striking feature of n-3 PUFA therapy in HCV was discovered when the efficacy of EPA was evaluated against ribavirin (RBV) associated hemolytic anemia. When EPA was supplemented with the standard combination therapy of peginterferon (PEG-IFN) and ribavirin (RBV), patients receiving EPA, required reduced RBV doses compared to the non-EPA group, and also showed decreased RBV-induced hemolysis, although rates of virological response are yet to be elucidated [85]. Therefore, it is encouraging that EPA can lower dependency on drugs in even such morbid disorders. Some of the other adverse effects of RBV treatment which have been effectively countered by n-3 PUFA include lowering of the impairment of the filterability of erythrocytes of chronic HCV patients in whom erythrocyte filterability was caused due to oxidative membrane damage induced by RBV which led to hemolytic anemia [86]. This finding further ensures the maintenance of the lymphocyte levels and improvement in hemoglobin levels in the patients of PEG-IFN and RBV treatment [87]. Thus, n-3 PUFA and their mediators can be recommended for concurrent administration with the recommended standard interferon and antibiotic therapy, to facilitate recovery, ameliorate adverse effects, and prevent mortalities in chronic cases.

2.8 Rhinitis

Allergic rhinitis (AR) is a highly prevalent heterogeneous disorder, caused by pollens, molds, dust mites, and animal dander, but often goes undiagnosed as a condition of rhinitis. The symptoms of AR, which include sneezing, itching, nasal congestion, and rhinorrhea are mostly IgE mediated and triggered through mucosal infiltration and action on eosinophils, plasma cells, and mast cells, [7]. Some other discomforts experienced by the patients in chronic rhinitis include sinusitis, eustachian-tube dysfunction, sleep disturbances, and forces the patient to breathe through the mouth. AR may not appear as a serious ailment but is clinically relevant as it triggers numerous complications, is a potential risk factor for asthma, deteriorates the quality of life and efficiency at work. The range of pharmaceutical options available to manage a rhinitis episode includes intranasal corticosteroids, which are effective and safe. The most common drug type is that of first-generation antihistamines, but they are well-known to cause sedation, psychomotor retardation, and decreased academic performance. AR is tackled most effectively from its natural history, using immunotherapy by targeting allergens singly [88]. Various dietary factors have been associated with allergic rhinitis, such as dietary antioxidants in vegetables may reduce wheezing symptoms, but fats such as butter and margarine tend to aggravate the symptoms [89]. The exact mechanism by which saturated fats alter airway inflammation has neither been fully understood nor has it received much attention. Yet, reports suggest that accumulation of saturated fatty acids is associated with changes in levels of serum cholesterol and arachidonic acid content in the cell membranes, which together modify the lymphocyte function. These observations and other epidemiological data led to the conclusion that saturated fats aggravate rhinitis, asthma and related conditions [90]. Since saturated fats were implicated in worsening rhinitis, the effect of unsaturated fats on allergic sensitization and allergic rhinitis was also investigated by many researchers. It was found that the presence of unsaturated fatty acids in membranes of red blood cells (RBC) in the form of EPA and the diet as ALA, lower sensitivity and incidence of rhinitis [8, 91]. Fish consumption during pregnancy as well as childhood was associated with a lowered risk of any respiratory illnesses—allergic rhinitis, wheezing or asthma in childhood and later, and long chain n-3 PUFA supplementation was also not related to the risk of postpartum hemorrhages in mothers as well [92, 93]. An insight into the mechanism by which this effect is achieved suggests that n-3 PUFA supplementation during pregnancy boosts levels of placental DHA and specialized pro-resolving lipid mediators (SPM) precursors without aggravating inflammatory gene expression [94]. Also, adjunctive supplementation of fish quickens the effects of routine pharmacotherapy in AR subjects by lowering AR related symptoms and serum levels of Ig E, which is crucial since IL-5 and Ig E are found in higher concentrations in nasal secretion and sputum of AR patients [95]. With such abundant information pointing towards the preventive and prophylactic effects of polyunsaturated fatty acids on the rhinitis, the dietary incorporation of n-3 PUFA, as well as their use as adjuvant therapy for lowering the incidence and severity of a rhinitis episode, can be strongly recommended. These fatty acids have also demonstrated a potential to slow down the progression of chronic rhinitis, thereby improving quality of life and productivity at work.

3. Summary and conclusion

Taken together, findings from in-vitro and in-vivo studies of inflammatory diseases suggest that n-3 PUFA could be potential therapeutic molecules to combat inflammatory diseases as described in this chapter. The results from mechanistic studies reveal the effects of LC-PUFAs (n-6 and n-3) on human health are highly favorable. However, the relation between the two families of LC-PUFAs on the mechanisms of action needs to be further understood as the presence of these two fatty acids in the diet play a significant effect on the metabolism of other. The inflammatory mechanism involved in the pathogenesis provides novel candidature targets for cost-effective pharmacological and nutraceutical drugs. Despite a better understanding of the beneficial effects of LC-PUFAs, there is a lack of data on attempts to replace or supplement with currently available drugs. Investigating the synergistic potential of various food bioactive molecules with existing clinically proven drugs could be the most effective therapy to ameliorate many of these inflammatory diseases.

Author details

Vinayak Uppin, Pooja Acharya and Ramaprasad Ravichandra Talahalli*
Department of Biochemistry, CSIR-Central Food Technological Research Institute, Mysore, Karnataka, India

*Address all correspondence to: ramaprasad@cftri.res.in

References

[1] Saetta M et al. Airway eosinophilia in chronic bronchitis during exacerbations. American Journal of Respiratory and Critical Care Medicine. 1994;**150**(6):1646-1652

[2] Dartt DA et al. Conjunctival goblet cell secretion stimulated by leukotrienes is reduced by resolvins D1 and E1 to promote resolution of inflammation. The Journal of Immunology. 2011;**186**(7):4455-4466

[3] Barnes PJ. Molecular genetics of chronic obstructive pulmonary disease. Thorax. 1999;**54**(3):245-252

[4] Sharon P, Stenson WF. Enhanced synthesis of leukotriene B4 by colonic mucosa in inflammatory bowel disease. Gastroenterology. 1984;**86**(3):453-460

[5] Ryder SD, Beckingham IJ. Acute hepatitis. BMJ. 2001;**322**(7279):151-153

[6] Hawkes JE, Chan TC, Krueger JG. Psoriasis pathogenesis and the development of novel targeted immune therapies. Journal of Allergy and Clinical Immunology. 2017;**140**(3):645-653

[7] Centers for Disease Control and Prevention (CDC). Arthritis: Disabilities and Limitations. Retrieved from: https://www.cdc.gov/arthritis/data_statistics/arthritis-disabilities-limitations-tables.htm

[8] Skoner DP. Allergic rhinitis: Definition, epidemiology, pathophysiology, detection, and diagnosis. Journal of Allergy and Clinical Immunology. 2001;**108**(1):S2-S8

[9] Maroon JC, Bost JW. ω-3 fatty acids (fish oil) as an anti-inflammatory: An alternative to nonsteroidal anti-inflammatory drugs for discogenic pain. Surgical Neurology. 2006;**65**(4):326-331

[10] Gerasimidis K et al. Dietary modifications, nutritional supplements and alternative medicine in paediatric patients with inflammatory bowel disease. Alimentary Pharmacology & Therapeutics. 2008;**27**(2):155-165

[11] Wall R et al. Fatty acids from fish: The anti-inflammatory potential of long-chain omega-3 fatty acids. Nutrition Reviews. 2010;**68**(5):280-289

[12] Simopoulos AP. The importance of the omega-6/omega-3 fatty acid ratio in cardiovascular disease and other chronic diseases. Experimental Biology and Medicine. 2008;**233**(6):674-688

[13] Simopoulos AP. Omega-6/omega-3 essential fatty acids: Biological effects. World Review of Nutrition and Dietetics. 2009;**99**(1):1-16

[14] Simopoulos AP. Importance of the omega-6/omega-3 balance in health and disease: Evolutionary aspects of diet. In: Healthy Agriculture, Healthy Nutrition, Healthy People. Vol. 102. Karger Publishers; 2011. pp. 10-21

[15] Das UN. Folic acid and polyunsaturated fatty acids improve cognitive function and prevent depression, dementia, and Alzheimer's disease—But how and why? Prostaglandins, Leukotrienes and Essential Fatty Acids. 2008;**78**(1):11-19

[16] Yang X et al. Effects of fatty acid unsaturation numbers on membrane fluidity and α-secretase-dependent amyloid precursor protein processing. Neurochemistry International. 2011;**58**(3):321-329

[17] Endres S et al. n-3 polyunsaturated fatty acids: Update 1995. European Journal of Clinical Investigation. 1995;**25**(9):629-638

[18] Menezes AM, Victora CG, Rigatto M. Prevalence and risk factors for

chronic bronchitis in Pelotas, RS, Brazil: A population-based study. Thorax. 1994;**49**(12):1217-1221

[19] Gibson PG, Fujimura M, Niimi A. Eosinophilic bronchitis: Clinical manifestations and implications for treatment. Thorax. 2002;**57**(2):178-182

[20] Smith SM et al. Antibiotics for acute bronchitis. Cochrane Database of Systematic Reviews. 2014;**3**

[21] Staykova T et al. Prophylactic antibiotic therapy for chronic bronchitis. Cochrane Database of Systematic Reviews. 2001;(2)

[22] Biltagi MA et al. Omega-3 fatty acids, vitamin C and Zn supplementation in asthmatic children: A randomized self-controlled study. Acta Paediatrica. 2009;**98**(4):737-742

[23] Villani F et al. Effect of dietary supplementation with polyunsaturated fatty acids on bronchial hyperreactivity in subjects with seasonal asthma. Respiration. 1998;**65**(4):265

[24] Novgorodtseva TP et al. Modification of the fatty acid composition of the erythrocyte membrane in patients with chronic respiratory diseases. Lipids in Health and Disease. 2013;**12**(1):117

[25] Okamoto M et al. Effects of dietary supplementation with n-3 fatty acids compared with n-6 fatty acids on bronchial asthma. Internal Medicine. 2000;**39**(2):107-111

[26] Hsiao H-M et al. A novel anti-inflammatory and pro-resolving role for resolvin D1 in acute cigarette smoke-induced lung inflammation. PLoS One. 2013;**8**(3):e58258

[27] Uddin M, Levy BD. Resolvins: Natural agonists for resolution of pulmonary inflammation. Progress in Lipid Research. 2011;**50**(1):75-88

[28] Podolsky DK. The current future understanding of inflammatory bowel disease. Best Practice & Research Clinical Gastroenterology. 2002;**16**(6):933-943

[29] Xavier RJ, Podolsky DK. Unravelling the pathogenesis of inflammatory bowel disease. Nature. 2007;**448**(7152):427

[30] McGovern DPB, Kugathasan S, Cho JH. Genetics of inflammatory bowel diseases. Gastroenterology. 2015;**149**(5):1163-1176

[31] Baumgart DC, Sandborn WJ. Inflammatory bowel disease: Clinical aspects and established and evolving therapies. The Lancet. 2007;**369**(9573):1641-1657

[32] Safroneeva E et al. Impact of the early use of immunomodulators or TNF antagonists on bowel damage and surgery in Crohn's disease. Alimentary Pharmacology & Therapeutics. 2015;**42**(8):977-989

[33] Shoda R et al. Epidemiologic analysis of Crohn disease in Japan: Increased dietary intake of n-6 polyunsaturated fatty acids and animal protein relates to the increased incidence of Crohn disease in Japan. The American Journal of Clinical Nutrition. 1996;**63**(5):741-745

[34] Costea I et al. Interactions between the dietary polyunsaturated fatty acid ratio and genetic factors determine susceptibility to pediatric Crohn's disease. Gastroenterology. 2014;**146**(4):929-931

[35] Ananthakrishnan AN et al. Long-term intake of dietary fat and risk of ulcerative colitis and Crohn's disease. Gut. 2014;**63**(5):776-784

[36] Hawthorne AB et al. Treatment of ulcerative colitis with fish oil supplementation: A prospective 12 month randomized controlled trial. Gut. 1992;**33**(7):922-928

[37] Stenson WF et al. Dietary supplementation with fish oil in ulcerative colitis. Annals of Internal Medicine. 1992;**116**(8):609-614

[38] Gil A. Polyunsaturated fatty acids and inflammatory diseases. Biomedicine & Pharmacotherapy. 2002;**56**(8):388-396

[39] Nieto N et al. Dietary polyunsaturated fatty acids improve histological and biochemical alterations in rats with experimental ulcerative colitis. The Journal of Nutrition. 2002;**132**(1):11-19

[40] MacLean CH et al. Systematic review of the effects of n−3 fatty acids in inflammatory bowel disease. The American Journal of Clinical Nutrition. 2005;**82**(3):611-619

[41] Uchiyama K et al. N-3 polyunsaturated fatty acid diet therapy for patients with inflammatory bowel disease. Inflammatory Bowel Diseases. 2010;**16**(10):1696-1707

[42] Calder PC, Kew S. The immune system: A target for functional foods? British Journal of Nutrition. 2002;**88**(S2):S165-S176

[43] Bassaganya-Riera J, Hontecillas R. Dietary CLA and n-3 PUFA in inflammatory bowel disease. Current Opinion in Clinical Nutrition and Metabolic Care. 2010;**13**(5):569

[44] Serhan CN et al. Novel functional sets of lipid-derived mediators with antiinflammatory actions generated from omega-3 fatty acids via cyclooxygenase 2–nonsteroidal antiinflammatory drugs and transcellular processing. Journal of Experimental Medicine. 2000;**192**(8):1197-1204

[45] Mayser P, Grimm H, Grimminger F. n-3 fatty acids in psoriasis. British Journal of Nutrition. 2002;**87**(S1):S77-S82

[46] Hawkes JE, Chan TC, Krueger JG. Psoriasis pathogenesis and the development of novel targeted immune therapies. Journal of Allergy and Clinical Immunology. 2017;**140**(3):645-653

[47] Paul C et al. Evidence-based recommendations on conventional systemic treatments in psoriasis: Systematic review and expert opinion of a panel of dermatologists. Journal of the European Academy of Dermatology and Venereology. 2011;**25**:2-11

[48] McCusker MM, Grant-Kels JM. Healing fats of the skin: The structural and immunologic roles of the ω-6 and ω-3 fatty acids. Clinics in Dermatology. 2010;**28**(4):440-451

[49] Guida B et al. Energy-restricted, n-3 polyunsaturated fatty acids-rich diet improves the clinical response to immuno-modulating drugs in obese patients with plaque-type psoriasis: A randomized control clinical trial. Clinical Nutrition. 2014;**33**(3):399-405

[50] Elisabeth S et al. Effect of dietary supplementation with very-long-chain n-3 fatty acids in patients with psoriasis. New England Journal of Medicine. 1993;**328**(25):1812-1816

[51] Sawada Y et al. Resolvin E1 attenuates murine psoriatic dermatitis. Scientific Reports. 2018;**8**(1):11873

[52] Rabe KF et al. Global strategy for the diagnosis, management, and prevention of chronic obstructive pulmonary disease: GOLD executive summary. American Journal of Respiratory and Critical Care Medicine. 2007;**176**(6):532-555

[53] Barnes PJ. Mechanisms in COPD: Differences from asthma. Chest. 2000;**117**(2):10S-14S

[54] Chung KF. Inflammatory mediators in chronic obstructive

pulmonary disease. Current Drug Targets-Inflammation & Allergy. 2005;**4**(6):619-625

[55] Wood LG et al. Innate immune mechanisms linking non-esterified fatty acids and respiratory disease. Progress in Lipid Research. 2009;**48**(1):27-43

[56] Shahar E et al. Docosahexaenoic acid and smoking-related chronic obstructive pulmonary disease. American Journal of Respiratory and Critical Care Medicine. 1999;**159**(6):1780-1785

[57] Hirayama F et al. Dietary intake of isoflavones and polyunsaturated fatty acids associated with lung function, breathlessness and the prevalence of chronic obstructive pulmonary disease: Possible protective effect of the traditional Japanese diet. Molecular Nutrition & Food Research. 2010;**54**(7):909-917

[58] Broekhuizen R et al. Polyunsaturated fatty acids improve exercise capacity in chronic obstructive pulmonary disease. Thorax. 2005;**60**(5):376-382

[59] Sugawara K et al. Effects of nutritional supplementation combined with low-intensity exercise in malnourished patients with COPD. Respiratory Medicine. 2010;**104**(12):1883-1889

[60] Sugawara K et al. Effect of anti-inflammatory supplementation with whey peptide and exercise therapy in patients with COPD. Respiratory Medicine. 2012;**106**(11):1526-1534

[61] Seki H et al. The anti-inflammatory and pro-resolving mediator resolvin E1 protects mice from bacterial pneumonia and acute lung injury. The Journal of Immunology. 2010;**184**(2):836-843

[62] Zvaifler NJ. The immunopathology of joint inflammation in rheumatoid arthritis. Advances in Immunology. 1973;**16**:265-336

[63] Han Z et al. AP-1 and NF-kB regulation in rheumatoid arthritis and murine collagen-induced arthritis. Autoimmunity. 1998;**28**(4):197-208

[64] Simopoulos AP. Omega-3 fatty acids in inflammation and autoimmune diseases. Journal of the American College of Nutrition. 2002;**21**(6):495-505

[65] Kremer JM et al. Dietary fish oil and olive oil supplementation in patients with rheumatoid arthritis clinical and immunologic effects. Arthritis & Rheumatism: Official Journal of the American College of Rheumatology. 1990;**33**(6):810-820

[66] Leventhal LJ, Boyce EG, Zurier RB. Treatment of rheumatoid arthritis with gammalinolenic acid. Annals of Internal Medicine. 1993;**119**(9):867-873

[67] Kolahi S et al. Fish oil supplementation decreases serum soluble receptor activator of nuclear factor-kappa B ligand/osteoprotegerin ratio in female patients with rheumatoid arthritis. Clinical Biochemistry. 2010;**43**(6):576-580

[68] Calder PC, Bevan SJ, Newsholme EA. The inhibition of T-lymphocyte proliferation by fatty acids is via an eicosanoid-independent mechanism. Immunology. 1992;**75**(1):108

[69] Volker DH, FitzGerald PEB, Garg ML. The eicosapentaenoic to docosahexaenoic acid ratio of diets affects the pathogenesis of arthritis in Lew/SSN rats. The Journal of Nutrition. 2000;**130**(3):559-565

[70] Calder PC. Marine omega-3 fatty acids and inflammatory processes: Effects, mechanisms and clinical relevance. Biochimica et Biophysica Acta (BBA)-Molecular and Cell Biology of Lipids. 2015;**1851**(4):469-484

[71] Morrow GL, Abbott RL. Conjunctivitis. American Family Physician. 1998;**57**(4):735-746

[72] Alfonso SA, Fawley JD, Alexa Lu X. Conjunctivitis. Primary Care: Clinics in Office Practice. 2015;**42**(3):325-345

[73] Ono SJ, Abelson MB. Allergic conjunctivitis: Update on pathophysiology and prospects for future treatment. Journal of Allergy and Clinical Immunology. 2005;**115**(1):118-122

[74] Bielory L, Friedlaender MH. Allergic conjunctivitis. Immunology and Allergy Clinics of North America. 2008;**28**(1):43-58

[75] Miyake Y et al. Dietary meat and fat intake and prevalence of rhinoconjunctivitis in pregnant Japanese women: Baseline data from the Kyushu Okinawa maternal and child health study. Nutrition Journal. 2012;**11**(1):19

[76] Li D et al. Resolvin D1 and aspirin-triggered resolvin D1 regulate histamine-stimulated conjunctival goblet cell secretion. Mucosal Immunology. 2013;**6**(6):1119

[77] Hodges RR et al. Lipoxin A 4 activates ALX/FPR2 receptor to regulate conjunctival goblet cell secretion. Mucosal Immunology. 2017;**10**(1):46

[78] Brignole-Baudouin F et al. A multicentre, double-masked, randomized, controlled trial assessing the effect of oral supplementation of omega-3 and omega-6 fatty acids on a conjunctival inflammatory marker in dry eye patients. Acta Ophthalmologica.

[79] Silva DA et al. Oral omega 3 in different proportions of EPA, DHA, and antioxidants as adjuvant in treatment of keratoconjunctivitis sicca in dogs. Arquivos Brasileiros de Oftalmologia. 2018;**81**(5):421-428

[80] Cortina MS, Bazan HEP. Docosahexaenoic acid, protectins and dry eye. Current Opinion in Clinical Nutrition and Metabolic Care. 2011;**14**(2):132

[81] Aleo D, Barabino S, Mangiafico S, Rolando M, Saita MGA. U.S. Patent No. 8,957,110. Washington, DC: U.S. Patent and Trademark Office; 2015

[82] Li Y et al. Endogenous n-3 polyunsaturated fatty acids attenuate T cell-mediated hepatitis via autophagy activation. Frontiers in Immunology. 2016;**7**:350

[83] Kuang H et al. Resolvin D1 and E1 alleviate the progress of hepatitis toward liver cancer in long-term concanavalin A-induced mice through inhibition of NF-κB activity. Oncology Reports. 2016;**35**(1):307-317

[84] Schmöcker C et al. Omega-3 fatty acids alleviate chemically induced acute hepatitis by suppression of cytokines. Hepatology. 2007;**45**(4):864-869

[85] Morsy KH, Zaghloul A, Mahmoud M. Can eicosapentaenoic acid maintain the original ribavirin dose or affect the response during the treatment course of chronic hepatitis C virus (HCV) patients? The Turkish Journal of Gastroenterology. 2016;**27**:55-61

[86] Seki R et al. Impaired filterability of erythrocytes from patients with chronic hepatitis C and effects of eicosapentaenoic acid on the filterability. The Journal of Physiological Sciences. 2007;**57**(1):43-49

[87] Kawashima A et al. Eicosapentaenoic acid supplementation for chronic hepatitis C patients during combination therapy of pegylated interferon α-2b and ribavirin. Lipids. 2008;**43**(4):325-333

[88] Greiner AN et al. Allergic rhinitis. The Lancet. 2011;**378**(9809):2112-2122

[89] Farchi S et al. Dietary factors associated with wheezing and

allergic rhinitis in children. European Respiratory Journal. 2003;**22**(5):772-780

[90] Huang S-L, Pan W-H. Dietary fats and asthma in teenagers: Analyses of the first nutrition and health survey in Taiwan (NAHSIT). Clinical & Experimental Allergy. 2001;***31***(12):1875-1880

[91] Hoff S et al. Allergic sensitisation and allergic rhinitis are associated with n-3 polyunsaturated fatty acids in the diet and in red blood cell membranes. European Journal of Clinical Nutrition. 2005;**59**(9):1071

[92] Gunaratne AW, Makrides M, Collins CT. Maternal prenatal and/or postnatal n-3 long chain polyunsaturated fatty acids (LCPUFA) supplementation for preventing allergies in early childhood. Cochrane Database of Systematic Reviews. 2015;(7)

[93] Stratakis N et al. Fish and seafood consumption during pregnancy and the risk of asthma and allergic rhinitis in childhood: A pooled analysis of 18 European and US birth cohorts. International Journal of Epidemiology. 2017;**46**(5):1465-1477

[94] Keelan JA et al. Effects of maternal n-3 fatty acid supplementation on placental cytokines, pro-resolving lipid mediators and their precursors. Reproduction. 2015;**149**(2):171-178

[95] Susibalan BD et al. Efficacy of snakehead fish (*Channa striatus*) in subjects with allergic rhinitis: A randomized controlled trial. Oriental Pharmacy and Experimental Medicine. 2018;**18**(3):209-215

Chapter 13

Beneficial Effect of Omega-3 Fatty Acids on Immune and Reproductive Endometrial Function

Maria A. Hidalgo, Marcelo Ratto and Rafael A. Burgos

Abstract

Omega-3 polyunsaturated fatty acids, such as docosahexaenoic acid (DHA) and eicosapentaenoic acid (EPA), are known by their anti-inflammatory effects through mechanisms such as formation of specialized pro-resolving mediators (SPM), and more recently a new mechanism dependent on the free fatty acid (FFA) receptors has been studied. DHA and EPA have shown an effect on the release of prostaglandins (PGs) E2 and F2α in endometrial cells, two PGs that have key function in fertility. In addition, other molecules such as cyclooxygenase-2, IL-1β, NF-κB, and intracellular signaling pathways are also affected by omega-3 fatty acids in endometrial cells. In this chapter, we will expose the following issues: eicosanoids in fertility and immune function in the uterus, effect of omega-3 fatty acids on endometrial function: in vivo and in vitro studies, mechanisms of action of omega-3 fatty acids in endometrial cells, and perspectives in health and diseases.

Keywords: omega-3 fatty acids, prostaglandin, endometrial cells, fertility, immune function

1. Introduction

Uterine function is key for a suitable reproductive performance and fertility. Endocrine and immune response play important roles for keeping the hormonal levels and the fetal-maternal interface. Hormones such as estrogen and progesterone are released from the ovaries; estrogen release is triggered by the hypothalamic-pituitary axis, and progesterone is secreted from the corpus luteum (CL) after ovulation. Estrogen is essential for uterine growth and cell proliferation and progesterone for endometrial receptivity and successful establishment of pregnancy [1, 2]. Estrogen and progesterone are also considered as regulators of innate immunity and inflammation in the endometrium [3]. Endometrial immune homeostasis plays an important role in the success of implantation and pregnancy, with complex interactions between the innate and adaptive immune system, through cells such as natural killer, antigen-presenting cells (macrophages and dendritic cells), and subtypes of T cells [4].

Prostaglandins (PGs), also known as prostanoids, are bioactive lipids with an important function as regulators of reproductive processes, including ovulation, fertilization, and implantation. PGs are synthesized from arachidonic acid by different cells, and five types of PGs have been described, with specific roles and mechanism in the female reproductive system [5, 6]. Prostaglandin F2α (PGF2α) has a luteolytic effect, whereas prostaglandin E2 (PGE2) is central in ovulation, fertilization, embryo development, and implantation [7, 8]. In addition, PGE2 plays important roles in inflammatory processes, being increased at first phases of inflammation [9].

Inflammation is a complex process with two differentiated steps or conditions: acute and chronic inflammation. A number of lipid mediators act as pro-inflammatory (i.e., leukotriene and prostaglandins) or anti-inflammatory and pro-resolving (lipoxins, resolvins, maresins, and protectins) mediators. Lipid mediators derived from polyunsaturated fatty acids (PUFA) have potent anti-inflammatory effect and promote the resolution of inflammation, through specialized pro-resolving lipid mediators (SPM) [10]. Omega-3 fatty acids are a type of PUFA with known beneficial effect, which, in addition to its pro-resolving mechanism, have shown two additional anti-inflammatory mechanisms: activation of the free fatty acid (FFA)-4 receptor and inflammasome inhibition. Recent evidences have suggested an anti-inflammatory effect of docosahexaenoic acid (DHA) and eicosapentaenoic acid (EPA) in the endometrium. Furthermore, FFA4 receptor was detected in the human, mouse, and bovine uterus [11, 12]. The following sections describe the effect and potential mechanisms of omega-3 fatty acids in the immune cells and endometrium and perspectives of these fatty acids in health and disease.

2. Eicosanoids in fertility and immune function in the uterus

Eicosanoids, which include prostaglandins and leukotrienes, are members of a large family of compounds that are synthesized from arachidonic acid through the cyclooxygenase and lipoxygenase pathways [13]. PGF2α and PGE2 exert opposite actions on the corpus luteum (CL); therefore, control over their synthesis and secretion is critical either for the initiation of luteolysis or maintenance of pregnancy [7, 8].

PGF2α is considered a pro-inflammatory molecule, and it may stimulate the synthesis of pro-inflammatory cytokines that enhance phagocytosis and lymphocyte functions [14]. PGF2α can increase IL1β, IL6, CCL2, and CXCL8 via ERK1/ERK2, PI3K, NFAT, and NF-κB pathways in the myometrial cells from term pregnant women, suggesting that PGF2α induces an inflammatory environment during the late stage of human pregnancy [15]. PGF2α in vitro enhanced neutrophil chemotaxis and the ability of neutrophils to ingest bacteria, and anti-PGF2α antibody blocked the chemotactic effects of PGF2α [16, 17]. Exogenous PGF2α increases uterine secretion of PGF2α through the activation of phospholipase A2 (PLA2) and cyclooxygenase 2 [18–20]. Also, it has been proposed that exogenous PGF2α increases luteal leukotriene B4 (LTB4) production [21]. LTB4 can stimulate chemotaxis, random migration, and antibody-independent cell-mediated cytotoxicity and may reduce the risk of uterine infections in cows [22].

Most studies about PGF2α and fertility have been performed in production animals. PGF2α and its analogs have been used to resolve uterine infections in livestock; however, its mechanism of action is not known. Moreover, it is unclear if modulation of sexual steroids levels induced by PGF2α directly alters the immune response in postpartum. Cattle are resistant to uterine infections when progesterone concentrations are basal, and they are susceptible to uterine infections when progesterone concentrations are increased [23, 24]. It has been proposed that exogenous PGF2α is an effective luteolytic factor to reduce progesterone levels and subsequent estrus, with increased estrogen level and myometrial contractions, and would be favorable for clearance of uterine infection [25, 26]. However, other authors report that PGF2α upregulate immune functions reducing vaginal discharge, uterine inflammation, endometrium fibrosis, and infection, which could be independent of progesterone levels [27]. Also, it has been proposed that PGF2α is more effective when progesterone levels are high or a corpus luteum is palpable [28, 29]. Thus, a direct effect of PGF2α on immune system has been proposed. The in vivo effect of exogenous PGF2α suggests that immune functions do not seem entirely independent of progesterone [30]. Fenprostalene, a long-acting PGF2α analog, injected between days 7 and 10 postpartum, when progesterone concentrations are basal, reduced the incidence of endometritis in dairy cows with dystocia and/or retained fetal membranes and reduced the interval from parturition to conception [31]. The studies with cloprostenol and fenprostalene indicate that increased PGF2α during the postpartum period in dairy cattle improves uterine health [27, 31]. Indeed, jugular concentrations of 13,14-dihydro-15-keto-PGF2α, which is a metabolite that seems to reflect uterine production of PGF2α postpartum, were less in postpartum dairy cows that subsequently developed uterine infections, than in cows that did not develop uterine infections [31, 32]. Despite the above antecedents, the evidence is contradictory if exogenous PGF2α can be useful as endometritis therapy in cows [33–36].

PGE2 is the most abundant eicosanoid lipid in the inflammatory environment. Thus, PGE2 plays a pivotal role in endometriosis-associated inflammation and pain, and its production is augmented in lesions and in the peritoneal cavity [37, 38]. Exogenous PGE2 pretreatment also modulates the innate immune response, increasing the Pam3CSK4-induced inflammatory responses through Toll-like receptor (TLR)-2 signaling in bovine endometrial epithelial cells [39]. PGE2 can increase the lipopolysaccharide (LPS)-induced response on PKA, ERK1/ERK2, and IκBα phosphorylation, as well as COX-2 and IL-6 expression, and downregulate the PGE2 receptor 4 (EP4) and TLR4 in bovine endometrial cells [40]. PGE2 via EP2 and EP4 receptors can reduce the expression of CXCL8, CCL2, and granulocyte macrophage colony-stimulating factor (GM-CSF) induced by IL-1β in primary human myometrial cells [41]. In human uterine epithelial cells, misoprostol, an analog of PGE2, increases cAMP levels via EP4 and reduces the expression of antimicrobial peptides such as β-defensins [42].

3. Effect of omega-3 fatty acids on endometrial function: in vivo and in vitro studies

Several studies have suggested that supplementation of omega-3 fatty acids during pregnancy is beneficial for establishment and maintenance of pregnancy, maintains gestation length and fetal growth, prevents preterm birth, and decreases the rate of gestational diabetes [43]. These effects of omega-3 fatty acids have been mainly studied in animals such as bovine and ovine; however, some recent studies have begun to be performed to demonstrate the beneficial effect of these fatty acids in humans and mice. Consumption of omega-3 fatty acids has been associated with a reduction of the symptoms and lower risk of developing endometriosis in women, a hormone-dependent chronic inflammatory condition [44, 45]. In wild-type mice, the administration of EPA reduced the number of endometriotic lesions, similarly as was observed in a transgenic mouse model with high levels of omega-3 fatty acids [46]. In a rat model of endometriosis, the EPA supplementation reduced the endometriotic lesions and expression of pro-inflammatory gene, suggesting that the EPA supplementation might be a strategy for the treatment of endometriosis [47].

Some studies in vitro have evidenced that the supplementation of mice with omega-3 fatty acids increased implantation markers such as laminin and leukemia inhibitory factor in endometrial epithelium and stroma, which would encourage the endometrium for a favorable environment of implantation [48]. Through an abortion mouse model and human stromal cells, it was suggested that omega-3 fatty acids activate the signaling pathways ERK1/ERK2 and AMPK, which increase FOXO1 and GLUT 1 expression, and the increased glucose uptake would be important for the maintenance of pregnancy [11].

Several studies in bovines have proposed that omega-3-rich diet improve the reproductive performance. It has been described that incorporation of fatty acids of omega-3, and also omega-6, in the bovine diet influences some of the reproductive process involved in the follicular development [49] and progesterone and PGF2α production [50, 51] regulating embryo survival and implantation.

It has been shown that high intake of omega-6 fatty acid induces a change in membrane phospholipids, increasing the proportion of arachidonic acid, which would favor the synthesis of PG of the series 2, and eicosanoid, so it would turn into a pro-inflammatory environment [52]. By contrast, the dietary increase of the omega-3 fatty acid especially EPA and DHA would increase the proportion of these phospholipids in the cell membranes, which would ultimately result in the decrease of the synthesis of the PG series 2, whereby they would act as an anti-inflammatory mechanism [52]. Based on the effect of fatty acids on PG secretion, several studies conducted in dairy cattle have been addressed to attenuate the endometrium secretion of PG at the time of the embryo-maternal recognition of pregnancy in order to improve embryo survival and pregnancy rate [53–55]. Dairy cows supplemented with conjugated linoleic acid (CLA) have higher pregnancy rates than their non-supplemented control group, and the probability of pregnancy increases by up to 26% and that the interval of first postpartum ovulation was reduced by 8 days [56].

Also, the supplementation of dairy cows with polyunsaturated omega-3 fatty acids as EPA can inhibit the synthesis of PGF2α through competition with arachidonic acid by COX-1 and COX-2 enzymes or in the case of DHA competing with arachidonic acid with the phospholipase A2 enzymes [57]. For this reason, fish meal included in the bovine diet could reduce PGF2α and delay regression of the CL, improving embryonic survival and female fertility [51]. The supplementation of cows with omega-3 fatty acid from the fish meal not only reduces the endometrial concentration of arachidonic acid but also increases the concentration of both EPA and omega-3 fatty acids in the endometrium [58]. When fish meal was included in the diet in a study conducted with beef cows (Angus), an increase in EPA and DHA in luteal tissue and a reduction of arachidonic acid in the endometrium resulting in an increase in the fertility of cows were observed [59]. However, in addition to its effect on PG secretion, some studies have concluded that diets rich in EPA and DHA can have a direct effect on the growth of the conceptus per se [60]. Others speculate that the delay on CL regression would allow not well-developed embryos to reach their competent size to initiate a maternal dialog before the luteolytic secretion of PG [50, 61].

The roles of omega-3 and omega-6 fatty acids on prostaglandin secretion have been well documented in in vitro and in vivo studies. The production of PGF2 α was suppressed in an endometrial cell culture when the culture medium was supplemented with omega-3 fatty acids [62]. However, when the medium was supplemented with omega-6 fatty acids, the increase in the ratio of omega-6 to omega-3 produced an increase of PGF2α [63]. Similarly, in studies conducted with dairy cows, the supplementation with different ratio of fatty acids from omega-6 to omega-3 altered the secretion on PGF2 α induced by either oxytocin [64] or spontaneous [49]. The production of PGE2 induced by LPS also was inhibited in the cellular line of bovine endometrium BEND treated with DHA [12].

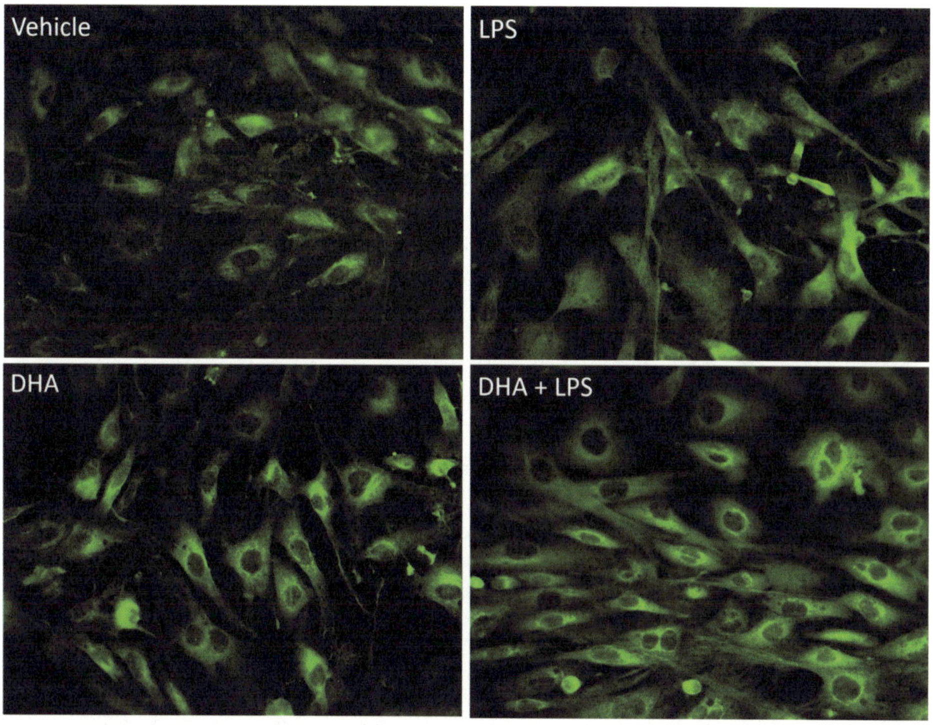

Figure 1.
Localization of NF-κB in BEND cells treated with DHA and stimulated with LPS. BEND cells were treated with 50 μM DHA for 15 min, and then 1 μg/ml LPS was added and incubated for 30 min. NF-κB was detected by immunocytochemistry and epifluorescence microscopy. Magnification 40X [65].

Additionally, recent evidences show an inhibition of the translocation of the transcription factor NF-κB induced by LPS in BEND cells treated with DHA (**Figure 1**; unpublished data).

4. Mechanisms of action of omega-3 fatty acids in endometrial cells

The first known anti-inflammatory mechanism of omega-3 fatty acids was the formation of specialized pro-resolving mediators (SPMs) derived from DHA. The enzymatic oxygenation of DHA via 12–/15-lipoxygenase (LOX) and 5-LOX leads to the formation of the D-series resolvins (RvD1, RvD2, RvD3, RvD4, RvD5, and RvD6), neuroprotectins/protectins, and maresins in different cells [10, 66], and resolvins have a potent effect on leukocyte migration and also reduce production of pro-inflammatory cytokines [67]. All those evidences have been obtained in different cellular types, but in the uterus or endometrial cells, there are not yet studies about formation of SPM.

Two more recent mechanisms have been described in macrophages and endothelial cells: (1) binding of DHA to FFA4 receptor/β-arrestin and inhibition of TAK1/NF-κB, thus reducing synthesis of pro-inflammatory factors, and (2) inhibition of NLRP3 inflammasome. FFA4 receptor is a G-protein-coupled receptor with high affinity by DHA described first in the intestine and macrophages. Recent studies evidenced the presence of FFA4 receptor in the human, mouse, and bovine endometrium [11, 12]. After ligand binding, FFA4 receptor couples to β-arrestin2, which is followed by receptor endocytosis and inhibition of TAB1-mediated activation of TAK1, a protein activated after inflammatory stimuli such as LPS, which induce signaling through the NF-κB pathway, thus reducing TNF-α, IL-6, and MCP-1 [68, 69]. Other studies have proposed omega-3 fatty acids to reduce the NLRP3 inflammasome activation [70–73]. The first evidence proposed two mechanisms dependent on FFA4 receptor to the reduction of inflammasome activation: first, DHA stimulation caused FFA4 receptor internalization through β-arrestin2, which reduced the initial inflammasome priming step by suppressing the nuclear translocation of NF-κB, and second, DHA enhanced autophagy, thereby reducing inflammasome complex formation or presenting inflammasome components for destruction [73]. Then, it was demonstrated that DHA reduced NLRP3 inflammasome expression in hepatocytes [70].

In the endometrium, only two lines of evidences about potential mechanisms of action of omega-3 have been studied. In human stromal cells, FFA4 receptor promoted decidualization through the upregulation of the GLUT1-mediated glucose uptake and glucose-6-phosphate dehydrogenase-mediated pentose-phosphate pathway [11]. In mice, FFA4 receptor protects LPS or RU486-induced abortion [11]. In summary, omega-3 fatty acids via FFA4 receptor increase ERK1/ERK2 and AMPK signaling and upregulate FOXO1 and GLUT1 expression, which increases glucose uptake and activates the pentose-phosphate pathway, promoting decidualization and maintenance of pregnancy. In addition, it was also shown that FFA4 receptor upregulates the expression of chemokines and cytokines such as CXCL12, TGFβ, and IL-15 [11]. In bovine endometrial cells, it was evidenced the presence of mRNA and protein of FFA4 receptor as well as an increase of intracellular calcium mobilization induced by DHA or a synthetic agonist (TUG891) of FFA4 receptor, which was inhibited by AH7614, a FFA4 receptor antagonist [12]. Also, DHA reduced NF-κB activation and PGE2 production induced by LPS; however, AH7614 did not modify these effects, suggesting that other mechanisms would be involved in the anti-inflammatory effect of DHA, which should be studied [12].

5. Perspectives in health and diseases

Until now, omega-3 fatty acids have been only used as dietary supplements, or DHA-rich diet has been recommended by their beneficial effects for health. However, although the mechanism of action of DHA has begun to be elucidated, it has not been recommended yet as an anti-inflammatory drug. The recent studies have described several possible anti-inflammatory mechanisms and propose omega-3 fatty acids as potential treatment for spontaneous abortion for its effect on decidualization and the maintenance of pregnancy [11]. Also, omega-3 fatty acids would be useful for the prevention and treatment of endometriosis because this disorder is characterized by a chronic inflammation [44, 46, 47]. In veterinary medicine, omega-3 fatty acids have potential use in fertility of dairy cows. Omega-3- rich supplements have been associated with improved reproductive performance, and the recent evidence of the presence of FFA4 receptor in the endometrium [12] could contribute to understand the mechanism as omega-3 fatty acids exert its effects, and open new possibilities for the prevention and treatment of the endometrial inflammation associated with infectious diseases, such as metritis or endometritis.

6. Conclusions

Omega-3 fatty acids have anti-inflammatory effects through different mechanisms, described in macrophages and endothelial cells: formation of SPMs, activation of the FFA4 receptor, inhibition of TAK1/NF-κB activation, and inflammasome inhibition. These mechanisms have not yet been demonstrated in the endometrium, but the presence of the FFA4 receptor and the inhibition of NF-κB, PGE2, and PGF2α suggest that similar anti-inflammatory mechanism could occur in the endometrium. Furthermore, omega-3 fatty acids could be useful for the treatment of disorders such as endometriosis or metritis/endometritis, as well as the prevention of spontaneous abortion and improvement of fertility.

Acknowledgements

Funded by Fondo Nacional de Desarrollo Científico y Tecnológico (Fondecyt No. 1151047).

Conflict of interest

The authors declare no conflict of interest.

Author details

Maria A. Hidalgo[1]*, Marcelo Ratto[2] and Rafael A. Burgos[1]

1 Institute of Pharmacology and Morphophysiology, University Austral of Chile, Valdivia, Chile

2 Institute of Animal Science, University Austral of Chile, Valdivia, Chile

*Address all correspondence to: mahidalgo@uach.cl

References

[1] Groothuis PG, Dassen HH, Romano A, Punyadeera C. Estrogen and the endometrium: Lessons learned from gene expression profiling in rodents and human. Human Reproduction Update. 2007;**13**(4):405-417. DOI: 10.1093/humupd/dmm009

[2] Young SL, Lessey BA. Progesterone function in human endometrium: Clinical perspectives. Seminars in Reproductive Medicine. 2010;**28**(1):5-16. DOI: 10.1055/s-0029-1242988

[3] Turner ML, Healey GD, Sheldon IM. Immunity and inflammation in the uterus. Reproduction in Domestic Animals. 2012;**47**(Suppl 4):402-409. DOI: 10.1111/j.1439-0531.2012.02104.x

[4] Liu S, Diao L, Huang C, Li Y, Zeng Y, Kwak-Kim JYH. The role of decidual immune cells on human pregnancy. Journal of Reproductive Immunology. 2017;**124**:44-53. DOI: 10.1016/j.jri.2017.10.045

[5] Niringiyumukiza JD, Cai H, Xiang W. Prostaglandin E2 involvement in mammalian female fertility: Ovulation, fertilization, embryo development and early implantation. Reproductive Biology and Endocrinology. 2018;**16**(1):43. DOI: 10.1186/s12958-018-0359-5

[6] Jabbour HN, Sales KJ. Prostaglandin receptor signalling and function in human endometrial pathology. Trends in Endocrinology and Metabolism. 2004;**15**(8):398-404. DOI: 10.1016/j.tem.2004.08.006

[7] Bennegard B, Hahlin M, Wennberg E, Noren H. Local luteolytic effect of prostaglandin F2 alpha in the human corpus luteum. Fertility and Sterility. 1991;**56**(6):1070-1076

[8] Hahlin M, Dennefors B, Johanson C, Hamberger L. Luteotropic effects of prostaglandin E2 on the human corpus luteum of the menstrual cycle and early pregnancy. The Journal of Clinical Endocrinology and Metabolism. 1988;**66**(5):909-914. DOI: 10.1210/jcem-66-5-909

[9] Jabbour HN, Sales KJ, Catalano RD, Norman JE. Inflammatory pathways in female reproductive health and disease. Reproduction. 2009;**138**(6):903-919. DOI: 10.1530/REP-09-0247

[10] Serhan CN, Petasis NA. Resolvins and protectins in inflammation resolution. Chemical Reviews. 2011;**111**(10):5922-5943. DOI: 10.1021/cr100396c

[11] Huang J, Xue M, Zhang J, Yu H, Gu Y, Du M, et al. Protective role of GPR120 in the maintenance of pregnancy by promoting decidualization via regulation of glucose metabolism. eBioMedicine. 2019;**39**:540-551. DOI: 10.1016/j.ebiom.2018.12.019

[12] Valenzuela P, Teuber S, Manosalva C, Alarcon P, Figueroa CD, Ratto M, et al. Functional expression of the free fatty acids receptor-1 and -4 (FFA1/GPR40 and FFA4/GPR120) in bovine endometrial cells. Veterinary Research Communications. 2019;**43**(3):179-186. DOI: 10.1007/s11259-019-09758-8

[13] Higgins AJ. The biology, pathophysiology and control of eicosanoids in inflammation. Journal of Veterinary Pharmacology and Therapeutics. 1985;**8**(1):1-18

[14] Lewis GS. Steroidal regulation of uterine immune defenses. Animal Reproduction Science. 2004;**82-83**:281-294. DOI: 10.1016/j.anireprosci.2004.04.026

[15] Xu C, Liu W, You X, Leimert K, Popowycz K, Fang X, et al. PGF2alpha

modulates the output of chemokines and pro-inflammatory cytokines in myometrial cells from term pregnant women through divergent signaling pathways. Molecular Human Reproduction. 2015;**21**(7):603-614. DOI: 10.1093/molehr/gav018

[16] Arnould T, Thibaut-Vercruyssen R, Bouaziz N, Dieu M, Remacle J, Michiels C. PGF(2alpha), a prostanoid released by endothelial cells activated by hypoxia, is a chemoattractant candidate for neutrophil recruitment. The American Journal of Pathology. 2001;**159**(1):345-357. DOI: 10.1016/s0002-9440(10)61701-4

[17] Hoedemaker M, Lund LA, Wagner WC. Influence of arachidonic acid metabolites and steroids on function of bovine polymorphonuclear neutrophils. American Journal of Veterinary Research. 1992;**53**(9):1534-1539

[18] Binelli M, Guzeloglu A, Badinga L, Arnold DR, Sirois J, Hansen TR, et al. Interferon-tau modulates phorbol ester-induced production of prostaglandin and expression of cyclooxygenase-2 and phospholipase-a(2) from bovine endometrial cells. Biology of Reproduction. 2000;**63**(2):417-424. DOI: 10.1095/biolreprod63.2.417

[19] Narayansingh RM, Senchyna M, Carlson JC. Treatment with prostaglandin F2alpha increases expression of prostaglandin synthase-2 in the rat corpus luteum. Prostaglandins & Other Lipid Mediators. 2002;**70**(1-2):145-160

[20] Wade DE, Lewis GS. Exogenous prostaglandin F2 alpha stimulates utero-ovarian release of prostaglandin F2 alpha in sheep: A possible component of the luteolytic mechanism of action of exogenous prostaglandin F2 alpha. Domestic Animal Endocrinology. 1996;**13**(5):383-398

[21] Steadman LE, Murdoch WJ. Production of leukotriene B4 by luteal tissues of sheep treated with prostaglandin F2 alpha. Prostaglandins. 1988;**36**(5):741-745

[22] Slama H, Vaillancourt D, Goff AK. Leukotriene B4 in cows with normal calving, and in cows with retained fetal membranes and/or uterine subinvolution. Canadian Journal of Veterinary Research. 1993;**57**(4):293-299

[23] Hawk HW, Brinsfield TH, Turner GD, Whitmore GW, Norcross MA. Effect of ovarian status on induced acute inflammatory responses in cattle uteri. American Journal of Veterinary Research. 1964;**25**:362-366

[24] Lewis GS. Role of ovarian progesterone and potential role of prostaglandin F2alpha and prostaglandin E2 in modulating the uterine response to infectious bacteria in postpartum ewes. Journal of Animal Science. 2003;**81**(1):285-293. DOI: 10.2527/2003.811285x

[25] LeBlanc SJ. Postpartum uterine disease and dairy herd reproductive performance: A review. Veterinary Journal. 2008;**176**(1):102-114. DOI: 10.1016/j.tvjl.2007.12.019

[26] McClary DG, Putnam MR, Wright JC, Sartin JL Jr. Effect of early postpartum treatment with prostaglandin F2alpha on subsequent fertility in the dairy cow. Theriogenology. 1989;**31**(3):565-570

[27] Bonnett BN, Etherington WG, Martin SW, Johnson WH. The effect of prostaglandin administration to Holstein-Friesian cows at day 26 postpartum on clinical findings, and histological and bacteriological results of endometrial biopsies at day 40. Theriogenology. 1990;**33**(4):877-890

[28] LeBlanc SJ, Duffield TF, Leslie KE, Bateman KG, Keefe GP,

Walton JS, et al. The effect of treatment of clinical endometritis on reproductive performance in dairy cows. Journal of Dairy Science. 2002;**85**(9):2237-2249. DOI: 10.3168/jds.S0022-0302(02)74303-8

[29] Sheldon IM, Noakes DE. Comparison of three treatments for bovine endometritis. The Veterinary Record. 1998;**142**(21):575-579. DOI: 10.1136/vr.142.21.575

[30] Lewis GS, Wulster-Radcliffe MC. Prostaglandin F2alpha upregulates uterine immune defenses in the presence of the immunosuppressive steroid progesterone. American Journal of Reproductive Immunology. 2006;**56**(2):102-111. DOI: 10.1111/j.1600-0897.2006.00391.x

[31] Nakao T, Gamal A, Osawa T, Nakada K, Moriyoshi M, Kawata K. Postpartum plasma PGF metabolite profile in cows with dystocia and/or retained placenta, and effect of fenprostalene on uterine involution and reproductive performance. The Journal of Veterinary Medical Science. 1997;**59**(9):791-794. DOI: 10.1292/jvms.59.791

[32] Seals RC, Matamoros I, Lewis GS. Relationship between postpartum changes in 13, 14-dihydro-15-keto-PGF2alpha concentrations in Holstein cows and their susceptibility to endometritis. Journal of Animal Science. 2002;**80**(4):1068-1073. DOI: 10.2527/2002.8041068x

[33] Haimerl P, Heuwieser W, Arlt S. Therapy of bovine endometritis with prostaglandin F2alpha: A meta-analysis. Journal of Dairy Science. 2013;**96**(5):2973-2987. DOI: 10.3168/jds.2012-6154

[34] Haimerl P, Heuwieser W, Arlt S. Short communication: Meta-analysis on therapy of bovine endometritis with prostaglandin F2alpha-an update. Journal of Dairy Science. 2018;**101**(11):10557-10564. DOI: 10.3168/jds.2018-14933

[35] Lima FS, Bisinotto RS, Ribeiro ES, Greco LF, Ayres H, Favoreto MG, et al. Effects of 1 or 2 treatments with prostaglandin F(2)alpha on subclinical endometritis and fertility in lactating dairy cows inseminated by timed artificial insemination. Journal of Dairy Science. 2013;**96**(10):6480-6488. DOI: 10.3168/jds.2013-6850

[36] Wagener K, Gabler C, Drillich M. A review of the ongoing discussion about definition, diagnosis and pathomechanism of subclinical endometritis in dairy cows. Theriogenology. 2017;**94**:21-30. DOI: 10.1016/j.theriogenology.2017.02.005

[37] Sacco K, Portelli M, Pollacco J, Schembri-Wismayer P, Calleja-Agius J. The role of prostaglandin E2 in endometriosis. Gynecological Endocrinology. 2012;**28**(2):134-138. DOI: 10.3109/09513590.2011.588753

[38] Wu MH, Lu CW, Chuang PC, Tsai SJ. Prostaglandin E2: The master of endometriosis? Experimental Biology and Medicine (Maywood, N.J.). 2010;**235**(6):668-677. DOI: 10.1258/ebm.2010.009321

[39] Shen Y, Feng S, Liu B, Mao W, Gao R, Wu J, et al. Prostaglandin E2 promotes Pam3CSK4-induced inflammation in endometrial epithelial cells of cattle. Animal Reproduction Science. 2019;**200**:51-59. DOI: 10.1016/j.anireprosci.2018.11.010

[40] Shen Y, Liu B, Mao W, Gao R, Feng S, Qian Y, et al. PGE2 downregulates LPS-induced inflammatory responses via the TLR4-NF-kappaB signaling pathway in bovine endometrial epithelial cells. Prostaglandins, Leukotrienes, and

Essential Fatty Acids. 2018;**129**:25-31. DOI: 10.1016/j.plefa.2018.01.004

[41] Mosher AA, Rainey KJ, Giembycz MA, Wood S, Slater DM. Prostaglandin E2 represses interleukin 1 beta-induced inflammatory mediator output from pregnant human myometrial cells through the EP2 and EP4 receptors. Biology of Reproduction. 2012;**87**(1):7, 1-10. DOI: 10.1095/biolreprod.112.100099

[42] Aronoff DM, Hao Y, Chung J, Coleman N, Lewis C, Peres CM, et al. Misoprostol impairs female reproductive tract innate immunity against Clostridium sordellii. Journal of Immunology. 2008;**180**(12):8222-8230. DOI: 10.4049/jimmunol.180.12.8222

[43] Saccone G, Berghella V, Maruotti GM, Sarno L, Martinelli P. Omega-3 supplementation during pregnancy to prevent recurrent intrauterine growth restriction: Systematic review and meta-analysis of randomized controlled trials. Ultrasound in Obstetrics & Gynecology. 2015;**46**(6):659-664. DOI: 10.1002/uog.14910

[44] Jurkiewicz-Przondziono J, Lemm M, Kwiatkowska-Pamula A, Ziolko E, Wojtowicz MK. Influence of diet on the risk of developing endometriosis. Ginekologia Polska. 2017;**88**(2):96-102. DOI: 10.5603/GP.a2017.0017

[45] Signorile PG, Viceconte R, Baldi A. Novel dietary supplement association reduces symptoms in endometriosis patients. Journal of Cellular Physiology. 2018;**233**(8):5920-5925. DOI: 10.1002/jcp.26401

[46] Tomio K, Kawana K, Taguchi A, Isobe Y, Iwamoto R, Yamashita A, et al. Omega-3 polyunsaturated fatty acids suppress the cystic lesion formation of peritoneal endometriosis in transgenic mouse models. PLoS One. 2013;**8**(9):e73085. DOI: 10.1371/journal.pone.0073085

[47] Netsu S, Konno R, Odagiri K, Soma M, Fujiwara H, Suzuki M. Oral eicosapentaenoic acid supplementation as possible therapy for endometriosis. Fertility and Sterility. 2008;**90**(4 Suppl):1496-1502. DOI: 10.1016/j.fertnstert.2007.08.014

[48] Sarsmaz K, Goker A, Micili SC, Ergur BU, Kuscu NK. Immunohistochemical and ultrastructural analysis of the effect of omega-3 on embryonic implantation in an experimental mouse model. Taiwanese Journal of Obstetrics & Gynecology. 2016;**55**(3):351-356. DOI: 10.1016/j.tjog.2016.04.011

[49] Greco LF, Neves Neto JT, Pedrico A, Lima FS, Bisinotto RS, Martinez N, et al. Effects of altering the ratio of dietary n-6 to n-3 fatty acids on spontaneous luteolysis in lactating dairy cows. Journal of Dairy Science. 2018;**101**(11):10536-10556. DOI: 10.3168/jds.2018-15065

[50] Mattos R, Staples CR, Thatcher WW. Effects of dietary fatty acids on reproduction in ruminants. Reviews of Reproduction. 2000;**5**(1):38-45

[51] Staples CR, Burke JM, Thatcher WW. Influence of supplemental fats on reproductive tissues and performance of lactating cows. Journal of Dairy Science. 1998;**81**(3):856-871. DOI: 10.3168/jds.S0022-0302(98)75644-9

[52] Calder PC. Long-chain fatty acids and inflammation. The Proceedings of the Nutrition Society. 2012;**71**(2):284-289. DOI: 10.1017/S0029665112000067

[53] Ambrose DJ, Kastelic JP, Corbett R, Pitney PA, Petit HV, Small JA, et al. Lower pregnancy losses in lactating

dairy cows fed a diet enriched in alpha-linolenic acid. Journal of Dairy Science. 2006;**89**(8):3066-3074. DOI: 10.3168/jds.S0022-0302(06)72581-4

[54] Silvestre FT, Carvalho TS, Francisco N, Santos JE, Staples CR, Jenkins TC, et al. Effects of differential supplementation of fatty acids during the peripartum and breeding periods of Holstein cows: I. uterine and metabolic responses, reproduction, and lactation. Journal of Dairy Science. 2011;**94**(1):189-204. DOI: 10.3168/jds.2010-3370

[55] Sinedino LD, Honda PM, Souza LR, Lock AL, Boland MP, Staples CR, et al. Effects of supplementation with docosahexaenoic acid on reproduction of dairy cows. Reproduction. 2017;**153**(5):707-723. DOI: 10.1530/REP-16-0642

[56] de Veth MJ, Bauman DE, Koch W, Mann GE, Pfeiffer AM, Butler WR. Efficacy of conjugated linoleic acid for improving reproduction: A multi-study analysis in early-lactation dairy cows. Journal of Dairy Science. 2009;**92**(6):2662-2669. DOI: 10.3168/jds.2008-1845

[57] Mattos R, Staples CR, Arteche A, Wiltbank MC, Diaz FJ, Jenkins TC, et al. The effects of feeding fish oil on uterine secretion of PGF2alpha, milk composition, and metabolic status of periparturient Holstein cows. Journal of Dairy Science. 2004;**87**(4):921-932. DOI: 10.3168/jds.S0022-0302(04)73236-1

[58] Bilby TR, Guzeloglu A, MacLaren LA, Staples CR, Thatcher WW. Pregnancy, bovine somatotropin, and dietary n-3 fatty acids in lactating dairy cows: II. Endometrial gene expression related to maintenance of pregnancy. Journal of Dairy Science. 2006;**89**(9):3375-3385. DOI: 10.3168/jds.S0022-0302(06)72374-8

[59] White NR, Burns PD, Cheatham RD, Romero RM, Nozykowski JP, Bruemmer JE, et al. Fish meal supplementation increases bovine plasma and luteal tissue omega-3 fatty acid composition. Journal of Animal Science. 2012;**90**(3):771-778. DOI: 10.2527/jas.2011-4208

[60] Ribeiro ES. Symposium review: Lipids as regulators of conceptus development: Implications for metabolic regulation of reproduction in dairy cattle. Journal of Dairy Science. 2018;**101**(4):3630-3641. DOI: 10.3168/jds.2017-13469

[61] Santos JE, Bilby TR, Thatcher WW, Staples CR, Silvestre FT. Long chain fatty acids of diet as factors influencing reproduction in cattle. Reproduction in Domestic Animals. 2008;**43**(Suppl 2):23-30. DOI: 10.1111/j.1439-0531.2008.01139.x

[62] Mattos R, Guzeloglu A, Badinga L, Staples CR, Thatcher WW. Polyunsaturated fatty acids and bovine interferon-tau modify phorbol ester-induced secretion of prostaglandin F2 alpha and expression of prostaglandin endoperoxide synthase-2 and phospholipase-A2 in bovine endometrial cells. Biology of Reproduction. 2003;**69**(3):780-787. DOI: 10.1095/biolreprod.102.015057

[63] Caldari-Torres C, Rodriguez-Sallaberry C, Greene ES, Badinga L. Differential effects of n-3 and n-6 fatty acids on prostaglandin F2alpha production by bovine endometrial cells. Journal of Dairy Science. 2006;**89**(3):971-977. DOI: 10.3168/jds.S0022-0302(06)72162-2

[64] Mattos R, Staples CR, Williams J, Amorocho A, McGuire MA, Thatcher WW. Uterine, ovarian, and production responses of lactating dairy cows to increasing dietary concentrations of menhaden fish

meal. Journal of Dairy Science. 2002;**85**(4):755-764. DOI: 10.3168/jds. S0022-0302(02)74133-7

[65] Valenzuela P. Modulación de la respuesta inmune mediante la activación de receptores de ácidos grasos de cadena larga en tejido endometrial bovino [thesis]. Valdivia: Universidad Austral de Chile; 2019

[66] Serhan CN, Yang R, Martinod K, Kasuga K, Pillai PS, Porter TF, et al. Maresins: Novel macrophage mediators with potent antiinflammatory and proresolving actions. The Journal of Experimental Medicine. 2009;**206**(1):15-23. DOI: 10.1084/jem.20081880

[67] Spite M, Norling LV, Summers L, Yang R, Cooper D, Petasis NA, et al. Resolvin D2 is a potent regulator of leukocytes and controls microbial sepsis. Nature. 2009;**461**(7268):1287-1291. DOI: 10.1038/nature08541

[68] Liu KL, Yang YC, Yao HT, Chia TW, Lu CY, Li CC, et al. Docosahexaenoic acid inhibits inflammation via free fatty acid receptor FFA4, disruption of TAB2 interaction with TAK1/TAB1 and downregulation of ERK-dependent Egr-1 expression in EA.hy926 cells. Molecular Nutrition & Food Research. 2016;**60**(2):430-443. DOI: 10.1002/mnfr.201500178

[69] Oh DY, Talukdar S, Bae EJ, Imamura T, Morinaga H, Fan W, et al. GPR120 is an omega-3 fatty acid receptor mediating potent anti-inflammatory and insulin-sensitizing effects. Cell. 2010;**142**(5):687-698. DOI: 10.1016/j.cell.2010.07.041

[70] Sui YH, Luo WJ, Xu QY, Hua J. Dietary saturated fatty acid and polyunsaturated fatty acid oppositely affect hepatic NOD-like receptor protein 3 inflammasome through regulating nuclear factor-kappa B activation. World Journal of Gastroenterology. 2016;**22**(8):2533-2544. DOI: 10.3748/wjg.v22.i8.2533

[71] Garay-Lugo N, Dominguez-Lopez A, Miliar Garcia A, Aguilar Barrera E, Gomez Lopez M, Gomez Alcala A, et al. n-3 fatty acids modulate the mRNA expression of the Nlrp3 inflammasome and Mtor in the liver of rats fed with high-fat or high-fat/fructose diets. Immunopharmacology and Immunotoxicology. 2016;**38**(5):353-363. DOI: 10.1080/08923973.2016.1208221

[72] Lin C, Chao H, Li Z, Xu X, Liu Y, Bao Z, et al. Omega-3 fatty acids regulate NLRP3 inflammasome activation and prevent behavior deficits after traumatic brain injury. Experimental Neurology. 2017;**290**:115-122. DOI: 10.1016/j.expneurol.2017.01.005

[73] Williams-Bey Y, Boularan C, Vural A, Huang NN, Hwang IY, Shan-Shi C, et al. Omega-3 free fatty acids suppress macrophage inflammasome activation by inhibiting NF-kappaB activation and enhancing autophagy. PLoS One. 2014;**9**(6):e97957. DOI: 10.1371/journal.pone.0097957